This is an authorized facsimile
of the original book, printed
by xerography on acid-free paper.

UNIVERSITY MICROFILMS INTERNATIONAL
Ann Arbor, Michigan, U.S.A.
London, England
1981

Solid State Chemistry of Energy Conversion and Storage

Solid State Chemistry of Energy Conversion and Storage

John B. Goodenough, EDITOR
University of Oxford

M. Stanley Whittingham, EDITOR
EXXON Research and Engineering Co.

A symposium sponsored by the Division of Inorganic Chemistry at the 171st Meeting of the American Chemical Society, New York, N.Y., April 5–8, 1976.

ADVANCES IN CHEMISTRY SERIES **163**

AMERICAN CHEMICAL SOCIETY
WASHINGTON, D. C.   1977

Library of Congress CIP Data

Solid state chemistry of energy conversion and storage.
(Advances in chemistry series; 163 ISSN 0065-2393)

Includes bibliographic references and index.

1. Solid state chemistry—Congresses. 2. Energy transfer—Congresses. 3. Energy storage—Congresses.
  I. Goodenough, John B., 1922-   . II. Whittingham,
Michael Stanley, 1941-   . III. American Chemical Society. Division of Inorganic Chemistry. IV. Series: Advances in chemistry series; 163.

QD1.A355   no. 163   (QD478)   540'.8s
(541'.042'1)   77-20011
ISBN 0-8412-0358-X   ADCSAJ 163 1–371 (1977)

Copyright © 1977

American Chemical Society

All Rights Reserved. No part of this book may be reproduced or transmitted in any form or by any means—graphic, electronic, including photocopying, recording, taping, or information storage and retrieval systems—without written permission from the American Chemical Society.

PRINTED IN THE UNITED STATES OF AMERICA

Advances in Chemistry Series
Robert F. Gould, *Editor*

Advisory Board

Donald G. Crosby

Jeremiah P. Freeman

E. Desmond Goddard

Robert A. Hofstader

John L. Margrave

Nina I. McClelland

John B. Pfeiffer

Joseph V. Rodricks

Alan C. Sartorelli

Raymond B. Seymour

Roy L. Whistler

Aaron Wold

FOREWORD

ADVANCES IN CHEMISTRY SERIES was founded in 1949 by the American Chemical Society as an outlet for symposia and collections of data in special areas of topical interest that could not be accommodated in the Society's journals. It provides a medium for symposia that would otherwise be fragmented, their papers distributed among several journals or not published at all. Papers are refereed critically according to ACS editorial standards and receive the careful attention and processing characteristic of ACS publications. Papers published in ADVANCES IN CHEMISTRY SERIES are original contributions not published elsewhere in whole or major part and include reports of research as well as reviews since symposia may embrace both types of presentation.

CONTENTS

Preface ix

1. Hydrogen as an Energy Carrier 1
   Paul Hagenmuller

2. Catalytic Synthesis of Hydrocarbons from Carbon Monoxide and Hydrogen 15
   M. A. Vannice

3. Photoelectrochemical Production of Hydrogen 33
   J. O'M. Bockris and K. Uosaki

4. Conversion of Visible Light to Electrical Energy: Stable Cadmium Selenide Photoelectrodes in Aqueous Electrolytes 71
   Mark S. Wrighton, Arthur B. Ellis, and Steven W. Kaiser

5. Solar Energy Conversion through Photosynthesis? 93
   Roderick K. Clayton

6. Photovoltaic Solar Cells 109
   Sigurd Wagner

7. Recrystallization of Semiconducting Polycrystalline Ribbons Using the Peltier Effect 134
   S. Vojdani and R. Hashemian

8. Wavelength-Selective Surfaces 149
   John C. C. Fan

9. Thermodynamic Studies of Some Electrode Materials 165
   P. G. Dickens

10. New Solid Electrolytes 179
    H. Y-P. Hong

11. Polarizability Enhancement of Ionic Conductivity for A^{1+} in $A^{1+}M_2X_6$ Series 195
    A. W. Sleight, J. E. Gulley, and T. Berzins

12. The Sodium–Sulfur Battery: Problems and Promises 205
    S. A. Weiner

13. Chemistry of Hot Corrosion 225
    John F. Elliott

14. Nonstoichiometry, Order, and Disorder in Fluorite-Related Materials for Energy Conversion 240
    Leroy Eyring

15. Solid Metal Hydrides: Properties Relating to Their Application in Solar Heating and Cooling 271
    G. G. Libowitz and Z. Blank

16. Storage of Hydrogen Isotopes in Intermetallic Compounds 284
    S. A. Steward, J. F. Lakner, and F. Uribe

17. Chemical Conversion Using Sheet-Silicate Intercalates 298
    John M. Thomas, John M. Adams, Samuel H. Graham, and D. Tilak B. Tennakoon

18. High Temperature Electrolysis/Fuel Cells: Materials Problems ... 316
    H. Obayashi and T. Kudo

Index 365

PREFACE

As the century of inexpensive oil and gas draws to a close, the need for more efficient utilization of fossil fuels and for the development of alternate energy sources is becoming urgent. Catalysis is the master key to more efficient processing of fossil fuels, whether in refineries or in the production of chemical feedstocks and synthetic fuels, and solid state design is fundamental to improved heterogeneous catalysis and catalytic electrodes. New materials and configurations are also required for the exploitation of alternate energy sources. Inevitably chemists will play a leading role in the technical innovations that must be forthcoming, and the introduction of solid state concepts into the old energy disciplines will prove to be an essential stimulus to creativity, as we are now witnessing in the battery field.

In the last thirty years solid state chemistry has been strongly associated with the electronics industry, but in a supportive role. Physicists provided the bridge between chemistry and electrical engineering; they also developed most of the theoretical constructs on which modern materials science is based. Today solid state chemistry has come of age, and we may anticipate its full participation not only in building bridges between chemistry and energy engineering but also in developing the theoretical structures in catalysis, electrochemistry, photochemistry, or biochemistry that will be needed for tomorrow's molecular engineering. Accordingly, a noticeable shift has already taken place in the solid state chemistry community, a shift from preoccupation with phase diagrams, the cataloging of new structures, and the preparation and characterization of high-purity single crystals to a greater emphasis on the design of materials having specified physical properties or selective chemical activity. This swing and its significance for tomorrow's energy technologies has not yet been fully appreciated either by the funding agencies or by the institutions of higher learning even though these organizations were the first to feel the impact of the government's shift in scientific priorities.

In this symposium we have chosen to eliminate topics relating to nuclear energy even though chemists are making critical contributions in some areas. We have also eliminated coal and geothermal energy, choosing to stress the challenge to the chemist of sunlight as an alternate energy source. In addition, we concentrated on catalysis, corrosion, and electrochemical cells, fields in which solid state concepts and techniques are making an important impact. In this endeavor we were fortunate to be able to bring a number of overseas scientists engaged in these problems, and we are pleased that all but three of the papers presented are included in this volume.

Linden, N.J.
August 18, 1977

JOHN B. GOODENOUGH
M. STANLEY WHITTINGHAM

Hydrogen as an Energy Carrier

A European Perspective of the Problem

PAUL HAGENMULLER

Laboratoire de Chimie du Solide du Centre National de la Recherche Scientifique, Universite de Bordeaux I, 351, cours de la Liberation, 33405 Talence, France

> *For countries with low fossil reserves the production of low-cost hydrogen may be useful both as energy storage of nuclear electric power plants and as feedstock for growing requests of the chemical industry. From a prospective point of view two methods of producing hydrogen from water and nuclear power have been considered: electrolysis at medium or high temperature, and thermochemical cycles. Transport, storage, and safety problems are discussed.*

The rising cost of fossil fuels and the prospect of depleted reserves in the 21st century have motivated nations—particularly those industrialized states dependent on foreign imports—to devise strategies that will lead to greater energy independence. These strategies differ from one country to another: the United States has coal and oil-shale reserves that are made economically attractive by the rising price of oil, West Germany has large coal deposits that can be transformed into liquid or gaseous hydrocarbons, and Britain is looking to its North Sea oil fields to carry it through the next 15–20 years. However, many countries—including France, Italy, India, and Japan—are poor in fossil-fuel reserves and therefore have chosen to increase their dependency on nuclear energy (*1, 2, 10, 11, 12*).

Choosing the nuclear option introduces a number of problems. The energy produced by a reactor should be distributed to the consumer in the most convenient form with the highest possible efficiency. An immediately obvious approach is to use the heat from a reactor to drive a conventional steam-turbine electric-power plant, but the theoretical con-

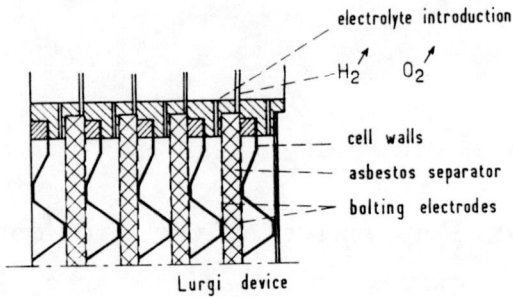

Figure 1. Example of a classical electrolytic cell

straint of the Carnot cycle and its practical low-temperature limit restrict the conversion efficiency of a present-day light-water reactor to 30–35%. Moreover, the nuclear reactor represents a constant-energy source, whereas electric-power demand fluctuates by a factor of ca. two in the day-night cycle and between weekdays and weekends. Since nuclear energy supplies more than the base load of electric power, a means of large-scale energy storage must be added to the system. Storage both decreases the system efficiency and increases its cost.

Energy storage in secondary batteries implies considerable cost, even if the high-specific-energy Na–S cells now under development replace the classical lead–acid batteries. The use of mechanical storage, whether as pumped water or as underground-stored compressed gas, would be cheaper, but it is not generally applicable.

An alternate approach is to electrolyze water, in a basic medium, during times of minimum demand (Figure 1) (3); the energy so stored could be converted back to electricity by burning the hydrogen—exhausting pure water—either in the combustion chamber of a conventional power plant or in a fuel cell. Unfortunately at present such hydrogen production is not efficient because of joule losses and electrode overvoltages in the electrolysis cell. In today's electrolysis cells, for example, about 5 kWh/Nm^3 (at 25°C and 2 bar) of hydrogen (4.6 kWh/Nm^3 in the best case) are used, whereas the theoretical value is only 3 kWh/Nm^3. Thus the overall efficiency for converting nuclear power to hydrogen is < 20%. This value is definitely too low for a practical storage scheme.

On the other hand, the efficient production of hydrogen is needed for other purposes besides load averaging in a power plant. Hydrogen is an industrial feedstock of growing importance. Present world consumption is ca. 200 billion Nm^3/year, and it is increasing at an average annual rate of 8%. Nearly half is used for the synthesis of ammonia, the cornerstone of the fertilizer industry. The petrochemical industry consumes another quarter for the hydrocracking and desulfurization of oil. In fact, developments in hydrocracking promise greater high-fraction yields in oil refinement, or an increase in the fraction of gasoline and a decrease in the fraction of residuals produced from a given volume of oil. Hydrogen is also used in the synthesis of methanol, cyclohexane, etc. In addition, hydrogen has potential uses, e.g., in the gasification or liquifaction of coal and in steel production. In iron-ore processing, an injection of hydrogen into the tank of the blast furnace could provide part of the required heat, thereby saving significant amounts of coke and fuel. Hydrogen could also save in the noncarburizing prereduction of iron ores before steel refining in the electrical furnace.

Hydrogen produced in reactors could be mixed eventually with domestic or industrial gas, even though it carries three times less thermal energy and has weak flame luminosity. Production of hydrogen rather than electric power with a reactor has an advantage that could make this mode of operation competitive. It is more expensive than natural gas but cheaper than electricity to transport. Moreover, the present natural-gas distribution net could be used without modification if the natural gas were enriched with 10–15% hydrogen (2).

However, cost will determine the choices made. The hydrogen now consumed by the chemical industry is produced either by steam reforming natural gas or naphtha, or by partial oxidation of heavy hydrocarbons. With a 50–60% yield, the present price is \$600/T, or \$5/10^6 Btu (the cost of fuel is about \$1.75/10^6 Btu). The cost of electrolytic hydrogen would now be about \$750/T in France for electricity at 1.2¢/kWh and a daily hydrogen production of 500 T. If the value of the co-produced oxygen is included and the effective yield from the electrolysis cell improved by 15%, electrolytic hydrogen may be almost competitive with hydrogen produced from hydrocarbons for use in the chemical industry as a reducing agent. However, it cannot compete with natural gas as a fuel in the near future.

In other words, the production of low-cost hydrogen may be useful both as energy storage for the development of nuclear electric-power plants and as feedstock for the growing demand of the chemical industry.

In the medium-to-long term, two methods of producing hydrogen more efficiently from nuclear power can be envisioned:

- Electrolysis at medium or high temperature.
- Thermal decomposition of water with thermochemical cycles.

Both methods are based on the decomposition of water, an inexhaustible raw material recovered during combustion.

Electrolysis at Medium and High Temperatures

Today's electrolysis plants require large capital investment and are relatively inefficient because of low current densities and high overvoltages. Optimum operating conditions correspond at present to voltages of ca. 2 V and current densities of only 0.2 A/cm².

To reduce the size, and hence the capital investment, of an electrolysis plant, the kinetics of the reactions at the electrodes must be increased and hence the temperature. If liquid electrolytes are used, the reaction must take place under pressure. Higher temperature decreases the overvoltages; it also increases the heat/electricity ratio of input energy because of the larger entropy term $T\Delta S$: the enthalpy of water formation, $\Delta H = T\Delta S + \Delta G$, is essentially temperature-independent, and the Gibbs free energy ΔG must be supplied as electrical energy. The Allis-Chalmers Co., for example, achieved a current density of 0.86 A/cm² at 120°C, a pressure of 21 kbar, and a voltage of 1.7 V, which corresponds to 4 kWh/Nm³H₂ (3). These conditions can be improved by raising the temperature and pressure, but such improvements will require better porous electrodes (lower overvoltages caused by better catalytic activity, higher corrosion resistance, and total elimination of the costly element platinum) and long-life separators. Moreover, the production of hydrogen under pressure has the added advantage of being more suitable to high-pressure storage, the form generally used.

A more radical process would utilize a solid electrolyte and water vapor above 800°C. Stabilized zirconia, for example, is a fast O^{2-} con-

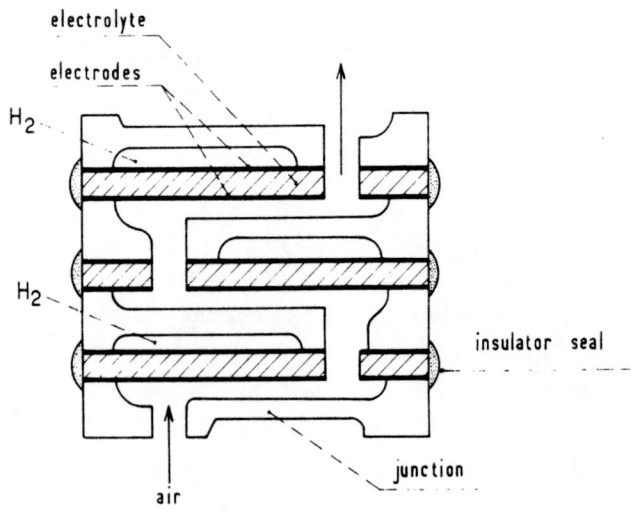

Figure 2. *Example of a proposed high temperature electrolytic cell*

ductor and an electronic insulator at these higher temperatures. The anodes would be porous electronic conductors that are resistant to the action of oxygen (doped SnO_2 or In_2O_3, for example) or mixed electronic and O^{2-} conductors such as $LaMn_{3+x}$ or UO_{2-x}. At 1000°C the voltage would be less than 1 V, energy consumption would be 2.3 kWh/Nm^3H_2 (compared with a theoretical value under standard conditions of 3 kWh/Nm^3H_2), and the current density would be a few A/cm^2. The electrolysis efficiency would then approach 80–90%, less than 100% essentially because of the resistance heating in the solid electrolyte (Figure 2). However, such a process requires a serious research effort in ceramic fabrication and cell design if it is to get beyond the present laboratory stage (3).

Decomposition of Water by Thermochemical Cycles

The production of hydrogen by electrolysis involves a fundamental objection: the production of fuel from a nuclear reactor is subject to the Carnot cycle, and the imposition of electrolysis as an intermediate step not only limits the cycle's upper temperature to that acceptable to the electric-power generator but also introduces losses associated with the electrolysis itself. Since some of the hydrogen produced by electrolysis will be used later to generate electric power, to this objection must be added the question of whether the electrolytic generation of hydrogen that is subsequently burned to regenerate electricity represents the most efficient way to store eletcrical energy in the future (4).

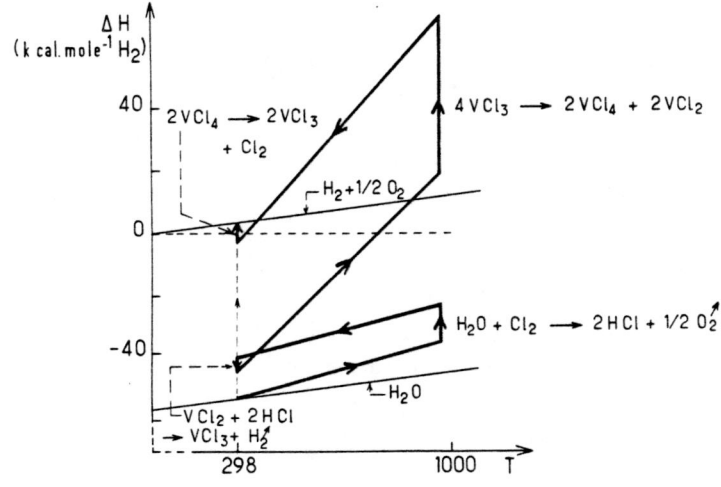

Figure 3. Thermodynamic diagram of a typical thermochemical cycle

The thermal dissociation of pure water requires very high temperatures ($\sim 4000°C$). To transform nuclear energy into chemical energy ($H_2 + \frac{1}{2}O_2$) by the thermal decomposition of water, the dissociation temperature must be reduced by the introduction of intermediate thermodynamic cycles. Several laboratories (Euratom, GDF, Aachen, Los Alamos, General Electric, etc.) have proposed such specific cycles for this purpose; however, these cycles all require a high-temperature point greater than the limiting temperatures available in light-water reactors. Temperatures greater than 650°C would be attainable today with an HTGR having a refractory uranium oxide core and the inert gas helium as the heat carrier. Studies made in West Germany have shown that HTGRs could be pushed for more than two years to provide a high-temperature heat source for the cycles above 900°C; this offers a Cornot efficiency greater than 50%.

The temperature–enthalpy diagram in Figure 3 illustrates the following four-stage cycle:

$$H_2O + Cl_2 \rightarrow 2HCl + \tfrac{1}{2}O_2\uparrow \qquad 700°C$$
$$2VCl_2 + 2HCl \rightarrow 2VCl_3 + H_2\uparrow \qquad 25°C$$
$$4VCl_3 \rightarrow 2VCl_4 + 2VCl_2 \qquad 700°C$$
$$2VCl_4 \rightarrow 2VCl_3 + Cl_2 \qquad 25°C$$

The goal of such thermodynamic cycles is to supply the enthalpy, 57.8 kcal/mol, needed to dissociate water at 25°C. The reactive masses should not be too great; the phase separations should be simple and few in number. For an efficient cycle a large number of reactions are necessary, but the number of reactions should be small enough to prevent the cycle from becoming too complex. A four-stage cycle is probably the best compromise (5). The standard free enthalpy of formation of hydrogen should be negative and have a high absolute value to give a high hydrogen pressure.

There are five different classifications of thermodynamic cycles:

(a) Cycles utilizing the Deacon equilibrium to liberate oxygen from water vapor with chlorine. However, at 700°C the yield of this reaction is only 60%. Figure 3 illustrates such a cycle using vanadium chlorides; according to Frank and Reinstrom, the total yield would be only 18%, mainly because of the high energy needed to disproportionate VCl_3 and the complexity of the phase-separation problems (6).

(b) Cycles utilizing the oxidation of carbon or carbon monoxide by water vapor (steam–carbon processes):

$$H_2O + C \rightarrow CO + H_2\uparrow$$
$$H_2O + CO \rightarrow CO_2 + H_2\uparrow$$

Carbon or carbon monoxide is regenerated by the reduction of CO_2. Mercury would be an attractive reducing agent since the oxidation product HgO dissociates at 600°C, but the toxicity of mercury creates a problem.

(c) Cycles utilizing transition-metal oxides with different oxidation states. The most reduced oxide is oxidized by water vapor with the formation of hydrogen. The resulting oxide dissociates at a higher temperature with the liberation of oxygen. FeO–Fe_3O_4, Mn_2O_3–MnO_2, and SnO–SnO_2 couples are examples. In general, these complex cycles have an acceptable theoretical yield only if the upper temperature of the cycle is high.

(d) Cycles utilizing halogenides. The hydrolysis of chlorides or bromides evolves HCl or HBr, which are subsequently dissociated to give hydrogen. The theoretical yields may be high, but the reactions generally involve large masses of reagents, which make recycling difficult. Moreover, the corrosive character of HCl and HBr is also a serious obstacle.

(e) A recent trend is the study of mixed cycles that combine classical chemical reactions with low-voltage, hence low-energy, electrolysis (7).

Presently a substantial effort is aimed at the electrolysis of SO_2 in aqueous solution, which produces hydrogen at the cathode and sulfuric acid at the anode. The acid is reduced back to SO_2 by chemical means (Westinghouse). Figure 4 summarizes parcticularly significant examples of the five types of cycles.

This brief overview shows that the development of a favorable cycle, even from purely theoretical considerations, is a rather difficult problem. In practice, the need to obtain high hydrogen pressure imposes not only the thermodynamic constraint already mentioned but also a kinetic problem that becomes more serious as the number of stages and the flow of reagents increases; the interface problem becomes more delicate. Moreover, corrosion problems complicate the choice of container materials, e.g., the choice of metals or oxides resistant to the action of alkali oxides or hydrogen halides (Figure 4). A joint investigation by CEA and GDF in France has recently shown that the reactor helium flow, used as a heat carrier for the chemical cycle, may significantly modify the yield: an increase from 500°C to 700°C of the low temperature point of the helium circuit would increase the yield by 50%, while variations of the high temperature have little influence. The thermodynamic cycle should be carefully coupled to the reactor.

In summary, the use of thermodynamic cycles to produce hydrogen from the thermal dissociation of water in HTGRs poses both a technological and a theoretical challenge of considerable difficulty. No cycle has yet reached the pilot stage. However, if a satisfactory cycle could be proposed and made to work, the efficiency of conversion from nuclear to

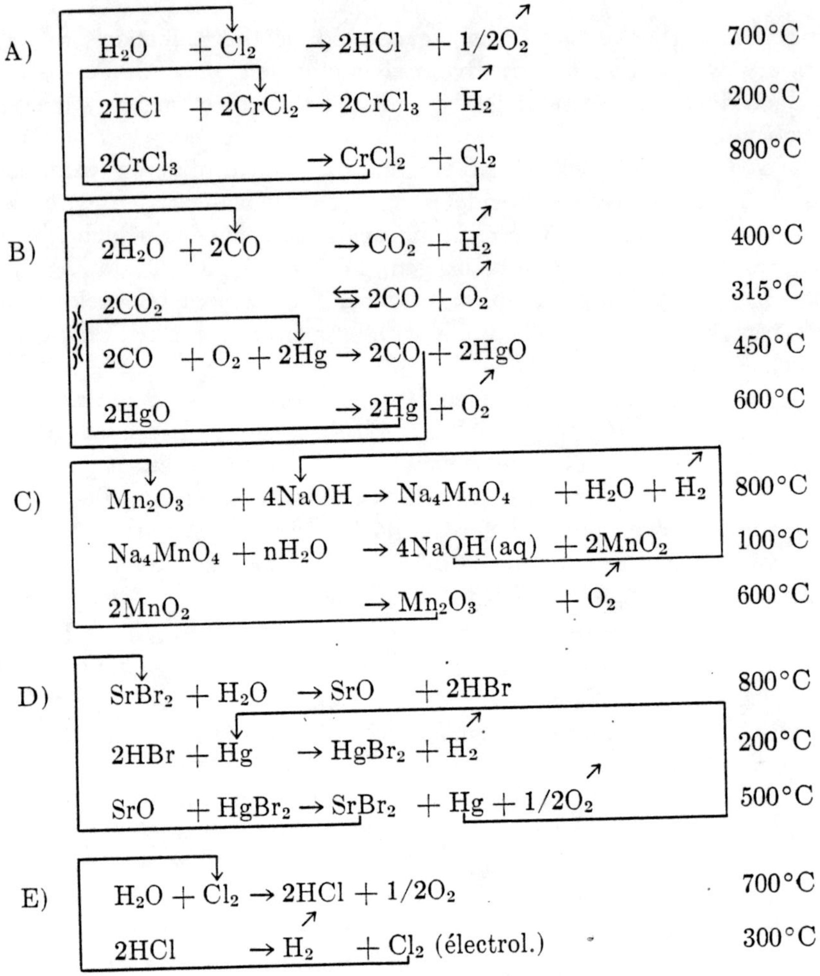

A)
$H_2O + Cl_2 \rightarrow 2HCl + 1/2O_2$	700°C
$2HCl + 2CrCl_2 \rightarrow 2CrCl_3 + H_2$	200°C
$2CrCl_3 \rightarrow CrCl_2 + Cl_2$	800°C

B)
$2H_2O + 2CO \rightarrow CO_2 + H_2$	400°C
$2CO_2 \rightleftarrows 2CO + O_2$	315°C
$2CO + O_2 + 2Hg \rightarrow 2CO + 2HgO$	450°C
$2HgO \rightarrow 2Hg + O_2$	600°C

C)
$Mn_2O_3 + 4NaOH \rightarrow Na_4MnO_4 + H_2O + H_2$	800°C
$Na_4MnO_4 + nH_2O \rightarrow 4NaOH(aq) + 2MnO_2$	100°C
$2MnO_2 \rightarrow Mn_2O_3 + O_2$	600°C

D)
$SrBr_2 + H_2O \rightarrow SrO + 2HBr$	800°C
$2HBr + Hg \rightarrow HgBr_2 + H_2$	200°C
$SrO + HgBr_2 \rightarrow SrBr_2 + Hg + 1/2O_2$	500°C

E)
| $H_2O + Cl_2 \rightarrow 2HCl + 1/2O_2$ | 700°C |
| $2HCl \rightarrow H_2 + Cl_2$ (électrol.) | 300°C |

Figure 4. *Examples of various redox cycles*

chemical energy would be significantly higher than any available electrolytic process and competitive with high-temperature electrolysis.

The production of hydrogen via thermodynamic cycles, as well as that via high-temperature electrolysis, does not require the use of a nuclear reactor; any high-temperature heat source will do.

Figure 5 compares the solar energy required to produce, by three different processes, a mass of hydrogen equivalent to 1 kWh of chemical energy: photovoltaic conversion followed by electrolysis, a high-temperature solar furnace using concentrators followed by either medium-temperature electrolysis or a thermodynamic cycle. The hypothetical yields used are the most optimistic (8). Not shown are the production of hydrogen by photoelectrolysis or by a solar furnace followed by high-temperature electrolysis. It should also be noted that the concentrators of the solar furnace use only direct sunlight.

Of the three production units considered in Figure 5 the first is by far the most expensive: for an average insolation of 350 W/m² the solar panel should have a surface area of ca. 25 m². If the panel is made of single-crystal silicon, the capital cost at today's prices would be $180,000. While this price would be significantly reduced, for instance, by the use of thin-film silicon on graphite, the smaller efficiency of the thin-film cells would increase the area required. With an electrolysis yield of 75% the second process would need an investment of only $1,000, most of it assigned to the solar collector. Replacement of the electrolysis cell by a thermodynamic cycle having a 45% yield would lower the capital cost to ca. $600 since the size of the solar collector would be decreased significantly. These cost estimates are, of course, intimately tied to the yields of the electrolytic process and the thermochemical cycle. The use of solar energy involves numerous problems not discussed here. However, two-axis concentrators would allow much higher temperatures than those available from HTGRs and would therefore permit more efficient cycles.

Distribution and Storage

The transport of pure hydrogen in pipes under 80–100 bar pressure does not create serious metallurgical problems. Although the lower density and viscosity of hydrogen gives it a faster flux than natural gas,

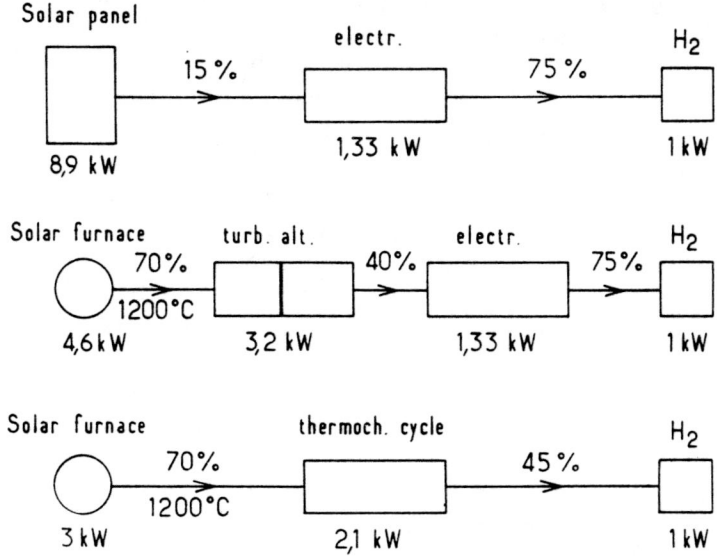

Figure 5. Production of hydrogen from solar energy

	Energy capacities	
	kWh/kg	kWh/l
H_2 200 bar light alloy reserv.	0,9	0,55
liq. H_2 20 K	7,3	2,1
$LaNi_5H_6$	0,5	5,3
MgH_2	3	4,2
Hydrocarbons	~8	~9

Figure 6. Mass and volume capacities of some energy sources

hydrogen carries only one-third the heating power; thus, compressors 5–6 times as powerful as those used for natural-gas systems are needed to attain the same rate of energy delivery. Nevertheless, for long distances hydrogen transport is much cheaper than electrical transport over power lines (2).

Large quantities of hydrogen, like natural gas, can be stored in underground water bedded cavities or in salt cavities purged of their salt by water injection. The simplest solution is to store hydrogen as a compressed gas, but the mass/volume ratio relative to that of liquid hydrocarbons is disastrous for mobile units because of the large mass of the high-pressure tanks (Figure 6). The specific energy of liquid hydrogen is much higher, but safety problems still seem to preclude, at this time, its use in cars, planes, or even trains. A proposed solution to this problem is the use of reversible hydrides that desorb their hydrogen with modest heating and/or lowering of the hydrogen pressure. The basic metal store can be recharged with hydrogen when it is not in use.

Among the reversible hydrides MgH_2 is a priori the most attractive because of its low mass and cost. Unfortunately, this hydride saturates only at relatively high temperatures and pressures.

Alloys of the rare-earth and transition elements offer alternative possibilities, but the mass/H_2 ratio and the cost are both high, as is also the dissociation pressure at room temperature. The best of these materials seems to be the hydride $LaNi_5H_6$.

$FeTiH_2$ is more stable and less expensive, but the mass/H_2 ratio is still quite high. This ratio is better in VH_2, but vanadium is expensive and the mass/H_2 ratio is higher (by about 50%) than that of MgH_2.

Recent research has led to an improved hydrogen diffusion in magnesium hydride. The approach used was to mix into sintered pellets of magnesium a dispersed phase of another metal or alloy that fixes hydrogen easily. Figures 7 and 8 show that the quantities of hydrogen fixed by

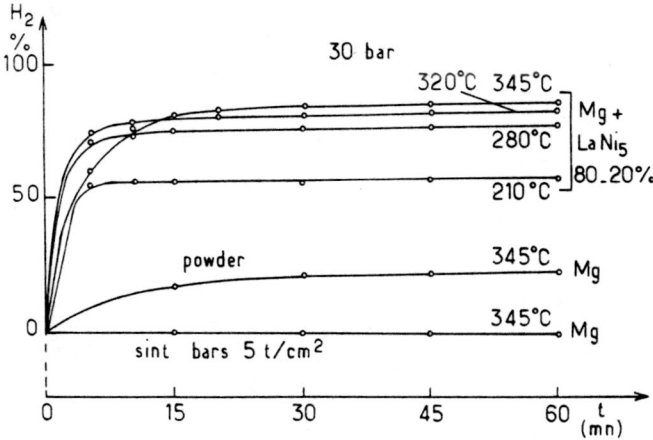

Figure 7. Absorption of hydrogen on pure and doped magnesium (p = 30 bar)

an 80:20 mixture of Mg and LaNi$_5$ at a given temperature, pressure, and time interval are much higher than those fixed by pure magnesium. Thus the temperature and pressure for fixing hydrogen can be reduced to tolerable values without excessive additional cost and mass. The origin of this synergetic effect is not yet well understood (9).

The reversible hydrides have a serious reliability problem. They tend to absorb oxygen and water vapor more easily than hydrogen, so that with time the amount of hydrogen absorbed and desorbed decreases. Little progress has been made on this problem.

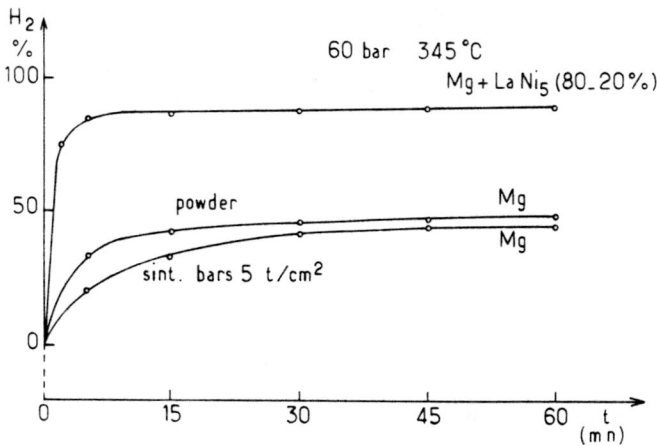

Figure 8. Absorption of hydrogen on pure and doped magnesium (p = 60 bar)

It appears that in the next 10–20 years the hydrides may be used for stationary hydrogen storage but not as mobile fuel sources for vehicles. This is a consequence of their chemical reactivity and low specific energy. For transport, hydrogen will increase the gasoline produced in oil refinement or produce synthetic liquid fuels. Nevertheless, it should be noted that the specific energy of MgH_2 is significantly higher than that of the Na–S batteries presently under development.

Metals capable of forming hydrides might be used as absorbing cathodes for load averaging in the electrolytic production of hydrogen. The gas absorbed at night would be liberated during the daytime hours of peak demand. Another possible use for hydrides is the isotopic separation of light hydrogen and deuterium (1).

Safety Problems

Hydrogen has had a bad reputation since the burning of the German zeppelin before World War II. Perhaps there is also an association in the public mind between its combustion and the explosion of the H-bomb.

In fact, gaseous hydrogen is no more dangerous than any other combustible gas; it ignites—and even explodes—in air, but the ignition and explosion thresholds for a mixture poor in air are higher than those of propane or butane (4% air instead of 2.1 and 1.9%. Compared with gaseous hydrocarbons, hydrogen ignites at a lower temperature (585°C), and its flame is practically invisible because of its low luminosity. However, because of its low density and viscosity, it diffuses more quickly in case of a leak, which reduces the risk of explosions or spontaneous ignition.

In West Germany and the USSR, pipelines transport hydrogen under pressure today, and the domestic gas used until recently in Europe contained 50% hydrogen. For many years now the chemical industry has used pressurized (several hundred bar) hydrogen in the synthesis of ammonia or for the Fischer–Tropsch process. On the other hand, the handling of liquid hydrogen is quite hazardous since, in the case of a leak, air would liquify on contact. Nevertheless, NASA has been transporting liquid hydrogen by highway and by rail without incident. Finally, neither hydrogen nor its combustion product water is toxic or polluting. The same cannot be said for fossil fuels.

In summary, as an alternate fuel of sufficient energy density, hydrogen will be competitive only when produced from water with a high enough efficiency to justify the large capital investment in production, storage, and distribution facilities. Figure 9 summarizes how hydrogen might eventually be distributed and used in the future.

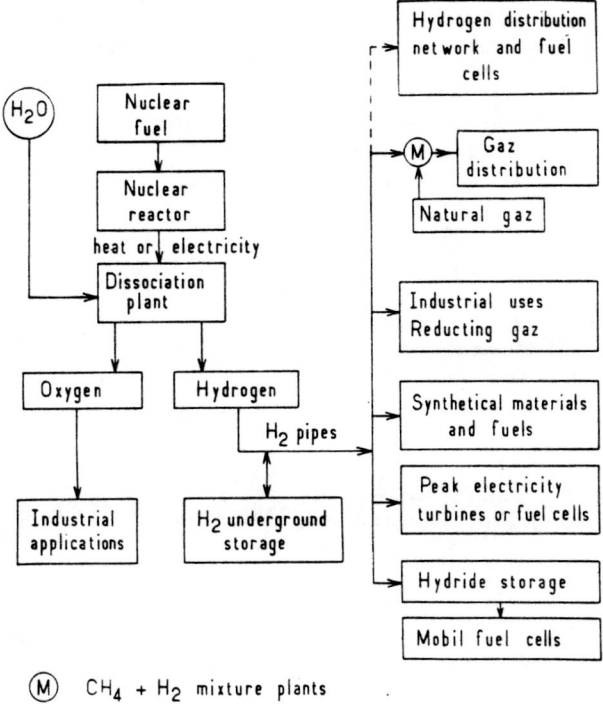

Figure 9. *Prospective distribution scheme of hydrogen*

Hydrogen produced from water by improved techniques, such as high-yield electrolysis or thermochemical cycles, would have a higher purity than that produced from hydrocarbons and could be an important industrial feedstock if its price became competitive. However, it is premature to consider hydrogen a real competitor of natural gas or oil for domestic and industrial heating (*10, 11, 12*).

Methanol is frequently proposed as the best medium-term energy carrier for automotive fuel. As a liquid at room temperature methanol is easier to store and distribute than hydrogen, and it requires little, if any, adjustments to present automobiles—particularly if methanol is used as an additive to gasoline. In fact, methanol is essentially an alternate way of storing hydrogen, which means that methanol has a lower specific energy than pure hydrogen—but a higher specific energy than MgH_2.

Acknowledgment

The author would like to thank the Petroleum Research Fund of the American Chemical Society for financial support at the New York meeting of the Society and John B. Goodenough for fruitful discussions.

Literature Cited

1. Hagenmuller, P., "Rapport du groupe HYDROGENE, prospective énergie," *C.N.R.S.*, 1974.
2. Massetti, C., "Hydrogène et énergie," *Entropie* (1974) **10**, 55.
3. Bonnemay, M., Bronoël, Lheritier, E., "Production de l'hydrogène par électrolyse," *C.N.R.S.* (1975) Feb.
4. Pottier, J., Rastouin, J., Souriau, D., Courvoisier, P., "Un nouveau support de l'Energie Nucléaire, l'hydrogène produit à partir de l'eau," *Rev. Energ.* (1974) **10**, 265.
5. Esteve, A., Leconanet, A., Roncato, J. P., GDF, *Entropie* (1975) **11**, 61.
6. Funk, J. E., Reinstrom, R. M., Allison Div., G. M. Report TID 20441, Washington, D.C., June, 1964.
7. "Hydrogen Production from Water Using Nuclear Heat," *Commun. Europ.*, Rep. No. 1, Dec. 1970; Rep. No. 2, Dec. 1971; Rep. No. 3, Dec. 1972; Rep. No. 4, Dec. 1973.
8. "A Hydrogen Energy Carrier," Vol. II, "Systems Analysis," NASA-ASEE, 43, 1974.
9. Tanguy, B., Soubeyroux, J. L., Pezat, M., Portier, T., Hagenmuller, P., *Mater. Res. Bull.* (1976) **11**, 1441.
10. *Proc. World Hydrogen Energy Conf.*, Miami Beach, Fla., March 1974 and 1976.
11. Actes des Journées Internationales d'Etude sur l'Hydrogène et ses Perspectives, Liège, Nov., 1976.
12. "A Hydrogen Energy Career," NASA-ASEE, 1973.
13. Balaceanu, J. C., "L'Hydrogène," Revue de l'Institut Français du Pétrole, May-June, 1974.

RECEIVED July 27, 1976.

Catalytic Synthesis of Hydrocarbons from Carbon Monoxide and Hydrogen

M. A. VANNICE[1]

Corporate Research Laboratories, Exxon Research and Engineering Co., Linden, N.J. 07036

> *Different synthesis processes are reviewed briefly before recent research on the catalytic behavior of the Group VIII metals in carbon monoxide hydrogenation, in which specific activities have been determined for the first time, is presented. Chemisorption measurements used to define reduced metal surface area indicate that the ordering of specific activities for these supported metals is significantly different from relative activities determined in older studies of unsupported metals in which metal surface areas were not measured. Supported Pt and Pd catalysts have much higher specific activities than unsupported Pt and Pd, and these activity increases are attributed to a crystallite size effect and a metal-support interaction, respectively. Supported Ni catalysts show a similar, but less pronounced, activity enhancement compared with unsupported Ni, and in addition exhibit a shift in selectivity to higher molecular weight hydrocarbons.*

The increasing demand for energy, coupled with the uncertainty and expense of crude oil imports, has renewed interest in the production of fuels and chemicals from hydrogen-deficient materials. Energy sources such as coal, residua, oil shale, and tar sands can be gasified with steam or oxygen to produce a gas containing large quantities of carbon monoxide and hydrogen. Once methane is removed from this $CO-H_2$ mixture, it is purified to remove sulfur poisons and then allowed to over a catalyst

[1] Current address: Department of Chemical Engineering, The Pennsylvania State University, University Park, Pa. 16802.

to produce a variety of organic products. The synthesis of hydrocarbon products, with the exception of methane, is commonly referred to as the Fischer–Tropsch synthesis reaction.

Many organic products can be formed by these $CO-H_2$ synthesis reactions. Control of the product distribution is of major importance since specific products are required for different end uses. For instance, methane and other light hydrocarbons are necessary for the production of substitute natural gas (SNG). However, if a synthetic naphtha, which can be upgraded into high octane motor fuel, is desired, then C_4-C_{10} hydrocarbon liquids are preferred. Another situation of interest involves methane which is now being flared in the Near East. As an alternative to burning this fuel, it could be transported to the U.S. as liquid natural gas (LNG), methanol, or paraffinic liquids. The latter two alternatives involve steam reforming of methane followed by different $CO-H_2$ synthesis reactions. Finally, the production of paraffinic liquids may be an important factor in a combined coal gasification–electrical generation power plant since these paraffinic liquids (or CH_3OH) can be easily stored during off-peak hours for use during peak-load hours.

The benefits of understanding and controlling product selectivity in $CO-H_2$ reactions are apparent, and knowledge of the catalytic behavior of the Group VIII metals is an important step toward achievement of this goal. A short review of existing synthesis processes will describe the state of the art today in $CO-H_2$ catalysis. Recent research using well-characterized, supported metal catalysts is presented, and the significance of these results is discussed.

Different Synthesis Processes

The gasification of heavy, hydrogen-deficient materials provides one route to produce clean fuels and chemicals. This process generates large quantities of CO and H_2 which can react to form a wide variety of products. A simplified scheme of the overall process is shown in Figure 1. The $CO-H_2$ stream is purified to remove CO_2 and sulfur-containing poisons. The desired H_2/CO ratio is then obtained by using the water–gas shift reaction. Finally, by the appropriate choice of catalyst and reaction conditions, the product distribution is adjusted to maximize production of the desired compounds.

The thermodynamics for the formation of organic compounds have been calculated and discussed in detail elsewhere (1). With the exception of a few compounds such as formaldehyde and acetylene, the $\Delta G°$ values at 298°K are negative; therefore, production of an enormous variety of compounds at reasonable reaction temperatures is theoretically possible. The production of different products, however, can be favored by a judicious choice of catalyst and the appropriate range of tempera-

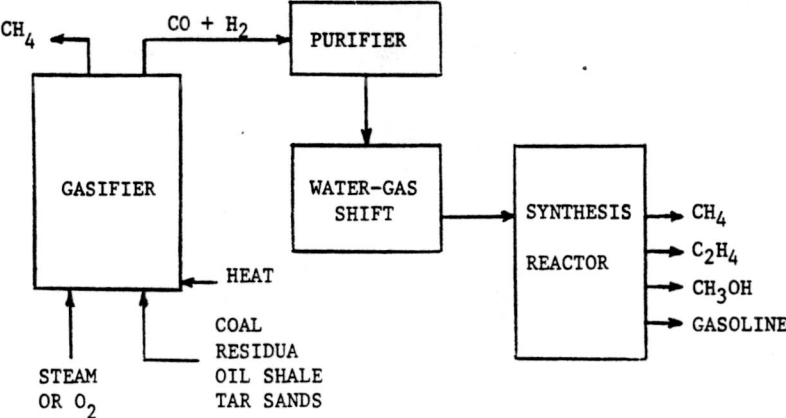

Figure 1. Gasification of hydrogen-deficient materials provides one route to produce clean fuels and chemicals

ture and pressure as illustrated in Figure 2. Metal catalysts favor the production of normal paraffins and olefins, whereas metal oxides such as ThO_2/Al_2O_3 can produce branched-chain hydrocarbons. Metal oxides and doped metal oxides are also required for the formation of alcohols. Aromatics, aldehydes, ketones, and acids can also be produced. A discussion of each of these processes in somewhat more detail will provide a familiarization with the catalysts used and the typical product distributions attained.

A number of complicating reactions can occur concurrently with the synthesis reaction; some of the most important reactions are shown in Table I. Since water is a primary product in most of the synthesis reactions, the water–gas shift reaction can occur between this water and carbon monoxide from the feed stream. This reaction changes the oxygen-containing by-products from H_2O to CO_2 and alters the usage ratio of hydrogen and carbon monoxide. The Boudouard reaction results in the disproportionation of CO to produce carbon on the catalyst surface and CO_2. Carbon deposition can also occur by the direct reaction of hydrogen and carbon monoxide. This is basically the reverse reaction of coal gasification if the latter is represented in a very simplified manner.

Table I. Complicating Reactions in CO Hydrogenation

Water gas shift	$CO + H_2O \rightarrow CO_2 + H_2$
Boudouard reaction	$2CO \rightarrow C + CO_2$
Coke deposition	$H_2 + CO \rightarrow C + H_2O$
Carbide formation	$xM + C \rightarrow M_xC$

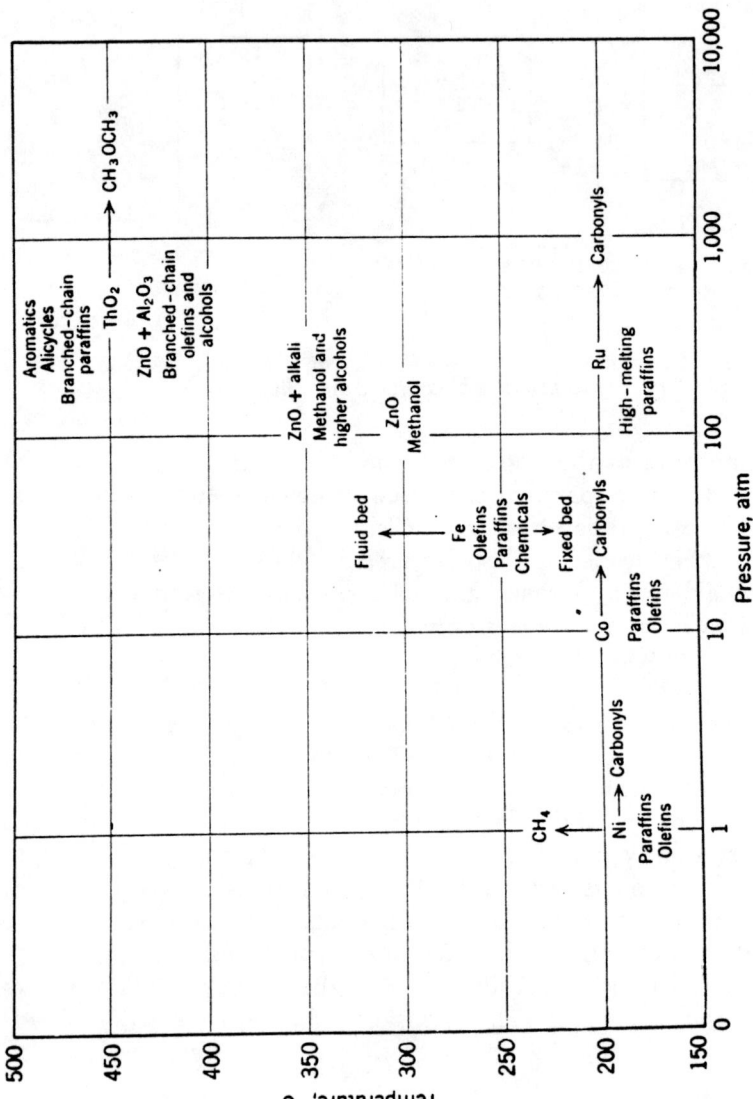

Figure 2. Many different carbon monoxide–hydrogen reactions can occur (6)

Kirk-Othmer Encyclopedia of Chemical Technology

Table II. Product Distribution from Medium-Pressure Synthesis Processes (1)

Catalyst	Mol %		
	Paraffins	Olefins	Alcohols
Fe (220°C, 10 atm)	46.4	33.1	20.5
Co (200°C, 7 atm)	79.0	20.0	1.0

Finally, when carbon exists on the surface of a metal catalyst, there is the possibility of metal carbide formation. The formation of these carbides, such as Fe_2C, Co_2C, and Ni_3C, can alter the catalytic behavior of the metal that was initially present.

Methanation has received the most attention during the past few years, primarily because it converts a low Btu syn-gas mixture into a high Btu substitute natural gas (SNG). Typically, Raney nickel or massive nickel/alumina catalysts are used in a temperature range of 523–673°K. Temperatures higher than 700°K can result in severe catalyst deactivation caused by metal sintering. A wide range of pressures can be employed, but operation at 3600–7200 kPa (500–1000 psi) allows a direct tie-in with pipeline natural gas. Under these operating conditions methane is the preponderant hydrocarbon product since only small amounts of ethane, propane, and butane are formed. This reaction is discussed in detail by Mills and Steffgen (2), Vlasenko and Yuzefovich (3), and Greyson (4).

The first commercial Fischer–Tropsch catalysts developed were co-precipitated Co/Kieselguhr and reduced, promoted iron oxide. There are substantial differences in the product distribution obtained in the medium-pressure synthesis process, which is normally run between 5 and 20 atm. Cobalt produces paraffins primarily, while promoted iron gives a larger percentage of olefins and oxygenated material which is mostly alcohols, as indicated in Table II. A look at just the hydrocarbon portion of the product reveals further differences in the catalytic behavior of Co and Fe. An examination of the molecular weight distributions shown in Figure 3 indicates that with Fe, a maximum occurs for the C_3 species on a wt % basis, whereas methane is the predominant product over Co with a second maximum occurring for the C_5 fraction. The enormous amount of data describing these synthesis processes has been discussed before (1).

In the late 1930s Pichler discovered that ruthenium produces very high molecular weight paraffinic waxes at low temperatures and very high pressures. This behavior is represented in Figure 4 where it can be seen that over 30 wt % of the product has an average molecular weight

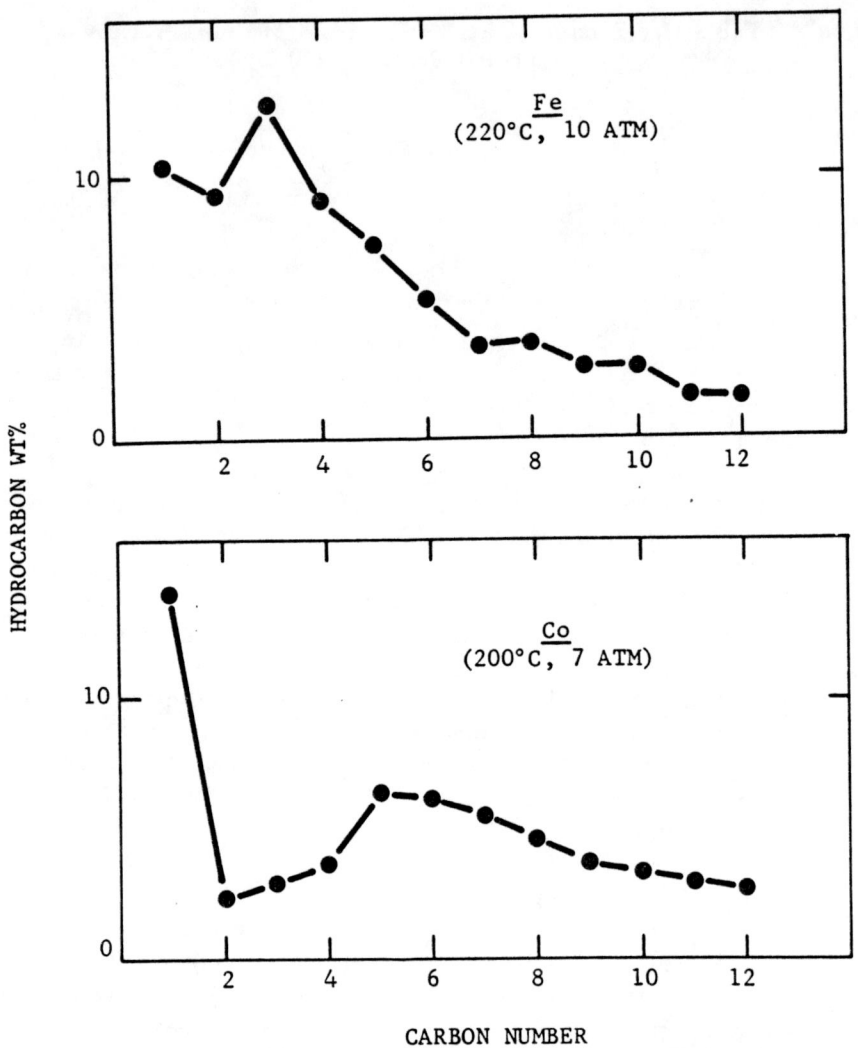

The Fischer-Tropsch and Related Syntheses

Figure 3. Molecular weight distribution of hydrocarbon products in the medium-pressure synthesis process (1)

> 200,000. This behavior has been discussed in greater detail by Pichler and co-workers (5, 6).

The metal oxides seem to be less active synthesis catalysts than the metals, thereby necessitating more demanding reaction conditions. They do produce a much different product spectrum, however. Methanol can be synthesized quite selectively with ZnO catalysts, and zinc oxide–chromia catalysts were first commercialized in the early 1920s. The low

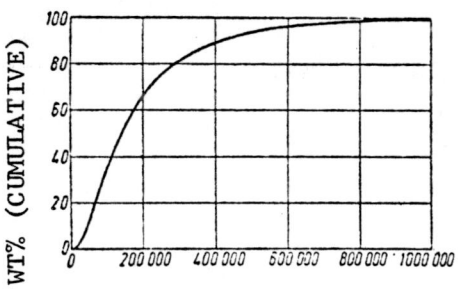

Figure 4. Ruthenium is unique in its production of high molecular weight paraffinic waxes (90°C, 2000 atm) (5).

activity of these catalysts required high operating temperatures of 550°–700°K, resulting in very unfavorable equilibrium conversions since $\Delta G°$ for the methanol synthesis reaction becomes positive at ca. 425°K. Very high reactor pressures of 100–600 atm were therefore necessary to achieve reasonable conversions because of this thermodynamic limitation. Catalysts developed recently which contain copper are more active than the zinc oxide–chromia catalysts and can be used at lower temperatures, thereby reducing the pressure required to achieve the same conversions. This change results in a more economical process since pressures of only 50–100 atm are now needed. A good review of the methanol synthesis reaction is provided by Natta (7).

The addition of alkali metals to zinc oxide imparts the capability to produce longer-chain alcohols, but these catalysts must be operated at high temperatures and high pressures. At these conditions 2-butanol is the primary product, exclusive of methanol, and comprises over 50 wt % of the product shown in Figure 5. The synthesis of higher molecular weight alcohols is discussed in greater detail by Natta, Colombo, and Pasquon (8).

At these high pressures and even higher temperatures, the use of ThO_2–Al_2O_3 catalysts results in the isosynthesis reaction in which the products are primarily branched paraffins rather than straight-chain hydrocarbons. At the conditions given in Figure 6, isobutane is the principal product on a wt % basis. The production of aromatics also has been observed in this system. A detailed description of the isosynthesis reaction is given by Cohn (9).

The newest development in CO–H_2 catalysis was disclosed recently by Union Carbide (10). By using a homogeneous Rh catalyst at moderate temperatures (525°K) and extremely high pressures of 20,000–

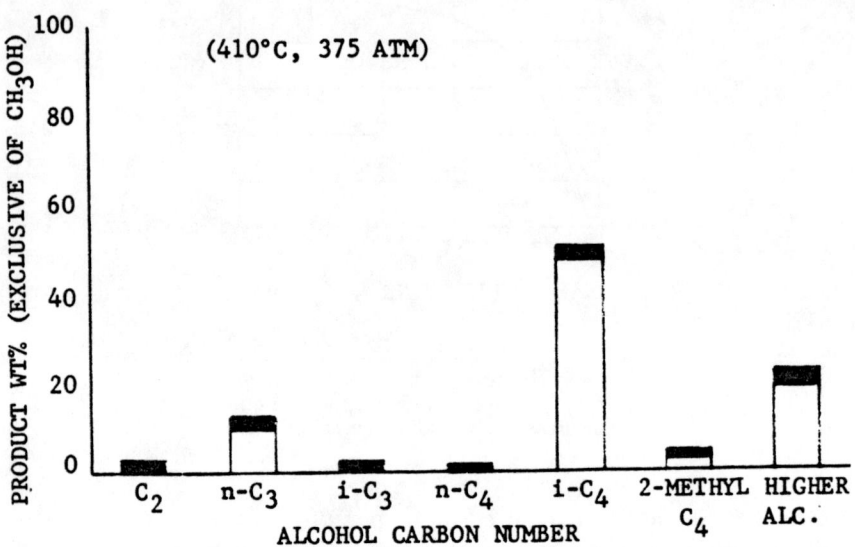

Figure 5. *Product distribution for the alcohol synthesis reaction. Catalyst: 14% K_2O/ZnO (8).*

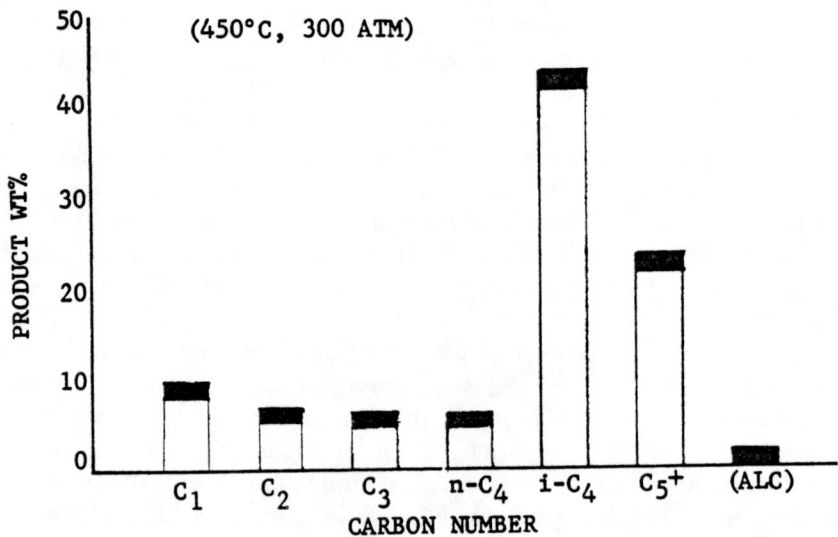

Figure 6. *The isosynthesis reaction favors isobutane formation. Catalyst: ThO_2/Al_2O_3 (9).*

50,000 psi, ethylene glycol can be synthesized directly from CO and H_2. About three-quarters of the polyhydroxy product is composed of ethylene glycol, with the balance composed of propylene glycol and glycerine. Methanol is also produced in this process. This result strongly suggests that new homogeneous catalysts may play an important future role in the catalysis of $CO-H_2$ synthesis reactions.

Although research concerning $CO-H_2$ synthesis reactions spans three-quarters of a century, a number of problems still remain to be solved. As typified by the figures presented, there is still a lack of product selectivity, especially in the synthesis of hydrocarbons other than methane. The capability of selectively forming a specified compound, i.e., ethylene, would be highly desirable. In addition to this problem, all the Group VIII metals are very sensitive to sulfur poisons such as H_2S, and a sulfur-tolerant catalyst would be a major improvement. For example, the methanation processes using nickel catalysts require that the H_2S level in the feed stream be reduced to 0.01–0.1 ppm to achieve satisfactory catalyst lifetimes. Because of this requirement, additional purification steps must be added to attain these very low sulfur levels.

Another problem is catalyst deactivation, which can occur because of sintering of metal particles, coke deposition, and metal carbide formation. Catalysts that are more sinter-resistant are particularly desirable. Also, some of the Group VIII metals form volatile carbonyls. This creates a problem not only because of the toxicity of these carbonyls but also because improper catalyst treatment can result in metal transport within the catalyst bed and even out of the reactor.

Although many studies involving heterogeneous catalysts for $CO-H_2$ reactions have been conducted, most of the data were obtained before the advent of the sensitive analytical techniques available today. Therefore, large product yields were required for product analyses, which were frequently represented in terms of distillation fractions. Detailed product distributions were not commonplace. Reactors were usually operated at high conversions, thereby providing kinetic data from integral reactors. These data are not so easily interpreted as data from differential reactors since heat and mass transfer effects, secondary reactions, and product inhibition can complicate kinetic analysis. In addition, no specific activity rate data had been determined in these earlier studies because chemisorption techniques had not been developed to the point where they were utilized as a routine catalyst characterization procedure. The comparison of rates on the basis of unit metal surface area or per metal surface site, i.e. turnover numbers, is the only meaningful way to compare the intrinsic activity of different metal catalysts. Because of these limitations still existing in the 1970s, a study was initiated to provide this necessary information for the first time.

Table III. Dispersion Variations for Different Metal Catalysts (11)

Catalyst	% Metal Dispersion
15% Fe/Al_2O_3	2[a]
5% Ru/Al_2O_3	6[b]
2% Co/Al_2O_3	8[a]
5% Ni/Al_2O_3	13[b]
2% Pd/Al_2O_3	22[a]
1% Rh/Al_2O_3	48[b]
1.75% Pt/Al_2O_3	88[a]
2% Ir/Al_2O_3	90[a]

[a] Assuming bridged bonding of CO.
[b] Assuming linear bonding of CO.

Supported Metal Catalysts

The behavior of Group VIII metals dispersed on a variety of typical metal oxide supports has been investigated in the CO–H_2 synthesis reaction (11). Dispersing a metal on a support is advantageous not only because it results in the formation of very small metal crystallites, thereby increasing the surface area per gram of the metal component, but also because the metal surface is stabilized under reaction conditions.

It was necessary to study these well-characterized Group VIII metal catalysts in order to make meaningful kinetic comparisons. Chemisorption of H_2 and CO measured the surface area of the reduced metal, thereby allowing the calculation of both specific activity, expressed as turnover numbers, and metal dispersion, which is the ratio of surface metal atoms to the total number of metal atoms in the catalyst. Table III shows the dispersion data for a series of alumina-supported metals. The more noble Group VIII metals are typically better dispersed. The wide

Table IV. New vs. Old Data for Methanation Activity (11)

Fischer et al. (1925) CH_4 Formation/g Metal	Vannice	
	Metal	N_{CH_4} @ 275°C (sec^{-1})
Ru	Ru	0.181
Ir	Fe	0.057
Rh	Ni	0.032
Ni	Co	0.020
Co	Rh	0.013
Os	Pd	0.012
Pt	Pt	0.0027
Fe	Ir	0.0018
Pd		

variation in dispersion values illustrates the importance of characterizing catalysts in this manner.

A differential, flow microreactor was operated at steady-state conditions, and conversion data free from heat and mass transfer limitations were obtained (11). Even at these low conversions, typically less than 5%, accurate product distributions could be determined by gas chromatography using sub-ambient temperature programming. Specific activities for the Group VIII metals are listed in Table IV. These results for the methanation reaction are compared with those from the only other kinetic study which encompassed all the Group VIII metals—that of Fischer in 1925 (12). Techniques to measure metal surface areas were not yet available, and these activities were compared on a gram-metal basis with no correction for surface area differences. When this correction is made, significant changes occur in the ordering of activity, especially for Fe and Ir. Another important feature is that a factor of only 100 in specific activity separates the least active and the most active metals in the methanation reaction. The same ordering of activity exists for total CO conversion (11).

Even at these low conversions, higher molecular weight products are easily detected when present, and these metals exhibit selectivity trends representative of their behavior under typical processing conditions. As expected, an increase in the H_2/CO ratio in the feed stream increases the relative formation of methane.

The kinetic parameters obtained for this series of alumina-supported metals are listed in Table V. By fitting data to a power rate law, activation energies and partial pressure dependencies were obtained. With the exception of Ru and Co, the methanation reaction is about first order in hydrogen and near zero order in carbon monoxide. These data reveal

Table V. Methanation Kinetics over Alumina-Supported Metals as Determined by a Power Rate Law (11)

$$N_{CH_4} = Ae^{E_m/RT} \cdot P_{H_2}^X \cdot P_{CO}^Y$$

Catalyst	N_{CH_4} (sec^{-1})	A (sec^{-1})	E_m (kcal/mol)	X	Y
Ru	0.181	5.7×10^8	24.2	1.6	−0.6
Fe	0.057	2.2×10^7	21.3	1.1	−0.1
Ni	0.032	2.3×10^8	25.0	0.8	−0.3
Co	0.020	9.0×10^7	27.0	1.2	−0.5
Rh	0.013	5.2×10^7	24.0	1.0	−0.2
Pd	0.012	1.2×10^6	19.7	1.0	0
Pt	0.0027	1.6×10^4	16.7	0.8	0
Ir	0.0018	1.4×10^4	16.9	1.0	0.1

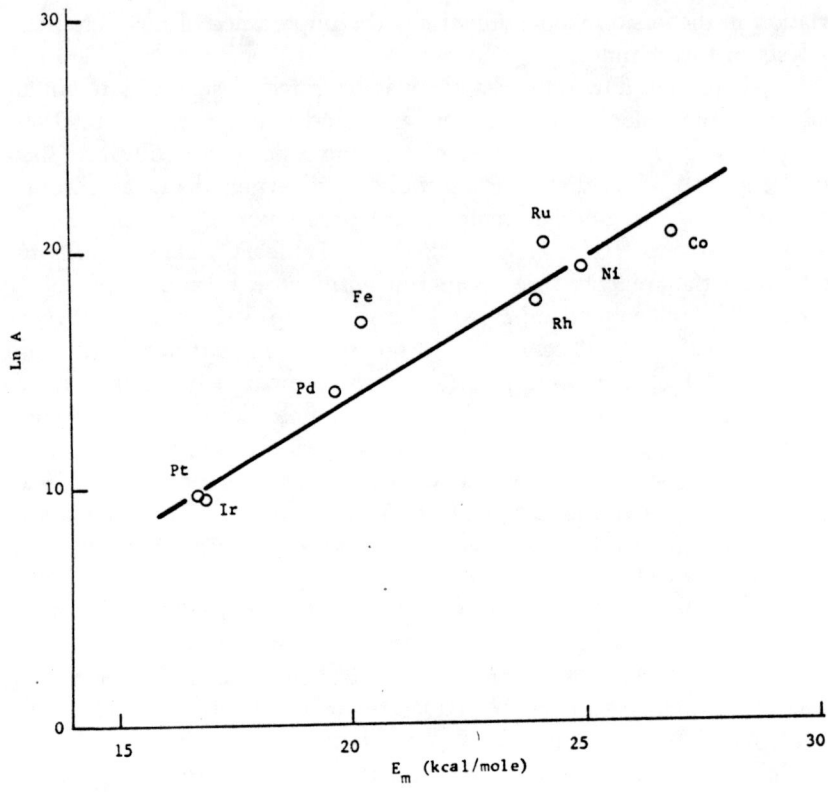

Figure 7. The compensation effect for the methanation reaction (11)

that a compensation effect exists for the methanation reaction as shown in Figure 7. Secondly, with adsorption data from the open literature, a strong correlation was found to occur between specific activity and the heat of adsorption of CO (*11*). This is represented in Figure 8. This correlation appears to be the right-hand portion of the well-known volcano plot. This relationship is an important result since it not only correlates a catalytic property with a physically measurable property, but it also tells us that weakening the metal–CO bond appears to result in higher activity.

Metal Crystallite Size Effects and Metal–Support Interactions

Metals were dispersed on a variety of materials that are typically used as supports, such as Al_2O_3, SiO_2, zeolites, and carbon. It was found that the support can play a very noticeable role in the catalysis of CO–H_2 reactions by influencing the behavior of the metal component (*13*). For

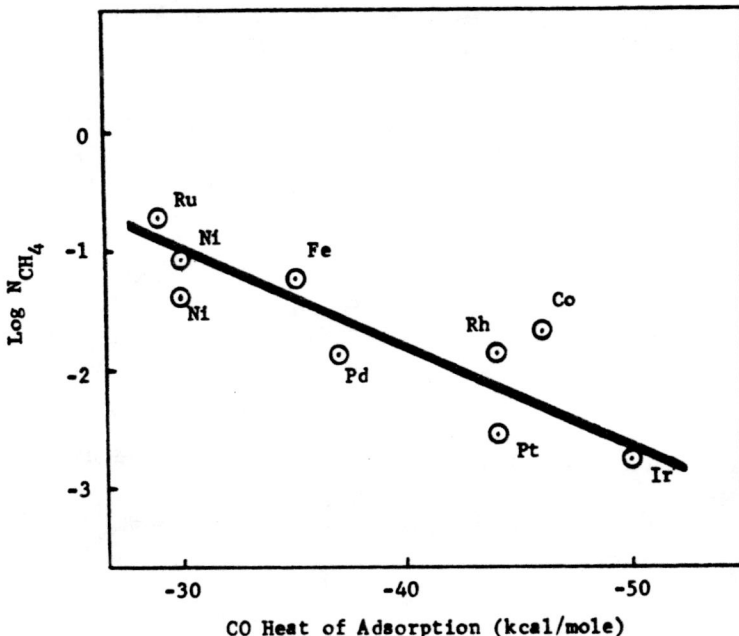

Figure 8. Correlation between methanation activity and CO heat of adsorption for alumina-supported metals. Both values for Ni/η-Al_2O_3 from Table VIII are included.

instance, supported Pt catalysts have a specific activity two orders of magnitude higher than unsupported Pt, as shown in Table VI. The major role of the support in this case appears to be the formation and stabilization of very small Pt crystallites since only small differences in turnover number exist between highly dispersed Pt on different supports. This

Table VI. Effect of Platinum Crystallite Size on Methanation Activity (14)

$H_2/CO = 3, P = 1$ atm

Catalyst	N_{CH_4} @ 275°C (sec^{-1} × 10^3)	Pt Crystallite Diameter (Å)
1.75% Pt/Al_2O_3	2.7	12
1.75% Pt/Al_2O_3 (sint.)	2.2	57
2.0% Pt/SiO_2	1.6	16
Pt Black	< 0.02	3600
25% Pt Black/Al_2O_3 (Physical mixture)	0.018	4300

Journal of Catalysis

Table VII. Methanation Activity Is Affected More by the Support than by Crystallite Size (14)

$H_2/CO = 3, P = 1$ atm

Catalyst	N_{CH_4} @ $275°C$ ($sec^{-1} \times 10^3$)	Pd Crystallite Diameter (Å)
0.5% Pd/H–Y Zeolite	5.9	31
2% Pd/Al$_2$O$_3$	12	48
2% Pd/Al$_2$O$_3$	7.4	82
9.5% Pd/Al$_2$O$_3$	10	120
4.75% Pd/SiO$_2$	0.32	28
Pd Black	0.15	2100

Journal of Catalysis

indicates that the methanation reaction may be structure-sensitive on Pt surfaces.

With Pd catalysts the support seems to play a more important role than just providing small Pd crystallites. Although smaller, supported Pd crystallites were always more active than larger, unsupported crystallites, the activity was much more sensitive to the support itself, as shown in Table VII. No pronounced correlation exists between activity and metal crystallite size, as observed for Pt, but the type of metal oxide support alters specific activity by a factor up to 80 compared with unsupported Pd. A trend does seem to exist, though, between the support acidity and specific activity. An increase in acidity increases the turnover number on Pd, as seen by comparing the two acidic supports, zeolite and alumina, with the nonacidic support, silica.

These effects can be explained by the assumption that any factor that weakens the Pt–CO or Pd–CO adsorption bond will result in an increase in catalytic activity. For Pt catalysts, it has been shown by the chemisorption studies of Freel (14) and the ir spectroscopy investigations of Eichens (15, 16) that the more weakly bound, linearly adsorbed form of CO is favored on small Pt crystallites. This trend then explains why small Pt crystallites are more active than large ones, since based on the correlation in Figure 8 a decrease in bond strength would be expected to enhance activity.

For Pd catalysts, van Harteveld and Hartog (17) have utilized ir spectroscopy to show that the support itself can greatly influence the form of CO adsorbed on the dispersed metal surface. They found that the more strongly bound bridged form of CO is favored on all but the most highly dispersed Pd/SiO$_2$ catalysts. When Pd is supported on Al$_2$O$_3$, however, the linear form of CO still predominates on crystallites as large as 11.5 nm. This alteration of CO adsorption again explains why Pd/Al$_2$O$_3$ catalysts are more active than Pd/SiO$_2$ catalysts. The exact manner

Table VIII. Activity of Nickel Catalysts in the Synthesis Reaction

$H_2/CO = 3, T = 275°C, P = 103$ kPa

Catalyst	N_{CH_4} (sec^{-1})	N_{CO} (sec^{-1})	Dispersion (H/Ni)
8.8% Ni/η-Al$_2$O$_3$	0.085	0.128	0.025
5% Ni/η-Al$_2$O$_3$	0.037	0.044	0.12
42% Ni/α-Al$_2$O$_3$	0.043	0.109	0.018
30% Ni/α-Al$_2$O$_3$	0.018	0.035	0.12
16.7% Ni/SiO$_2$	0.034	0.047	0.085
16.7% Ni/SiO$_2$	0.045	0.062	0.077
20% Ni/Graphite	0.051	0.079	0.031
Ni Metal	0.016	0.018	0.0035

in which the support influences the catalytic behavior of the metal is not clear, and further work is required to provide a better understanding of these metal-support interactions.

The catalytic properties of nickel catalysts also appear to be sensitive to the support, as shown in Table VIII. Although the differences in turnover number are less than an order of magnitude, they still indicate that nickel may be susceptible to the influence of the support or to crystallite size effects. Although they are included for completeness, the 42% Ni/α-Al$_2$O$_3$ and 30% Ni/α-Al$_2$O$_3$ catalysts should not be compared directly with the other alumina-supported catalysts since they were commercial samples. These commercial catalysts can contain promoters which can alter the catalytic behavior of the catalyst. Similar Ni-support effects were found in the ethane hydrogenolysis reaction by Sinfelt and co-workers (18). The comparison of product distributions given in Table IX also demonstrates the support effect. The formation of higher

Table IX. Supported Nickel Crystallites Favor the Production of Higher Molecular Weight Paraffins

$H_2/CO = 3, P = 103$ kPa

Catalyst	$T°C$	% CO Conversion	Mol % of Each Carbon Number				
			C_1	C_2	C_3	C_4	C_5^+
Bulk Ni	229	2.8	90	10	—	—	—
42% Ni/α-Al$_2$O$_3$	236	2.1	76	14	5	3	1
30% Ni/α-Al$_2$O$_3$	229	8.2	81	11	5	2	1
8.8% Ni/η-Al$_2$O$_3$	230	3.1	81	14	3	2	—
5% Ni/η-Al$_2$O$_3$	235	4.9	87	9	3	1	—
16.7% Ni/SiO$_2$	220	5.3	92	5	3	1	—
20% Ni/Graphite	218	7.0	88	9	2	1	0.5

Table X. Specific Activities Measured in Different Laboratories

$H_2/CO = 3, T = 275°C, P = 103$ kPa

Catalyst	N_{CH_4} (sec$^{-1} \times 10^3$)	N_{CO} (sec$^{-1} \times 10^3$)	Reference
5% Ni/Al$_2$O$_3$	37	44	11
15% Ni/Al$_2$O$_3$	35	43	20
Ni/Al$_2$O$_3$	35–75	—	21, 22
Raney Ni	45	115	23
8.8% Ni/Al$_2$O$_3$	85	128	This study
3% Ni/Al$_2$O$_3$	99	119	20
2% Ni/Al$_2$O$_3$	90	220	23
5% Ni/ZrO$_2$	91	170	24
16.7% Ni/SiO$_2$	0.89[a]	—	This study
12.2% Ni/SiO$_2$	0.61[a]	—	25

[a] Measured at 212°C.

molecular weight paraffins, particularly C_3^+ hydrocarbons, is favored by dispersing Ni on a variety of support materials. Unsupported Ni produces only methane and ethane at these conditions. It is encouraging to learn that selectivity can be altered by this synergism between metal and support. Such behavior is perhaps not unexpected, at least in retrospect, since the strength of CO adsorption on Ni has been found to be dependent on crystallite size (17). In addition, O'Neill and Yates (19) have shown by ir spectroscopy that the support can alter the state of CO adsorption on Ni surfaces. Should such effects weaken the Ni–CO bond, we might expect a rate enhancement on that Ni surface, and if all possible synthesis reactions are not altered to the same extent, an overall change in product distribution will occur.

For many years the absence of surface area measurements on catalysts made it impossible to compare directly intrinsic activity in the synthesis reaction. However, the increasing use of specific activity representations is welcome, and the results from five different laboratories for Ni catalysts are shown in Table X (20–25). Corrected to identical temperatures and pressures, turnover numbers for catalysts using the same support agree to within a factor of two. The consistency in these data again illustrates the importance and usefulness of catalyst characterization.

Conclusions

The specific activities of the Group VIII metals for CO hydrogenation have now been measured. This study of well-characterized, supported metals provided not only intrinsic rate data, but also a measure of metal dispersion. The specific activities of these metals were correlated with the heats of adsorption for CO on their surfaces. The conclusion that a decrease in metal–CO bond strength would increase activity appeared to be verified by the behavior of Pt and Pd catalysts. Factors such as metal crystallite size and metal-support interactions, both of which have been shown to alter the bonding state of adsorbed CO on the metal surface, were found to alter catalytic activity. Supported Ni catalysts showed this effect to a lesser extent. With Ni catalysts, however, changes in product distribution occurred which can be attributed to the presence of the support. This is particularly intriguing since alteration and control of product distribution are two of the primary goals in the synthesis reaction. Finally, a comparison of specific activities obtained for Ni catalysts in five different laboratories shows excellent agreement. The increasing trend to report specific activities for this reaction and other reactions is indeed welcome.

The behavior of these different supported catalysts is certainly interesting, but it is not well understood at this time. Regardless, these new findings increase our understanding of $CO-H_2$ reactions occurring on metal surfaces and encourage us in developing new catalysts to alter product selectivity in the synthesis reaction.

Literature Cited

1. Storch, H. H., Golumbic, H., Anderson, R. B., "The Fischer–Tropsch and Related Syntheses," Wiley, New York, 1951.
2. Mills, G. A., Steffgen, F. W., *Catal. Rev.* (1973) **8**, 159.
3. Vlasenko, V. M., Yuzefovich, G. E., *Russ. Chem. Rev.* (1969) **38**, 728.
4. Greyson, M., "Catalysis," Vol. IV, P. H. Emmett, Ed., Reinhold, New York, 1956.
5. Pichler, H., Schulz, H., *Chem. Ing. Tech.* (1970) **42**, 1162.
6. Pichler, H., Hector, A., *Kirk-Othmer Encycl. of Chem. Technol.* (1964) **4**, 446.
7. Natta, G., "Catalysis," Vol. III, P. H. Emmett, Ed., Reinhold, New York, 1955.
8. Natta, G., Colombo, U., Pasquon, I., "Catalysis," Vol. V, P. H. Emmett, Ed., Reinhold, New York, 1957.
9. Cohn, E. M., "Catalysis," Vol. IV, P. H. Emmett, Ed., Reinhold, New York, 1956.
10. Pruett, R. E., Walker, W. E., U.S. Patent No. **3,833,634**, 1974.
11. Vannice, M. A., *J. Catal.* (1975) **37**, 449, 462.
12. Fischer, F., Tropsch, H., Dilthey, P., *Brennst. Chem.* (1925) **6**, 265.
13. Vannice, M. A., *J. Catal.* (1975) **40**, 129.
14. Freel, J., *J. Catal.* (1972) **25**, 139, 149.
15. Eichens, R. P., *Adv. Catal.* (1958) **10**, 1.
16. Darensbourg, D. J., Eichens, R. P., *Proc. Int. Congr. Catal. 5th* (1973) **1**, 371.
17. Van Hardeveld, R., Hartog, F., *Adv. Catal.*, **22**, D. D. Eley, H. Pines, and P. B. Weisz, Eds., Academic, New York, 1972.
18. Taylor, W. F., Yates, D. J. C., Sinfelt, J. H., *J. Phys. Chem.* (1964) **68**, 2962.

19. O'Neill, C. E., Yates, D. J. C., *J. Phys. Chem.* (1961) **65,** 901.
20. Bartholomew, C. H., *Quarterly Tech. Prog. Report to ERDA,* Apr. 22–July 22, 1975 (FE-1790-1).
21. Bousquet, J. L., Teichner, S. J., *Bull. Soc. Chim. Fr.* (1969) 2963.
22. *Ibid.* (1972) 3689.
23. Dalla Betta, R. A., Piken, A. G., Shelef, M., *J. Catal.* (1975) **40,** 173.
24. Dalla Betta, R. A., Piken, A. G., Shelef, M., *J. Catal.* (1974) **35,** 54.
25. Fontaine, R., Ph.D. Dissertation, Cornell Univ., 1973.

RECEIVED October 4, 1976.

3

Photoelectrochemical Production of Hydrogen

J. O'M. BOCKRIS and K. UOSAKI

School of Physical Sciences, The Flinders University of South Australia, Bedford Park, South Australia 5042

The recent work on the photoelectrochemical production of hydrogen is surveyed briefly. Theoretical expressions for photocurrents at p- and n-type semiconductors are set up under the assumption that the rate-determining step is charge transfer. The photocurrents are estimated by using typical characteristics of semiconductor electrodes. Calculated quantum efficiencies are from 10^{-6} to 10% for the semiconductors considered. The theoretical results for TiO_2 can be made consistent with the observations if surface states are assumed and the rate-determining step is considered to be dependent on the supply of holes to the surface. Photocurrents and hence photo-hydrogen yields are calculated for a typical cell. Optimization conditions are deduced. Quantum efficiencies of TiO_2 single crystal are measured as functions of wavelength and electrode potential. TiO_2 films are made by heating Ti in air, and the quantum efficiencies of these films are compared with the results of TiO_2 single crystal. Iron titanate gives results similar to those of TiO_2. A protective TiO_2 film is made on CdS ($E_g = 2.5\ eV$) by chemical vapor deposition, and photoelectrochemical reactions are carried out on the specimens. Anodic dissolution of the substrate (CdS) is greatly reduced by the TiO_2 coating. The quantum efficiency was greater than that for TiO_2.

An efficient process for the photoelectrochemical production of hydrogen would contribute significantly to the solution of the problems of future energy requirements and their drain on the limited availability of fossil fuels.

The founding paper of semiconductor electrochemistry by Brattain and Garrett (1) in 1955 examined the electrochemical properties of germanium. The first detailed theoretical study of semiconductor electrochemistry made by Green (*2*) introduced electrode kinetics into the field. Since then many studies of semiconductor electrodes have been carried out. The main groups during the 1960s were Gerischer et al. (*3, 4, 5, 6*), Pleskov et al. (*7*), and Memming et al. (*8, 9*). Dewald (*10, 11*) also contributed significantly to the experimental and theoretical understanding of the semiconductor–electrolyte interface. Table I lists some of the semiconductors which have been investigated photoelectrochemically.

Table I. List of Semiconductors Investigated Photoelectrochemically

Semiconductor	Type	Energy Gap (eV)	Ref.
Ge	n,p	0.66	*1*
Si	n,p	1.1	*12–18*
GaAs	n,p	1.35–1.43	*5, 7, 19–21*
^{a}[CdTe	n,p	1.44–1.5	*22*]
CdSe	n	1.74	*23, 24*
[CuO		1.95	*22*]
[CdO		2.2	*22*]
Cu_2O	p	2.2	*23, 25, 26*
Fe_2O_3	n	2.2	*27*
V_2O_5	n	2.23	*28, 29*
ZnTe	p	2.26	*23, 30*
GaP	n,p	2.25–2.35	*15, 31*
CdS	n	2.4	*23, 32–34*
ZnSe	n	2.6	*35*
WO_3	n	2.7	*36*
α-SiC	n,p	2.75–3.1	*37*
TiO_2	n	3.0	*38–42*
ZnO	n	3.2	*43–46*
$SrTiO_3$	n	3.2	*47, 48*
$KTaO_3$	n	3.5	*49*
SnO_2	n	3.7	*50, 51*
[NiO	p	4.0	*52, 53*]
Ta_2O_5	n	4.6	*54*

^{a} Semiconductors shown in brackets were examined in dark only.

Fujishima and Honda (*40*) first pointed out the possible application of an irradiated semiconductor–electrolyte system when they reported the photo-evolution of hydrogen for a platinum cathode in a cell involving an irradiated TiO_2 (reduced) anode. Mavroides et al. (*55*) reported that the photocurrent observed by Fujishima and Honda (*40*) was caused by oxygen reduction at the cathode. Later Fujishima, Kohayakawa, and Honda (*56*) obtained hydrogen by introducing a pH gradient (pH = 14 near TiO_2, pH = 0 near platinum). They reported on the results of various methods of making photoactive titanium dioxide in addition to that of a single crystal. When they made the film anodically, the current was one-tenth that on the single crystal because the oxide was too thin to absorb light; however, with a thermally produced crystal, when the film was thicker than that formed anodically, the currents were the same as that on a single crystal. Hydrogen production in sunlight gave 0.4% efficiency.

Yoneyama et al. (57) used titanium dioxide and gallium phosphide as electrodes, the latter acting as the cathode. Both electrodes were irradiated. The experiments were carried out in homogeneous solutions, both in 1N sulfuric acid and 1N sodium hydroxide. They obtained hydrogen and oxygen, but the gallium phosphide deteriorated; current densities fell to one tenth of the original value in ca. one hour.

Hardee and Bard (27) compared these single-crystal results for photoeffect on TiO_2 with those for films of TiO_2 prepared from the vapor phase; the single-crystal results gave four times more current than those from the thinner films.

Wrighton et al. (58) set up a variety of TiO_2 (single crystal)/Pt cells with different pH gradients and confirmed the evolution of O_2 at the TiO_2 anode and H_2 at the Pt cathode using a mass spectroscope. Using a power supply (0.5–2.0 V) or a pH gradient, substantial currents (0.5 mA cm^{-2}) were obtained. Doping the TiO_2 with tungsten moved the response to the blue region (i.e., the wrong direction to increase the efficiency); neverthless the qualitative result is important.

Ohnichi et al. (59) repeated the work of Fujishima and Honda and reported a quantum yield of 10^{-3}. For the first time the importance of the position of the flat band potential was related to the overall quantum efficiency of the cell.

Nozik (60) investigated the quantum efficiencies of TiO_2 (single crystal)/Pt cells using external "biasing" and both homogeneous and heterogeneous [i.e., different pH at cathode (pH = 0) and anode (pH = 14)]. Conversion efficiencies of up to 10% were observed.

Mavroides et al. (55) compared the quantum efficiency with wavelength curves for TiO_2 single crystals, polycrystals, and thermally produced thin films and obtained similarly shaped curves for all anodes though the relative intensities of the absorption peaks for TiO_2 thin films depend on the film thickness and preparation procedure. Keeney et al. (61) have also reported photoelectrochemical sensitivity of thermally produced TiO_2 thin films.

Mavroides et al. (47) and Watanabe et al. (48) have reported that a $SrTiO_3$/Pt can act as a photo-self driven cell in a homogeneous medium without the application of an external bias. The energy gap of $SrTiO_3$ (3.2 eV) is too high to be useful under solar irradiation.

Gerischer (62) has pointed out that in principle all p- and n-type semiconductors can be used for a photocell electrode when a suitable reversible redox system is found which induces the formation of a depletion of a space–charge layer. The problems lie in finding a semiconductor with a suitable energy gap (1.3–1.8 eV) that is stable in solution and whose critical photopotential (i.e., potential at which the onset of a photocurrent is observed) is sufficiently negative for the electrode to be used in a photo-driven cell without the application of an external bias. For a comprehensive coverage of earlier works on semiconductor electrochemistry there are several books (62, 63) and reviews (2, 64, 65) available.

Theory of Hydrogen Production in a Photoelectrochemical Cell

Photoeffect on Electrode Reactions at a Semiconductor–Solution Interface. PHOTOEFFECT ON A CATHODIC CURRENT AT A p-TYPE SEMICONDUCTOR–SOLUTION INTERFACE. There are many holes in the valence band and few electrons in the conduction band of p-type semiconductors without illumination. Illumination with light, the energy per photon of which is larger than the energy gap of the semiconductor, creates electrons in the conduction band and holes in the valence band (Figure 1). However, since there are many holes in the valence band without illumination and usually the number of holes created by light is small compared with the

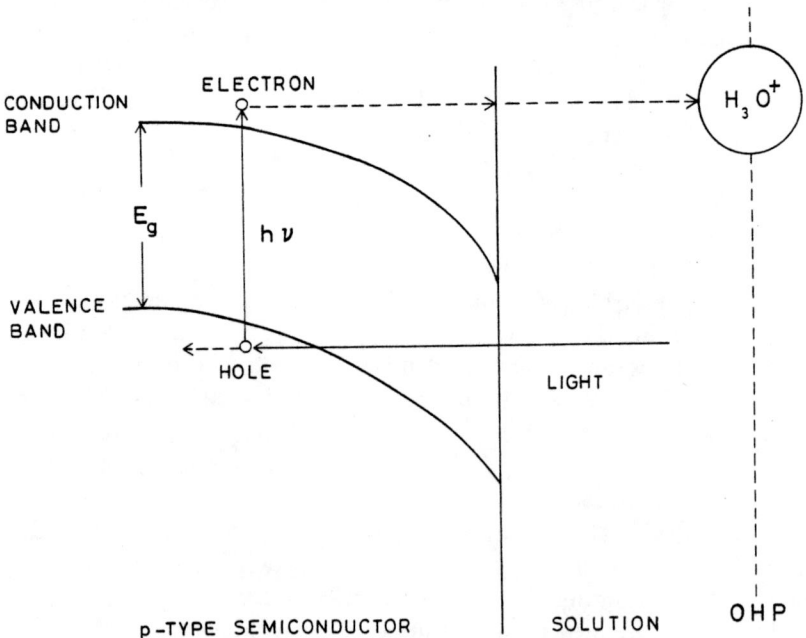

Figure 1. Schematic of the photoeffect on a cathodic current at a p-type semiconductor–solution interface

number of holes without light, the values of the anodic current via the valence band with and without illumination are nearly the same. On the other hand, there are a very few electrons in the conduction band and therefore a very low cathodic current via the conduction band without illumination. Consequently, a photoeffect would be expected via the conduction band for p-type semiconductors and would be cathodic.

In hydrogen evolution reaction in acidic solutions one assumes a rate-determining step of

$$SC + H_3O^+ \rightarrow SC - H + H_2O \tag{1}$$

The photocurrent corresponding to this reaction i_p, is given by (66):

$$i_p = e_0 \frac{C_A}{C_T} \int_0^\infty N_e(E) G_A(E) W_c(E) \, dE \tag{2}$$

where $N_e(E)$ is the number of electrons with energy E which strike the semiconductor surface per unit time and area; $G_A(E)$ is the distribution function for the vibrational-rotational states of an acceptor, H_3O^+, at energy E; $W_c(E)$ is the WKB tunneling probability of electrons through the potential barrier at energy, E; C_A is the number of electron acceptors per unit area in the outer Helmholtz plane (OHP); C_T is the total number of sites per unit area in the OHP; e_0 is the electronic charge. Energy levels are taken as zero at the bottom of the conduction band.

Enthalpy Change for Electron Transfer from Semiconductor to H_3O^+. The standard enthalpy change $\Delta H(e)$ for the electron transfer reaction corresponding to Equation 1 from the bottom of the conduction band of the semiconductor to the proton, when the proton–solvent system is in its ground state, can be obtained by using the following thermodynamic cycle:

$$\begin{array}{ccc}
& \Delta H(e) & \\
\text{P-S.C}(e) + H_3O^+ & \longrightarrow & \text{P-S.C--H--H}_2O \\
\uparrow L_0 & & \downarrow -R \\
\text{P-S.C}(e) + H^+ + H_2O & & \text{P-S.C--H} + H_2O \\
\uparrow -\chi & J & \downarrow -A \\
\text{P-S.C} + e + H^+ + H_2O & \longleftarrow & \text{P-S.C} + H + H_2O
\end{array}$$

where R, A, J, χ, and L_0 represent respectively the H–H$_2$O repulsive energy (R), the heat of adsorption of a hydrogen atom on the semiconductor (A), the ionization potential of the hydrogen atom (J), the electron affinity of the semiconductor (χ), and the hydration energy of a proton (L_0). Therefore,

$$\Delta H(e) = -L_0 + \chi - J + A + R \tag{3}$$

The values of L_o, R, and J are -261 kcal/mol (67), -3 kcal/mol [the value for the H atom in ice (68)], and 313 kcal/mol (67), respectively. The value of χ is 78–115 kcal/mol depending on the semiconductor (69). The value of A is -6 kcal/mol to -7 kcal/mol depending on the semiconductor. Therefore the estimated value of $\Delta H(e)$ is 16–54 kcal/mol.

Photon Absorption Step (66). The number of photons absorbed by the semiconductor from light normal to the surface between x and $x + dx$ from the surface, $N_{ph}(x)$, is given by:

$$N_{ph}(x) = I_o(1 - R)\alpha_{ph}e^{-\alpha_{ph}x}dx \quad (4)$$

where I_o is the total number of photons of incident light (cm^{-2} sec^{-1}), R is the reflectivity of the semiconductor, and α_{ph} is the absorption coefficient of the semiconductor at the wavelength λ (cm^{-1}).

Creation of Excited Electrons. Each absorbed photon, the energy of which is greater than the energy gap of the semiconductor, makes an excited electron in the conduction band and a hole in the valance band (Figure 1). The number of excited electrons, the energy of which is between E and $E + dE$ in the conduction band, between x and $x + dx$, $N_e(E,x)$ is (66):

$$N_e(E,x) = N_{ph}(x) \cdot \frac{\rho(E,x)\{1 - f(E)\}\rho(E - h\nu,x)f(E - h\nu)dE}{\int_{E_{c,x}}^{\infty} \rho(E,x)\{1 - f(E)\}\rho(E - h\nu,x)f(E - h\nu)dE}$$

i.e., from Equation 4

$$= I_o(1 - R)\alpha_{ph}e^{-\alpha_{ph}x}\frac{\rho(E,x)\{1 - f(E)\}\rho(E - h\nu,x)f(E - h\nu)dE}{\int_{E_{c,x}}^{\infty} \rho(E,x)\{1 - f(E)\}\rho(E - h\nu,x)f(E - h\nu)dE}dx \quad (5)$$

where $E_{c,x}$ is the energy of the bottom of the conduction band at x and $\rho(E,x)$ is the density of states at E at x and given by Equation 6 (70).

$$\rho(E,x) - (8\pi m^*/h^3)S(E,x) = \begin{cases} (8\pi m^*/h^3)\sqrt{E - E_{c,x}} & \text{when } (E \geq E_{c,x}) \\ 0 & (E_{c,x} \geq E \geq E_{v,x}) \\ (8\pi m^*/h^3)\sqrt{E_{v,x} - E} & (E \leq E_{v,x}) \end{cases} \quad (6)$$

Here $f(E)$ is the Fermi distribution function, and m^* is the effective mass of the electron in the semiconductor.

Potential Profile in a p-Type Semiconductor. The number of excited electrons which reach the electrode surface depends not only on the gradient of the chemical potential of the excited electrons conduction band but also on the internal electric field. The potential-distance profile inside a semiconductor may be obtained by solving Poisson's Equation 7.

$$\frac{d^2V_x}{dx^2} = \frac{4\pi e_0}{\epsilon}\left[p_0 \exp\left\{-\frac{e_0(V_x - V_B)}{kT}\right\} - n_0 \exp\left\{\frac{e_0(V_x - V_B)}{kT}\right\} + N_D - N_A\right] \quad (7)$$

where e_0 is the electronic charge; ϵ is the dielectric constant of the semiconductor; p_0, n_0, and V_B are, respectively, the concentration of holes and electrons and the potential in the bulk of the semiconductor in the region beyond the space charge (when $x \to \infty$). V_x is the potential at x; N_D and N_A are the concentrations of donor levels and acceptor levels, respectively (values are in the Gaussian system of units). The donors and the acceptors are assumed to be completely ionized and immobile.

Evaluation gives the following equation from Equation 7 (62, 71)

$$\frac{dV_x}{dx} = \pm\sqrt{\frac{8\pi kT}{\epsilon}\left[-(N_D - N_A)y + p_0(e^{-y} - 1) + n_0(e^y - 1)\right]} \quad (8)$$

where y represents

$$y = \frac{e_0(V_x - B_B)}{kT} \quad (9)$$

Numerical solutions have been found for V as a function of x. Results are shown in Figure 2. The parameters used in the calculation are $N_A = p_0 = 10^{15} - 10^{17}/cm^3$; $N_D = n_0 = 0/cm^3$; $\epsilon = 20.0$, and $V_B - V_S = 0-1.0$ V, where V_B and V_S are the potential in the bulk and at the surface of semiconductor, respectively.

Number of Excited Electrons Arriving at the Surface. The number of photo-excited electrons, originally expressed for the distance, x_1 in Equation 5, decreases to $N_{e,x\text{-}dx}(E,x)$ given by Equation 10, after travelling dx.

$$N_{e,x\text{-}dx}(E,x) = N_e(E,x) e^{-\frac{1}{L(x)}dx}, \quad (10)$$

where $L(x)$ is the mean free path of electrons at x and is given by Equation 11 (72).

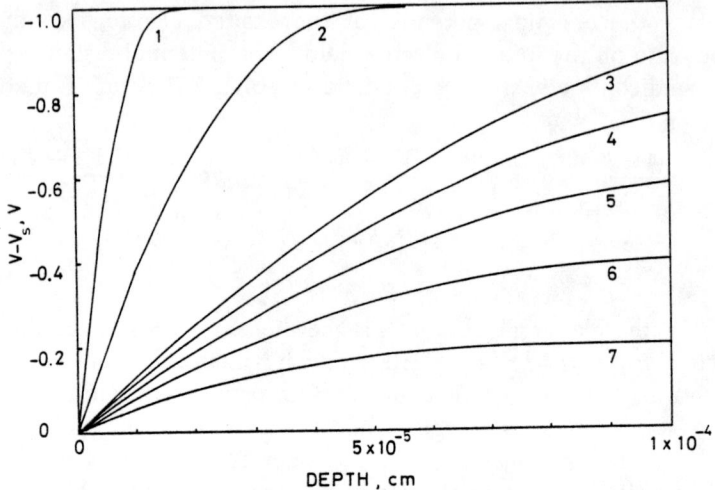

Figure 2. Potential-distance profile inside a p-type semiconductor. 1. $N_A = p_o = 10^{17}/cm^3$ $V_B - V_S = -1.0$ V. 2. $N_A = p_o = 10^{16}/cm^3$ $V_B - V_S = 1.0$ V. 3-7. $N_A = p_o = 10^{15}/cm^3$. 3. $V_B - V_S = -1.0$ V. 4. $V_B - V_S = -0.8$ V. 5. $V_B - V_S = -0.6$ V. 6. $V_B - V_S = -0.4$ V. 7. $V_B - V_S = -0.2$ V.

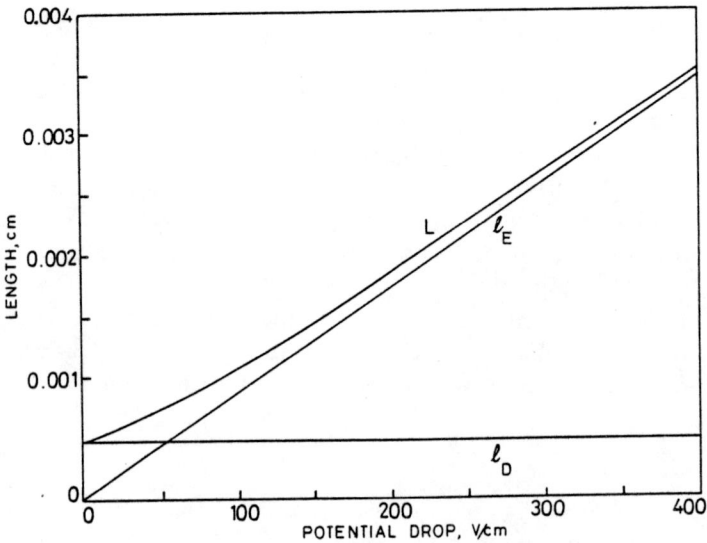

Figure 3. Dependence of mean free path, diffusion length, and drift length of electron on potential drop

$$L(x) = \frac{2l_D^2}{\sqrt{l_E^2 + 4l_D^2} - l_E} \tag{11}$$

where l_D is the diffusion length and l_E is the drift length. The terms l_D and l_E are given by:

$$l_D = \sqrt{D\tau_e} = \sqrt{300\mu_e \tau_e kT/e_o} \tag{12}$$

$$l_E = \tau_e \mu_e V'(x) \tag{13}$$

respectively, where μ_e is the mobility of the electron in the semiconductor (cm^2/V · sec), τ_e is the lifetime of the electron, and $V'(x)$ is the potential drop at $x[= (dV/dx)_x;$ V/cm]. The parameters k and e_o are in cgs units.

When $V'(x) = 0$, $L(x)$ becomes l_D and when $V'(x)$ is sufficiently high, $l_E \gg l_D$, and $L(x)$ becomes l_E (Figure 3). Here, L changes after travelling dx corresponding to a change of $V'(x)$.

Similarly,

$$N_{e,x=x-2dx}(E,x) = N_{e,x=x-dx}(E,x) e^{-\frac{1}{L(x-dx)} dx}$$
$$= N_e(E,x) e^{-\left(\frac{1}{L(x)} + \frac{1}{L(x-dx)}\right) dx} \tag{14}$$

$$N_{e,x=x-3dx}(E,x) = N_{e,x=x-2dx}(E,x) e^{-\frac{1}{L(x-2dx)} dx}$$
$$= N_e(E,x) e^{-\left(\frac{1}{L(x)} + \frac{1}{L(x-dx)} + \frac{1}{L(x-2dx)}\right) dx} \tag{15}$$

and so on.

After N steps ($N = x/dx$),

$$N_{e,x=0}(E,x) = N_{e,x=dx}(E,x) e^{-\frac{1}{L(dx)} dx}$$
$$= N_e(e,x) \exp\left\{-\left(\frac{1}{L(x)} + \frac{1}{L(x-dx)} + \cdots + \frac{1}{L(dx)}\right) dx\right\} \tag{16}$$

$N_{e,x=0}(E,x)$ gives the number of electrons at the surface which were excited between x and $x + dx$, the energy of which is between E and $E + dE$.

The total number of electrons, $N_e(E)$, which has an energy between E and $E + dE$ at the surface is given by

$$N_e(E) = \int_0^\infty N_{e,x=0}(E,x)\,dx \qquad (17)$$

Electron Transfer Process.

(a) Construction of the potential energy barrier.

(i) Interaction of an electron with the dipole potential of adsorbed water. The work required for the electron to cross the water layer is given as the negative value of the surface potential made by the waters, multiplied by e_o. The concentration of surface state is assumed to be less than $10^{14}/cm^2$; the charge on the electrode is almost zero (2). The surface potential caused by adsorbed water dipoles at the potential of zero charge was calculated (73) and is 0.03–0.04 at Hg, Cd, and Zn electrodes. The diameter of the water molecule is 2.76 Å.

(ii) Interaction with ions in the OHP and their images. When a photo-excited electron leaves the semiconductor surface, it interacts with all the ions in the OHP and their images in the semiconductor. The Coulombic force between this electron (x from electrode) and all ions in OHP and their images, $F(x)$ is shown in Equation 18 (74).

$$F(x) = -\frac{e_o^2}{(d-x)^2 \epsilon_{op}} + \frac{e_o^2}{(d+x)^2 \epsilon_{op}} \frac{\epsilon - \epsilon_{op}}{\epsilon + \epsilon_{op}}$$

$$-\frac{2\pi e_o^2}{\epsilon_{op}} \sum_{n=1}^{\infty} \left(\frac{1}{(d-x)^2 + n^2 R_i^2} - \frac{1}{(d+x)^2 + n^2 R_i^2} \frac{\epsilon - \epsilon_{op}}{\epsilon + \epsilon_{op}} \right) \qquad (18)$$

where d is the distance between the semiconductor surface and OHP, ϵ_{op} is the optical dielectric constant of water, $n = 1, 2, 3$ (and represents the succession of rings of ions around a given central ion), and R_i is the distance between two ions in the OHP depending on its coverage with ions and determined by $R_i = 4r_i/(\pi\theta)^{1/2}$, where θ is the coverage and r_i is the radius of the ions.

The potential energy of an electron at x caused by this force, $U_{ion}(x)$, is given by:

$$U_{ion}(x) = \int_0^x F(x)\,dx \qquad (19)$$

and this value is obtained by numerical integration.

(iii) Optical Born energy of the electron. In the region outside that of the water layer attached to the electrode it is difficult to calculate the electron interaction with the water upon a modelistic basis because of limited knowledge concerning the water structure. The energy of an electron with respect to vacuum can be estimated when it has entered this water layer by means of the Born equation, with an optical dielectric constant:

$$U_{opt.Born} = -\frac{e_0^2}{2r_c}\left[1 - \frac{1}{\epsilon_{op}}\right] \tag{20}$$

The value of r_c in this equation is taken from the S.C.F. calculations of Fueki et al. as 1 Å (75). The value of $U_{opt.Born}$ is then -3.2 eV.

(iv) Calculation of the barrier maximum. The potential energy of the electron with respect to the bottom of the conduction band is given by

$$U = \chi - U_{H_2O} + U_{ion} + U_{opt.Born} \tag{21}$$

where U_{H_2O} is the interaction energy between an electron and an adsorbed water dipole layer. A potential energy profile is given as a function of distance in Figure 4 assuming d is 6.5 Å (76) in Equation 18. The maxi-

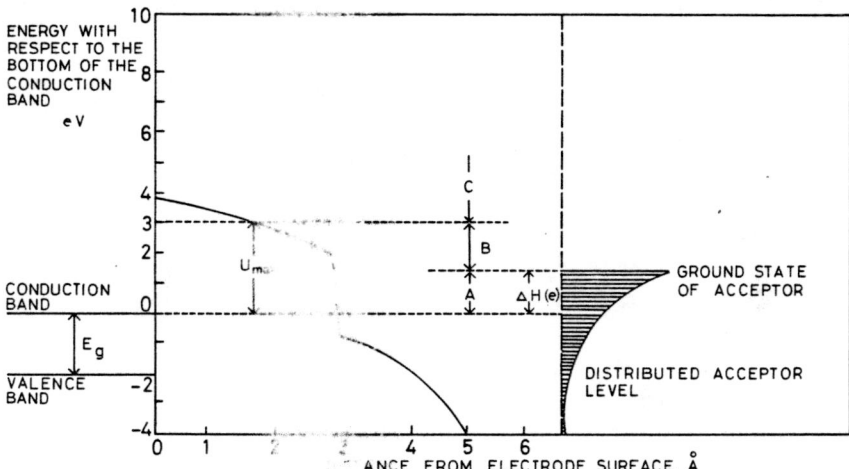

Figure 4. Schematic of the potential energy barrier for electron transfer from a p-type semiconductor

mum value of the barrier height, U_{max}, depends on the electron affinity of the semiconductor χ and is given (in eV) by:

$$U_{max} = \chi - 0.2 \tag{22}$$

(b) Equation for the photocurrent. The equation for the current density can then be divided into three parts. In section A of Figure 4 the photoelectrons pass through the barrier and are accepted by H_3O^+ ions in solution. The current for this section, $i_{p,A}$, is given by:

$$i_{p,A} = e_0 \frac{C_A}{C_T} \int_0^{\Delta H(e)} N_e(E)\, G_A(E)\, W_c(E)\, dE$$

$$= e_0 \left\{ \frac{C_A}{C_T} \int_0^{\Delta H(e)} N_e(E) \exp[-\beta(\Delta H(e) - E)/kT] \exp\left[-\frac{\pi^2 l}{h}(U_{max} - E) \right] dE \right\} \quad (23)$$

where $N_e(E)$ is given on p. 42, l is barrier width. In section B the photoelectrons do not find an acceptor state in H_3O^+ but in water. The current density for the B region is represented by:

$$i_{p,B} = e_0 \frac{C_A}{C_T} \left\{ \int_{\Delta H(e)}^{U_{max}} N_e(E)\, W_c(E)\, dE \right\} = e_0 \frac{C_A}{C_T} \left\{ \int_{\Delta H(e)}^{U_{max}} N_e(E) \exp\left\{ -\frac{\pi^2 l}{h}(U_{max} - E) \right\} dE \right\} \quad (24)$$

Finally, some photoelectrons may pass over the barrier and into the solvent water, and the current density from this contribution is given by:

$$i_{p,C} = e_0 \frac{C_A}{C_T} \left\{ \int_{U_{max}}^{\infty} N_e(E)\, dE \right\} \quad (25)$$

In total, the photocurrent is given by:

$$i_p = i_{p,A} + i_{p,B} + i_{p,C} \quad (26)$$

THE PHOTOEFFECT ON THE ANODIC CURRENT AT n-TYPE SEMICONDUCTORS. By analogy with the arguments on p. 37 anodic photocurrents can be considered to arise through the donation of electrons by ions in solution to holes in the valency band of n-type semiconductors. The mechanism of anodic reactions is less well known than that of cathodic reactions. One complication is that there may be two reactions corresponding to the anodic current. One is the anodic dissolution of the semiconductor itself, and another is the oxidation of a species (e.g., OH^-, water, halogen ion) in the solution.

Let us suppose that the anodic current corresponds to the oxidation of OH^- and water and that the rate-determining steps are:

$$n\text{-S.C(hole)} + OH_{aq}^- \rightarrow n\text{-S.C--OH--H}_2O \quad (27)$$

$$n\text{-S.C(hole)} + H_2O \rightarrow n\text{-S.C--OH--H}_3O^+ \quad (28)$$

Estimation of the Standard Enthalpy Change for Electron Transfer from the Donor to a Hole in the Valence Band. The standard enthalpy change $\Delta H_1'(e)$ for an electron transfer reaction corresponding to Equation 27 from the electron level of the hydroxyl ion to the top of the valence band of the semiconductor, when the hydroxyl ion-solvent system is in its ground state, is obtained using a thermodynamic cycle similar to that for electron transfer from the conduction band to an acceptor.

One finds

$$\Delta H'(e) = -L_o' + E.A. - \chi - E_g + A' + R' \qquad (29)$$

The values of R', L_o', and $E.A.$ are -3 kcal/mol (*see* p. 37), -87 kcal/mol (77), and 42 kcal/mol (78), respectively. The value of χ is 78–115 kcal/mol depending on the semiconductor (69). Although the value of E_g can vary over a wide range, 30–70 kcal/mol is a reasonable range of choice for this purpose because too large an E_g reduces photon absorption in the solar spectrum range. The value of A' ranges from -3 kcal/mol to -4 kcal/mol. Therefore the estimated value of $\Delta H_1'(e)$ is $-69- + 15$ kcal/mol.

Similarly, the standard enthalpy change $\Delta H_2'(e)$ for an electron transfer reaction corresponding to Equation 28 from the electron level of the water molecule to the top of the valence band of the semiconductor, when the H–OH bond system is in its ground state, can be obtained using a cycle as before, and is:

$$\Delta H_2'(e) = L_o + A' - \chi - E_g + J + E_{\text{Diss}} + E_{\text{vap}} \qquad (30)$$

The values of L_o, J, E_{Diss}, and E_{vap} are -261 kcal/mol (67), 313 kcal/mol (67), 119 kcal/mol (79), and 10.4 kcal/mol (80), respectively. The value of χ is 78–115 kcal/mol depending on the semiconductor (69). Although the value of E_g can vary in wide ranges, 25–69 kcal/mol is chosen because of the reason mentioned above. Also, the value from -3 kcal/mol to -4 kcal/mol taken for A' and $\Delta H_2'(e)$ then varies from -12 kcal/mol to 71 kcal/mol.

The Creation of Holes. Each photon, the energy of which is greater than the energy gap of the semiconductor, creates an electron in the conduction band and a hole in the valence band. The number of created holes, the energy of which is between $(E - h\nu)$ and $(E - h\nu) + dE$ in the valence band and at distances from the electrode surface between x and $x + dx$, $N_h(E - h\nu, x)$, is equal to the number of created electrons the energy of which is between E and $E + dE$ in the conduction band and at a distance between x and $x + dx$, $N_e(E, x)$. Hence, using Equation 5, $N_h(E, x)$ is given by:

$$N_h(E,x) = I_0(1-R)_{-\alpha_{ph}}e^{\alpha_{ph}} \frac{\rho(E+h\nu,x)\{1-f(E+h\nu)\}\rho(E,x)f(E)dE}{\int_{-\infty}^{E_{v,x}} \rho(E+h\nu,x)\{1-f(E+h\nu)\}\rho(E,x)f(E)dE} \quad (31)$$

where $E_{v,x}$ is the energy of the top of the valence band at x from the electrode surface.

Potential Energy Profiles in n-Type Semiconductor. Equations 8 and 9 are applicable to an *n*-type semiconductor.

Number of Holes Arriving at the Surface. The number of holes at the surface, the energy of which is between $E + dE$, is given in a similar way to that of electrons discussed on p. 41, with the changes mentioned below. Thus,

$$l_D = \sqrt{D\tau_h} = \sqrt{300\mu_h\tau_h kT/e_o} \quad (32)$$

$$l_E = -\tau_h\mu_h V'(x) \quad (33)$$

where μ_h is the mobility of the hole in the semiconductor and τ_h is the life time of the hole.

Electron Transfer Process. Electrons are transferred from a donor in solution to a hole in the valence band. Since $N_D(E)$ is an inverted Boltzman distribution for donors (*81*), a photocurrent corresponding to a direct electron transfer from a donor to a hole is expected only when the energy of ground state of the donor is lower than the top of the valence band at the surface (Figure 4), i.e., $\Delta H' < 0$.

(a) Construction of the potential energy barrier. The potential energy barrier for the electron from OH⁻ is constructed by considering the optical Born energy, Coulombic interaction, and interaction energy between electron and adsorbed water, analogous to *p*-type semiconductor case (*see* p. 42). The potential energy of the electron with respect to its value in the ground state of OH⁻, U', is given by

$$U' = E.A. + U_{\text{opt. Born}} + U'_{\text{ion}} + U_{H_2O} \quad (34)$$

where
$$U_{\text{ion}} = \int_0^x \left\{ -\frac{e_o^2}{(2d-x)^2\epsilon_{op}} \cdot \frac{\epsilon - \epsilon_{op}}{\epsilon + \epsilon_{op}} + \frac{e_o^2}{x^2\epsilon_{op}} + \frac{2\pi e_o^2}{\epsilon_{op}}\sum_{n=1}^{\infty} n\left(\frac{1}{(2d-x)^2 + n^2R_i^2} \cdot \frac{\epsilon - \epsilon_{op}}{\epsilon + \epsilon_{op}} - \frac{1}{x^2 + n^2R_i^2}\right) \right\}dx \quad (35)$$

Similarly, the potential energy of the electron from water with respect to its value in the ground state of water U'' is given by:

$$U'' = E_{Diss} - J + U_{opt.\,Born} + U'_{ion} + U_{H_2O} \tag{36}$$

A potential energy profile based on these considerations is shown as a function of distance in Figure 5 by assuming d is 6.5 Å in Equation 35. Other parameters taken in this calculation are $E_g = 1.5$ eV and $\chi = 3.5$ eV.

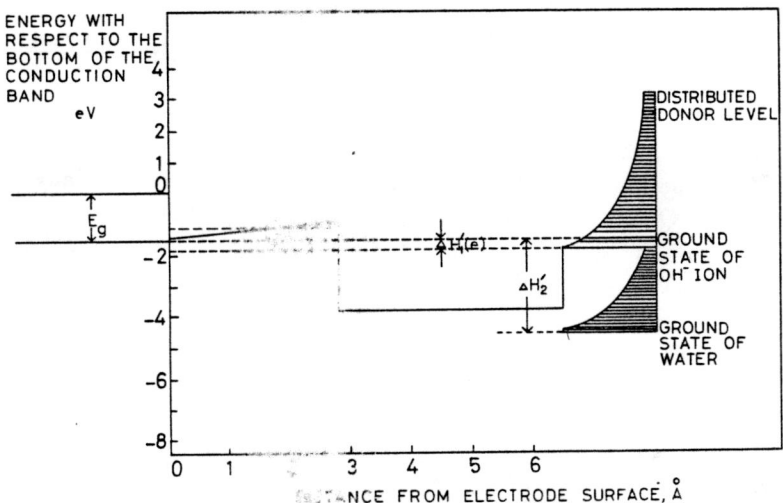

Figure 5. Schematic of the potential energy barrier for the electron transfer from donors (OH^- ion and water) in solution

(b) The equation for the photocurrent. From Equation 5 the photocurrent corresponding to an electron transfer from OH^- is given in general by (see also Figure 5):

$$i_{p,OH^-} = e_o \frac{C_L}{C_T} \int_{-\infty}^{-E_g} N_h(E)\, G_D(E)\, W_A(E)\, dE \tag{37}$$

where $N_h(E)$ is the number of holes with energy E which strike the semiconductor surface per unit time per unit area; $G_D(E)$ is the distribution of vibrational–rotational states of donor at energy E in the solution; $W_A(E)$ is the WKB tunneling probability of electrons through the potential barrier at energy E; C_D is the number of electron donors per unit area in the OHP.

Since $-E_g > -E_g - \Delta H_1'(e) > -E_g - \Delta H_2'(e)$, i_p' is:

$$i_p' = e_o \left[\frac{C_{H_2O}}{C_T} \int_{-E_g - \Delta H_2'(e)}^{-E_g} N_h(E) \exp\{\beta(-E_g - \Delta H_2'(e) - E/kT\} \exp \left\{ -\frac{\pi^2 l}{h} (-E_g - \Delta H_2'(e) + U''_{max} - E) \right\} dE + \frac{C_{OH^-}}{C_T} \int_{-E_g - \Delta H_1'(e)}^{-E_g} N_h(E) \exp\{\beta(-E_g - \Delta H_1'(e) - E)/kT\} \exp \left\{ -\frac{\pi^2 l}{h} (-E_g - \Delta H_1'(e) + U'_{max} - E) \right\} dE \right] \quad (38)$$

where C_{H_2O} and C_{OH^-} are the number of water molecules and OH⁻ ions per unit area in the OHP, respectively; U''_{max} and U'_{max} are the barrier maxima for the electrons from water and OH⁻ ion, respectively. $N_h(E)$ is given on p. 46.

The Computation of the Quantum Efficiency for an Individual Electrode. Quantum efficiencies are given by Equation 39. They have been computed using Equation 26 for a *p*-type semiconductor and Equation 38 for an *n*-type semiconductor.

$$\text{Quantum efficiency} = \frac{i_p}{e_o I_o} = \frac{\text{electrons transferred from or to a semiconductor}}{\text{incident photons}} \quad (39)$$

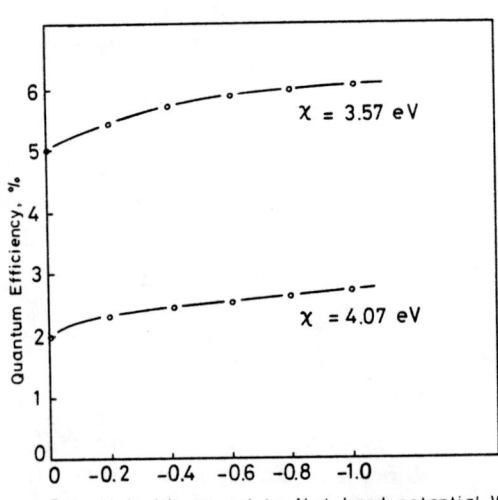

Figure 6. Dependence of quantum efficiency of p-type semiconductor on electrode potential for several values of electron affinities

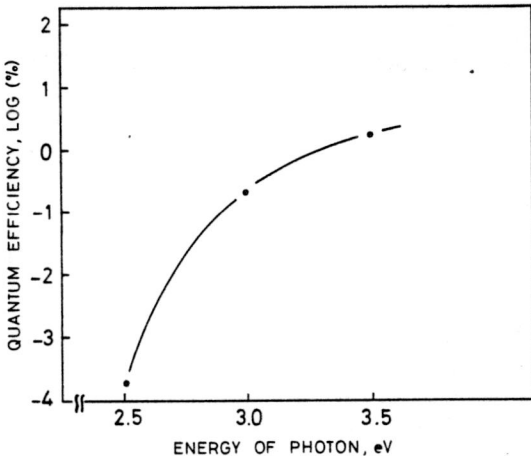

Figure 7. Dependence of quantum efficiency of p-type semiconductor on the energy of photon

PREDICTED RESULTS FOR CATHODIC PHOTOCURRENTS AT p-TYPE SEMICONDUCTOR ELECTRODES. *Effect of Potential.* In Figure 6 the quantum efficiency calculated by Equation 39 is shown as a function of potential. The parameters for computation are $h\nu = 3$ eV, $R = 0.47$ (82), $\alpha_{ph} = 4.4 \times 10^5$ cm^{-1} (82), $E_g = 1.40$ eV (69), $\chi = 4.07$ eV (69), $\mu_e = 8{,}600$ cm^2/V · sec (83), and $\tau = 10^{-12}$ sec. The parameters used are those for GaAs.

Effect of the Energy of the Photon. The optical constants of GaAs at several photon energies are available from Ref. 82. Quantum efficiencies at several photon energies calculated by Equation 39 are given in Figure 7 as a function of the energy of the photon. The parameters for computation are the same as those on p. 49 and $V = -1.0$ V with respect to the flat band potential.

Effect of Lifetime. The quantum efficiency (calculated by Equation 39) dependence on life time is shown in Figure 8. When $l_D \gg 1/\alpha_{ph}$, most of the excited electrons reach the surface; further increase of lifetime does not affect the photocurrent (*see* p. 46). Parameters for calculation are the same as those on p. 49 and $V = -1.0$ V with respect to the flat band potential.

Effect of Electron Affinity of the Semiconductor. When χ changes, the relative position of the acceptor changes, and this changes $N_e(E)$ and $G(E)$. The calculated quantum efficiencies for several values of χ are shown in Figure 6. The smaller the values of χ, the larger are the quantum efficiencies calculated.

Energy Gap Effect. The calculated quantum efficiency is shown for three values of the energy gap in Figure 9. The parameters for the calculation are the same as those on p. 49, except for E_g and V. V is

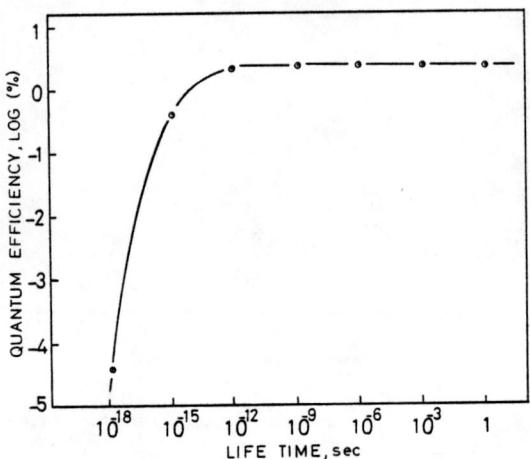

Figure 8. Dependence of quantum efficiency of p-type semiconductor on life time of electron

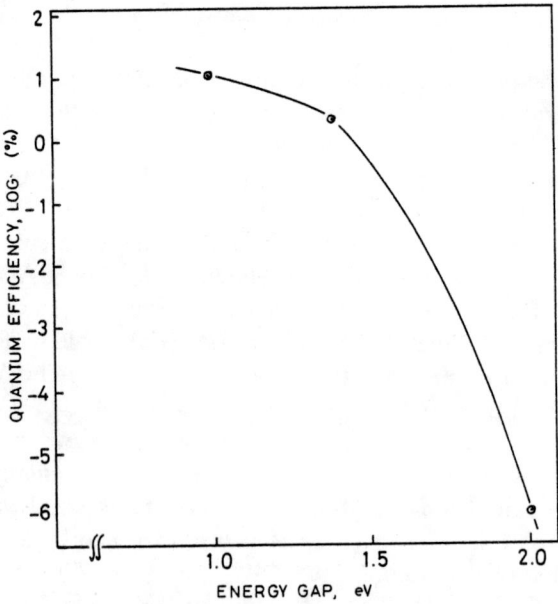

Figure 9. Dependence of quantum efficiency of p-type semiconductor on the energy gap

taken as -1.0 V with respect to the flat band potential. The smaller the value of E_g, the larger the quantum efficiency.

Effect of Electron Mobility. The quantum efficiency calculated from Equation 39 as a function of the electron mobility is shown in Figure 10. When the electron mobility is large enough to have $l_D \gg 1/\alpha_{ph}$, most of the excited electrons reach the surface and saturate the photocurrent; further increase in μ_e does not affect the photocurrent. The parameters for the calculation are the same as those on p. 49, except for μ_e and V. $V = -1.0$ V with respect to the flat band potential is used.

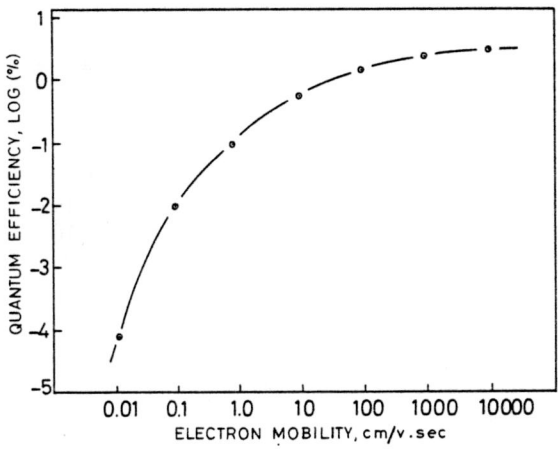

Figure 10. Dependence of quantum efficiency of p-type semiconductor on electron mobility

PREDICTED RESULTS FOR n-TYPE SEMICONDUCTORS. *Effect of Potential.* The potential dependence of the quantum efficiency calculated by Equation 39 is shown in Table II(a). The parameters for computation are the same as those of GaAs except that $\mu_h = 400$ cm^2/V · sec and $\chi = 3.5$ eV.

Effect of Electron Affinity of the Semiconductor. The quantum efficiencies calculated by Equation 39 for several values of electron affinity are shown in Table II(b). The parameters for the calculation are the same as those on p. 51 except χ and V. The value of $V = +1.0$ V with respect to flat band potential is used.

Effect of the Energy Gap. In Table II(c) the calculated quantum efficiencies for several E_g are shown while other parameters are kept constant. A value of $V = +1.0$ V with respect to flat band potential is used. Other parameters for the calculation are the same as those on p. 51.

Comparison with the Experimental Results of n-Type Semiconductors. The calculated results of quantum efficiencies for the n-type semi-

Table II. Computed Results for n-Type Semiconductor

Effect of	Quantum Efficiency (%)
(a) Potential (V with respect to the flat band potential)	
0	1.0×10^{-2}
0.2	1.8×10^{-2}
0.4	2.2×10^{-2}
0.6	2.7×10^{-2}
0.8	3.0×10^{-2}
1.0	3.2×10^{-2}
(b) Electron Affinity (eV)	
3.1	4.5×10^{-2}
3.3	4.1×10^{-2}
3.5	3.2×10^{-2}
(c) Energy Gap (eV)	
1.0	4.0×10^{-2}
1.2	3.5×10^{-2}
1.4	3.0×10^{-2}

conductors are low. The physical reasons for this, for the model taken, can be seen from Figure 4. Electrons are donated from distributed energy states of a donor. The presence of the distribution law in the n-type situation (see also the position of the donor in Figure 5) leads to a reduction of available states. Conversely, experimental results for TiO_2 (n-type) electrodes show high quantum efficiencies, i.e., up to 60% at 3.4 eV.

Surface States Involved in the TiO_2 Mechanisms. Hole consumption at the surface can occur by:

(1) direct electron transfer from a donor in solution;
(2) surface recombination with electrons in the conduction band;
(3) surface recombinations with electrons in surface states;
(4) oxidation of anions in crystal, e.g., $S^{-2} + 2p^+ \rightarrow S$ in CdS.

The reaction mechanism for oxygen evolution on TiO_2 without illumination was studied by Boddy (38): electron tunneling from a surface state to the bulk of the electrode is the rate-determining step. Correspondingly, the high quantum efficiency of the oxygen evolution reaction on TiO_2 can be explained by considering surface states. Since the energy levels of the donors are higher than that of the top of the valence band, electron transfer from the donor to a hole can occur easily if surface states exist. Thus most of the holes arriving at the surface from the bulk are expected to react with electrons of the surface states because path (1) cannot occur as explained before, (4) does not occur, as shown by Fujishima, Honda, and Kikuchi (84), and (3) may occur, but the contribution is small because of anodic polarization. Also, in the presence of a halide anion halogen production occurs instead of oxygen evolution, and this is explained by the reaction between holes and adsorbed halide anions (85).

In Figure 11 the quantum efficiency is shown as a function of potential (*see* p. 46). The parameters for the calculation, the results of which are shown in Figure 12, are $E_g = 3$ eV, $h\nu = 3.5$ eV, $R = 0.15$ (59), $\alpha = 5 \times 10^3$ cm^{-1} [the value of TiO$_2$ film (86)], $\tau = 10^{-7}$ sec, and $\mu_h = 1.0$ cm/V · sec (value for the electron), and $\epsilon = 89$ (value for the *a*-direction).

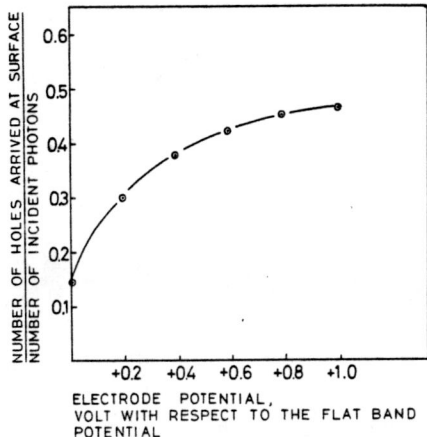

Figure 11. Dependence of the number of holes arriving at the surface per unit time per unit area divided by the number of incident photons per unit time per unit area on the electrode potential

Whole Cell System. Expressions for the quantum efficiency of individual electrodes (*n*- and *p*-type semiconductors) have been computed. The photocurrent of the electrochemical cell, a cathode and an anode without other external potential source, has the important practical meaning. In the evaluation of the cell carried out below, the photocurrent–energy relation at chosen electrode potential has been integrated over the whole solar spectrum.

RELATION BETWEEN THE POTENTIAL OF AN ELECTRODE AND A CELL. The potential of a self-driving cell at a current I, V_cell, is given by

$$V_\text{cell} = E_\text{so} - E_\text{si} - IR_c \qquad (40)$$

where E_so is the potential of the cathode with respect to a reference electrode, E_si is the potential of the anode with respect to the same reference electrode, and R_c is the inner cell resistance. Also, the potential of a driven cell at a current I, V_ext, is given by Equation 41 (87).

$$V_{\text{ext}} = E_{\text{si}} - E_{\text{so}} + IR_c \qquad (41)$$

Let V be:

$$V = E_{\text{so}} - E_{\text{si}} - IR_c \qquad (42)$$

If V is positive at current I, the cell is self-driven. If V is negative at I, the cell is driven. A cell potential, $-V$, is required to cause a current I to flow. This is shown in Figure 12. If two electrodes are dipped into an acidic solution and one of them is in contact with oxygen and another is in contact with hydrogen, the system works as a self-driving cell. The

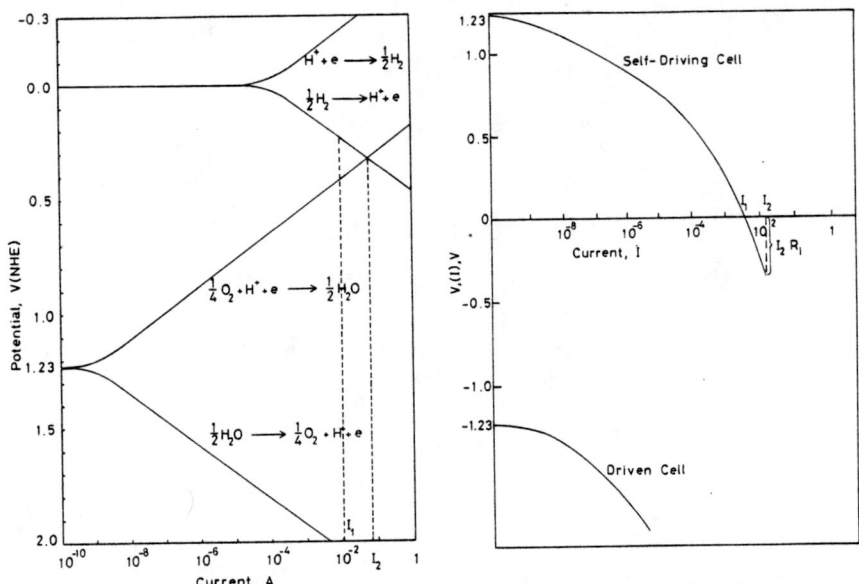

Figure 12. Potential–current relationships of individual electrodes and cell

oxygen electrode works as a cathode ($\frac{1}{4}O_2 + H^+ + e \rightarrow \frac{1}{2}H_2O$) and the hydrogen electrode as an anode ($\frac{1}{2}H_2 \rightarrow H^+ + e$) until I reaches I_1 (at I_1, $V = E_{\text{so}} - E_{\text{si}} - IR_c = 0$). To obtain a current I_2 an extra external potential $I_2 R_c$ ($= -V$) is needed. Correspondingly, to get hydrogen from one electrode and oxygen from another in the normal dark case one must supply an external potential, V_{ext} [$= -V = -(E_{\text{so}} - E_{\text{si}} - IR_c)$].

CALCULATED HYDROGEN PRODUCTION RATE FROM SOLAR ENERGY USING PHOTO-DRIVEN CELLS. Photocurrents from individual electrodes stimulated by solar energy at a certain potential have been calculated by integrating the photocurrent of the monochromatic light through the whole solar energy range. Thus one can get a current–potential relationship of an individual electrode by carrying out this process over some potential range. Then one can get a cell potential–current relationship if one knows the value of the flat band potential of the semiconductor electrode with respect to a reference electrode.

As examples, Equation 42 has been applied to cell 1, TiO_2 solution (pH = 14)/solution (pH = 0, 7, 14)/Pt, and cell 2, TiO_2/solution (pH = 0)/GaAs, with the assumption that all holes arriving at the surface react with electrons from the OOH⁻ ion (or H_2O) of the solution. The parameters used are as follows. The flat band potential of TiO_2 at pH = 0 and GaAs at pH = 0 are −0.8 V(NHE) (88), −0.05 V(NHE) (88) and 0.43 V(NHE) (89); R is 100 Ω; i_o of Pt electrode is 10^{-4} A/cm², and optical constants of TiO_2 were taken from Ref. 32. Results are shown in Figures 13 and 14. The maximum cell current of cell 1 is a function of pH gradient. When the pH of the cathodic compartment is 0, the maximum cell current is 0.14 mA/cm², and the maximum hydrogen production rate is 0.06 cc/hr cm². The efficiency of this cell is 0.3% of the total solar energy (0.0739 w/cm²).

In cell 2 in which both electrodes are illuminated, current flows even when the pH gradient is zero. The maximum current is determined by the cathodic reaction and is 0.03 mA/cm² (Figure 14).

OPTIMUM PROPERTIES OF ELECTRODES.

(a) Electron affinity. For p-type semiconductors the lower the electron affinity, the greater the quantum efficiency. The minimum value is 3.5 eV. For n-type semiconductors, since the energy level of donor must be higher than that of the top of the valence band of semiconductor and lower than that of the bottom of the conduction band of semiconductor, $-E_g < \Delta H_1' < 0$, i.e. $\chi + E_g > 5.3$ and $\chi < 5.3$.

(b) Energy gap. The smaller the energy gap, the larger the portion of solar energy absorbed by the semiconductors. However, if the energy gap is too small (e.g., < 1.23 eV), water decomposition cannot occur.

(c) Electron mobility and lifetime. When these values increase, the quantum efficiencies increase until saturation. If μ is large, most carriers arrive at the surface even when τ is small. Therefore the optimum condition is electron mobility × lifetime $> 10^{-8}$.

(d) Flat band potential. This property determines whether the cell is self-driven or not. Since the p-type semiconductor works as a photocathode, the more positive the flat band potential, the better. If this is more positive than 1.23 V(NHE) at pH = 0, even when the counter electrode (anode) is a metal, and there is no pH gradient, the cell works like a self-driven one. If this is more positive than 0.42 V(NHE) at pH = 0, the cell works like a self-driven one when the counter electrode is metal and at pH = 14. For n-type semiconductors the more negative the flat band potential, the better. If the value is more negative than −0.83

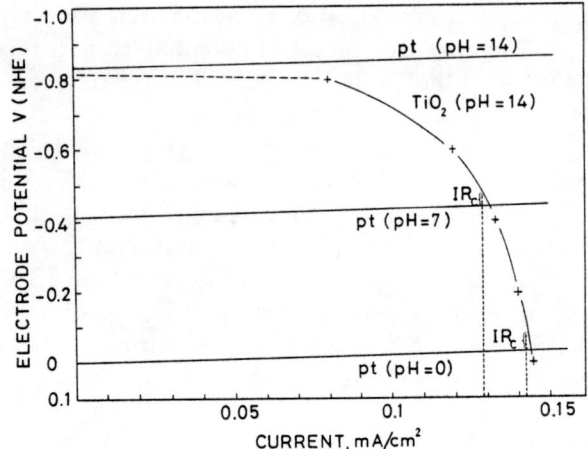

Figure 13. Potential–current relationship of the cell, TiO_2/solution (pH = 14)/solution (pH = 0, 7, 14)/Pt

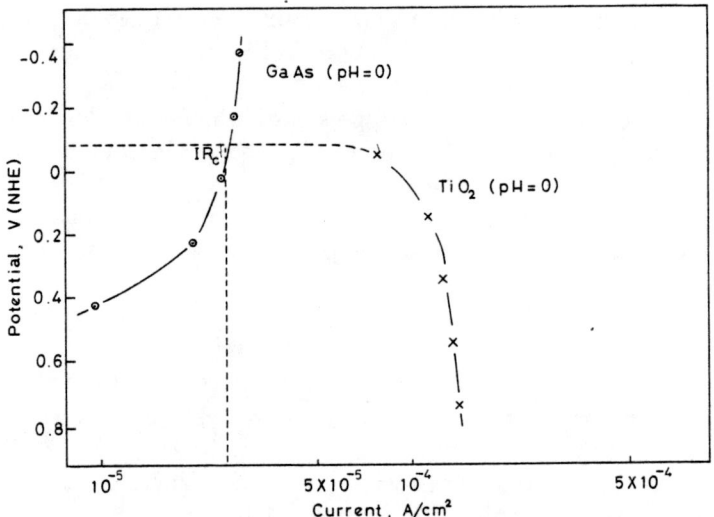

Figure 14. Potential–current relationship of the cell, TiO_2/solution (pH = 0)/GaAs

V(NHE) at pH = 14, even when the counter electrode (cathode) is a metal and there is no pH gradient, the cell works like a self-driven one. If this is more negative than 0.0 V(NHE) at pH = 14, the cell is self-driven when the counter electrode is metal and in pH = 0. If the flat band potential of a p-type semiconductor is more positive than that of an n-type semiconductor in the same solution, the cell works like a self-

Table III. Optimum Condition of Semiconductor Electrodes

	p-S.C	n-S.C
Electron affinity	The smaller, the better. Minimum ca. 3.5 eV	$\chi + E_g > 5.3$ eV $\chi < 5.3$ eV
Life time, mobility	$\mu \times \tau > 10^{-8}$ (cm²/V)	
Energy gap	The less, the better down to a minimum of about 1.23 eV	
Flat band potential	> 0.4 V (NHE)	< 0.0 V (NHE)
Surface state	—	very important

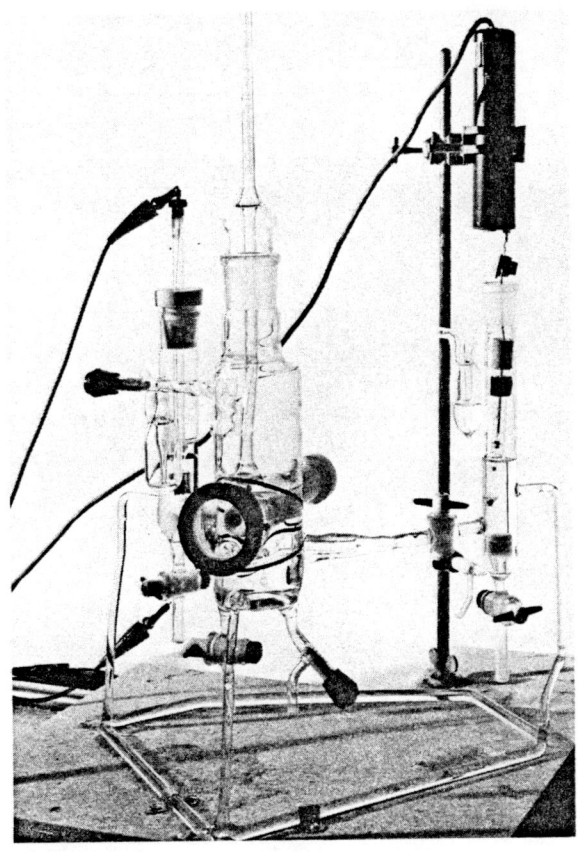

Figure 15. Photoelectrochemical cell

driven cell. A summary of the predicted optimum properties for a semiconductor to be used in a photo-driven cell is listed in Table III.

Photoelectrochemical Reaction on TiO_2 Single Crystal

Experimental. A TiO_2 single crystal was reduced in H_2 at 800°C. An ohmic contact was made and the specimen mounted in a Teflon electrode holder. A three-compartment cell was used (Figure 15). A 150-W Xe lamp served as the light source; a monochrometer was used.

Results. CURRENT–POTENTIAL RELATIONSHIP AT SEVERAL pH VALUES. Current–potential relationships for the anodic reaction on reduced TiO_2 under illumination were measured for several solutions (1N NaOH, 0.1N NaOH, 1N Na_2SO_4, 0.1N H_2SO_4, and 1N H_2SO_4) potentiostatically at room temperature. Results are in Figure 16. The shape of the curves is the same for all solutions although the curves are shifted along the

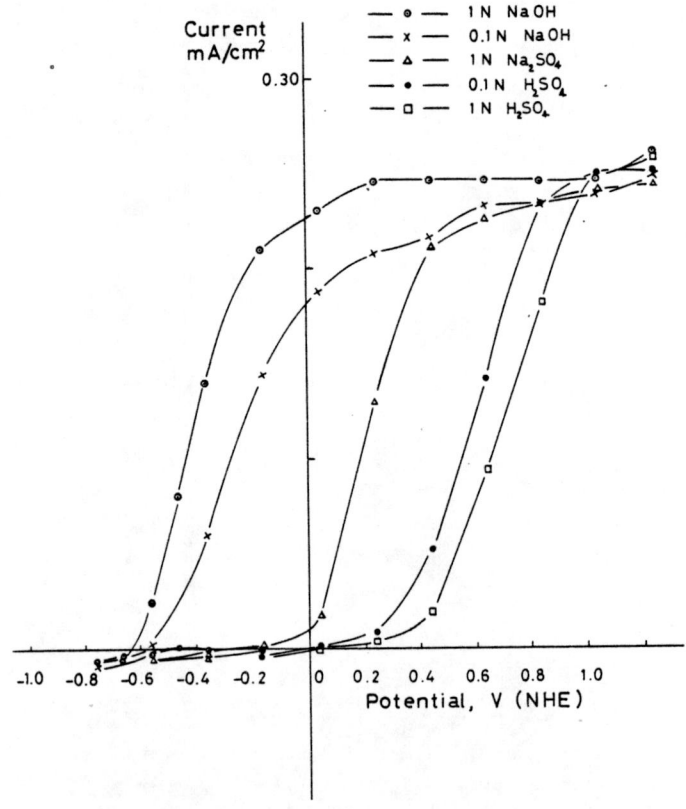

Figure 16. Photocurrent–potential relationships of TiO_2 single crystal at several pH solutions

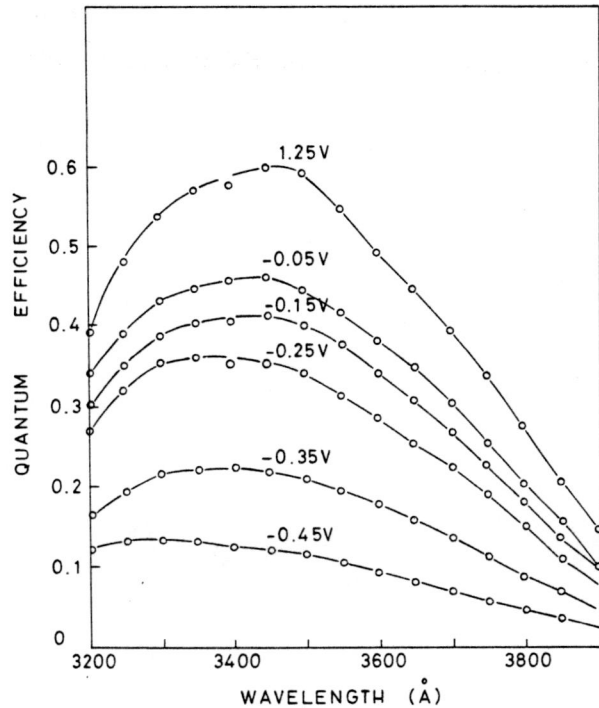

Figure 17. Quantum efficiencies–wavelength relation of TiO_2 single crystal at several electrode potentials

potential axis corresponding to the pH change. The potential at which the photocurrent can first be observed shifts by about 0.059 V/pH.

Quantum Efficiency—Photon Energy Relationships at Several Potentials. Current–photon energy relationships were measured at several potentials in 0.1N NaOH solution.

Quantum efficiencies were calculated from photocurrents and the photons per unit time (Figure 17). The more negative the electrode potential, the larger the ratio of the quantum efficiency at high, compared with that at low, energies of the incident light (Figure 18).

MECHANISM OF OXYGEN EVOLUTION REACTION ON TiO_2 UNDER ILLUMINATION. The photocurrents at the same potential with respect to the flat band potential are the same in all pH ranges; this suggests a reaction step inside the semiconductor as the rate-determining step.

One possible mechanism is shown in Figure 19(a). (A) shows the TiO_2 surface which becomes a cation center (B) after reaction with holes and gives oxygen (Step I). OH⁻ ions and water molecules are adsorbed on these centers (Step II), and (B) becomes (C) and (D). There is an equilibrium between (C) and (B) which causes the pH dependence of

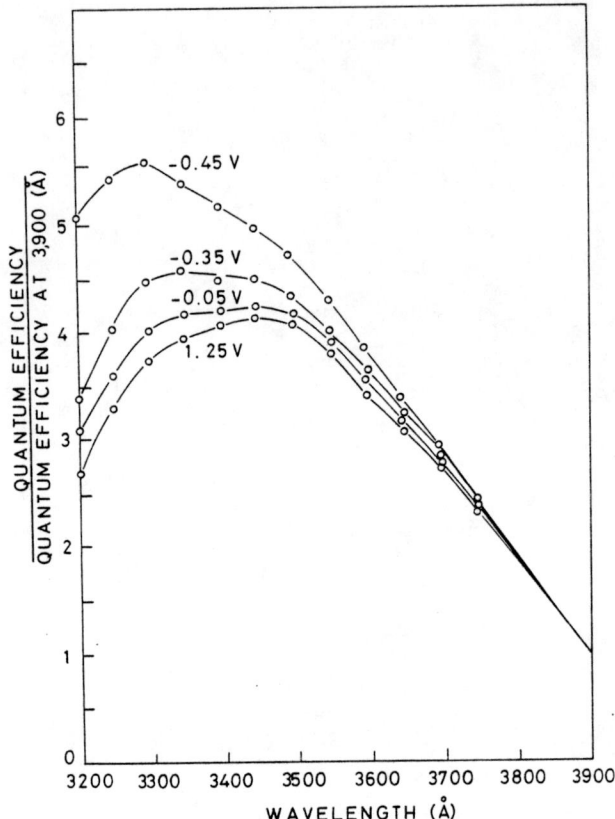

Figure 18. Relative values of quantum efficiencies (Q/Q at 3900 Å)—wavelength relation of TiO_2 single crystal at several electrode potentials

the flat band potential. Holes arriving at the surface react with these states (Steps IV, IV', and IV''), and the surface oxide is recovered (A). This is the rate-determining step. Then Step I occurs again. The processes repeat.

Another possible mechanism is in Figure 19(b). (A) shows oxygen vacancies at the surface, and these become cation centers (B) after donating electrons to the conduction band which can occur easily because an oxygen vacancy is an electron donor (Step I). OH^- ions and water molecules are adsorbed on these centers (Step II), and (B) becomes (C) and (D). There is an equilibrium between (C) and (D) which causes the pH dependence of the flat band potential as mentioned on p. 59. Holes arriving at the surface react with these states (Steps IV, IV', and IV''), and a surface oxide is made (E). This step is rate determining. State (E) may be Ti–O–O–Ti, the surface structure of TiO_2 without

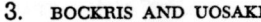

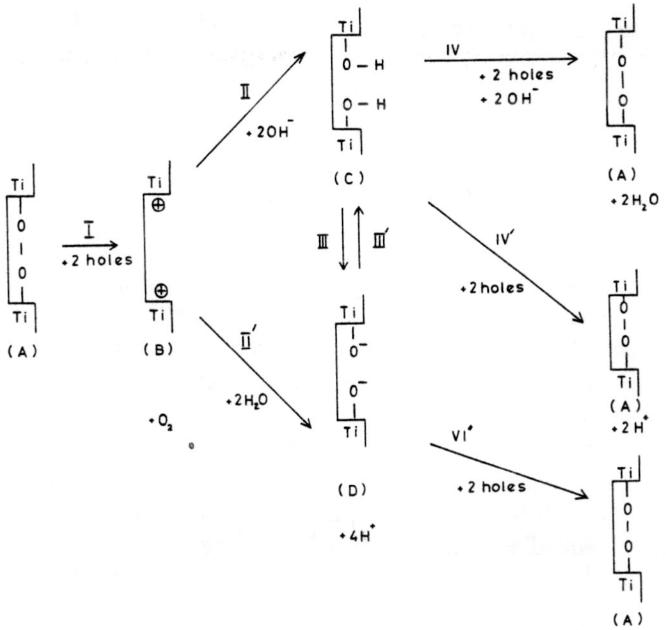

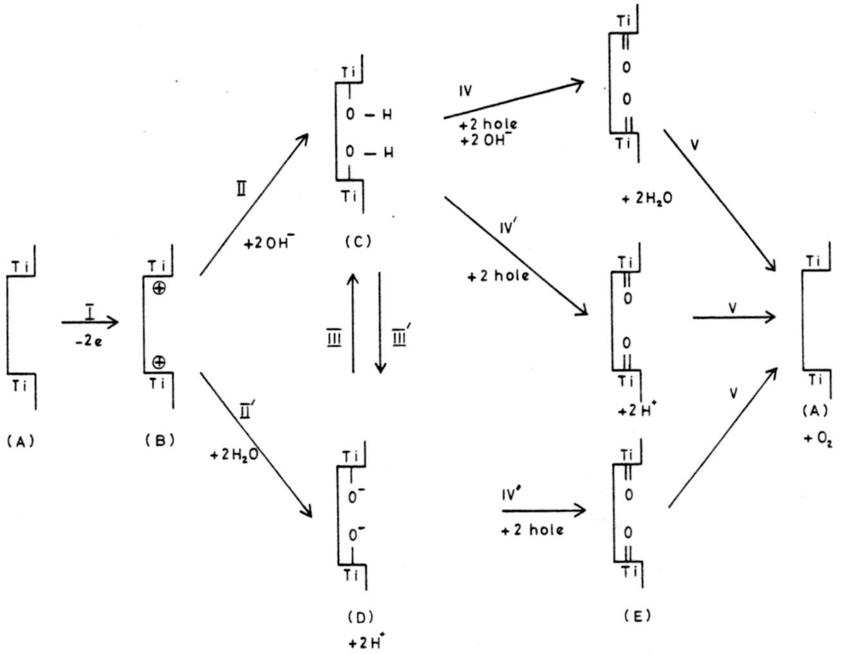

Figure 19. Proposed reaction mechanisms of oxygen evolution reaction on TiO_2 single crystal under illumination

defects. State (E) gives oxygen and state (A) (Step V). The processes repeat.

The difference between two mechanisms consists of the assumed differences in structure of the surface oxide.

Photoelectrochemical Properties of TiO₂ Film

Experimental. A section of titanium rod (0.62-cm diameter) about 5 mm thick was polished, degreased, and washed in distilled water. Specimens were heated in an electric furnace in air at 600°C for various periods up to three hours. The TiO_2 films formed were then reduced by heating in hydrogen. The same electrochemical measurements as those for TiO_2 single crystals were performed in $0.1N$ NaOH.

Results. POTENTIAL–CURRENT RELATIONSHIP. Figure 20 shows the potential–current relationships of several types of Ti electrodes. The photocurrents at 1.0 V(NHE) are plotted against heating time for unreduced and reduced samples (Figure 21). Photocurrents on oxidized Ti were 20% of those of TiO_2 single crystal (2.7 mA/cm²) and depend little on heating time. The photocurrents increase when the oxidation

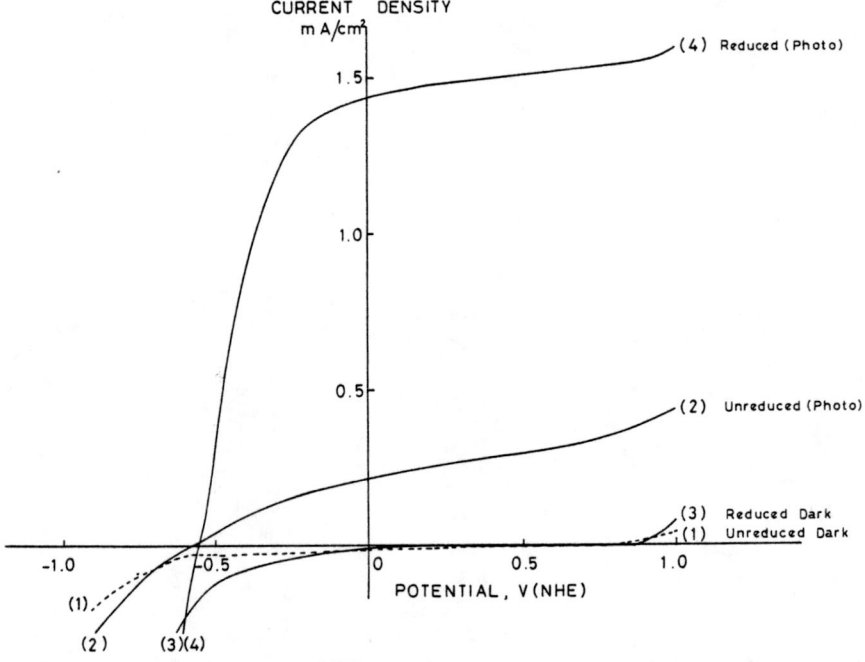

Figure 20. Potential–current relationships of TiO_2 films made by high temperature oxidation

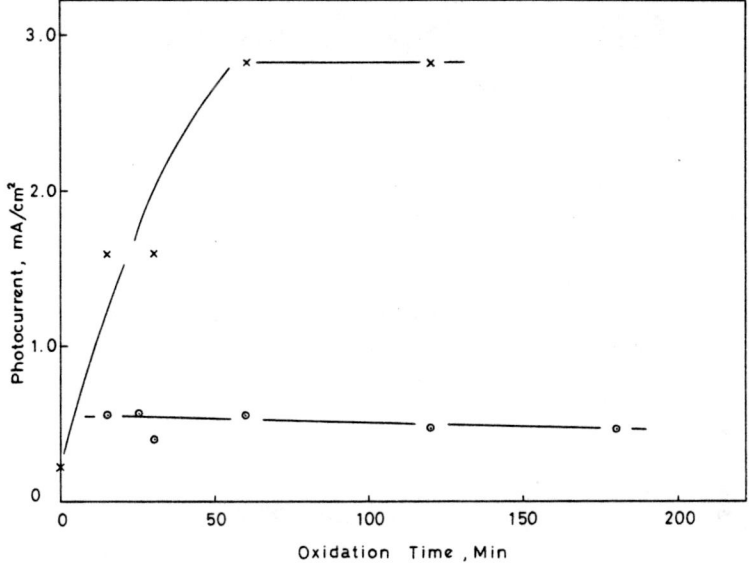

Figure 21. Dependence of photocurrent of TiO_2 films made by high temperature oxidation on oxidation time

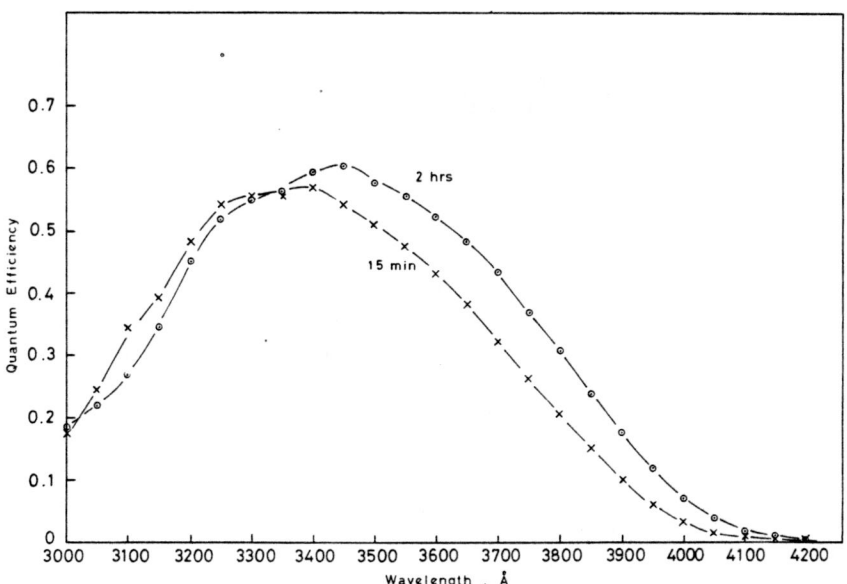

Figure 22. Quantum efficiencies–wavelength relations of TiO_2 films made by high temperature oxidation

time increases and saturate after one hour of oxidation. The saturated photocurrents are the same as those of TiO_2 single crystal.

QUANTUM EFFICIENCY–WAVELENGTH RELATIONSHIPS. Quantum efficiencies of the 15-min oxidized and 2-hr reduced Ti and 2-hr oxidized and 2-hr reduced Ti are shown in Figure 22.

Discussion. Results found in the potential–current relationship above agree with those of Fujishima et al. The low photocurrents before reduction (Figure 21) arise because of the resistance of the TiO_2 films. The longer the heating time, the thicker the film, and therefore the higher the resistance. However, the thicker the film, the larger the photon absorption. Consequently, before reduction the oxidation time does not affect

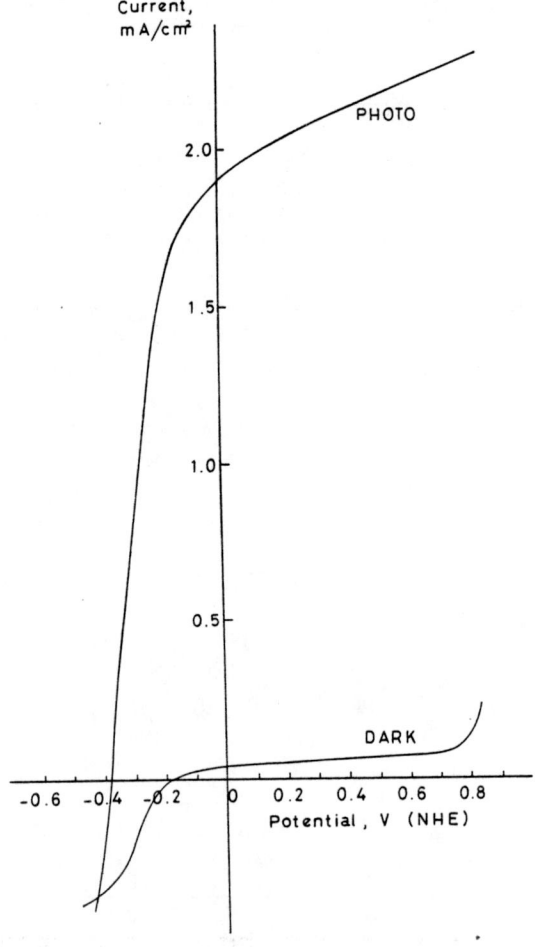

Figure 23. *Potential–current relationships of Fe_2TiO_5*

the photocurrent. After reduction resistances are reduced, and photocurrents increase. The thicker the film, the higher the photocurrent (Figure 21). The resistance effect on the photocurrent before reduction is seen in Figure 20.

The positions of the maximum point of the quantum efficiency–wavelength relationship depend upon the oxidation time and, therefore, the thickness of the film. (*See also* the results of Möllers et al. (*86*).)

The shorter the wavelength of light, the larger the absorption coefficient (*87*). The penetration depth is proportional to the reciprocal value of the absorption coefficient so that the larger the wavelength, the longer the penetration length. If the oxide layer is not thick enough, not all photons are absorbed by oxide layer when the wavelength of the light is large, but all of them are absorbed by oxide layer when the wavelength of the light is short. However, if the oxide layer is thick enough, all photons are absorbed by the oxide layer even when the wavelength is large. Consequently, the maxima of the qauntum efficiency–wavelength relationships appear on the shorter wavelength side when the oxide thickness is small or the oxidation time short.

Photoelectrochemical Reactions on Metal Titanates

Iron titanate was tried as an intended low-gap semiconductor.

Experimental. SYNTHESIS OF Fe_2TiO_5 (*88*). Equimolar amounts of Fe_2O_3 and TiO_2 powders were mixed and heated in an oxygen atmosphere at 1000°C for three days. The samples were reground and pelleted by pressing at 2000 kg/cm^2. These pellets were heated in an oxygen atmosphere at 1000°C for three days. The samples were partially reduced in hydrogen at 800°C for 30 min, and the resistances became 1.6 kΩ · cm.

Results. The potential–current relationships of the samples in 0.1N NaOH with and without illumination are shown in Figure 23. The photocurrents were almost the same as those of TiO_2 single crystals. The quantum efficiency vs. wavelength relationship at 0.85 V(NHE) is shown in Figure 24. The result suggests that the energy gap of Fe_2TiO_5 is near that of TiO_2.

Figure 25 shows the photocurrents at 0.85 V(NHE) as a function of time. The photocurrent is stable for over 35 hr. Bubbles on the electrode were observed. After 3.5 hr no Fe^{++} was detected in solution by adding KSCN.

Discussion. The photocurrent is oxygen evolution. The electrode satisfies the condition of no dissolution under illumination. Since the quantum efficiency–wavelength relationship is similar to that of TiO_2, a better photocurrent under solar illumination cannot be expected (*see* Figure 22).

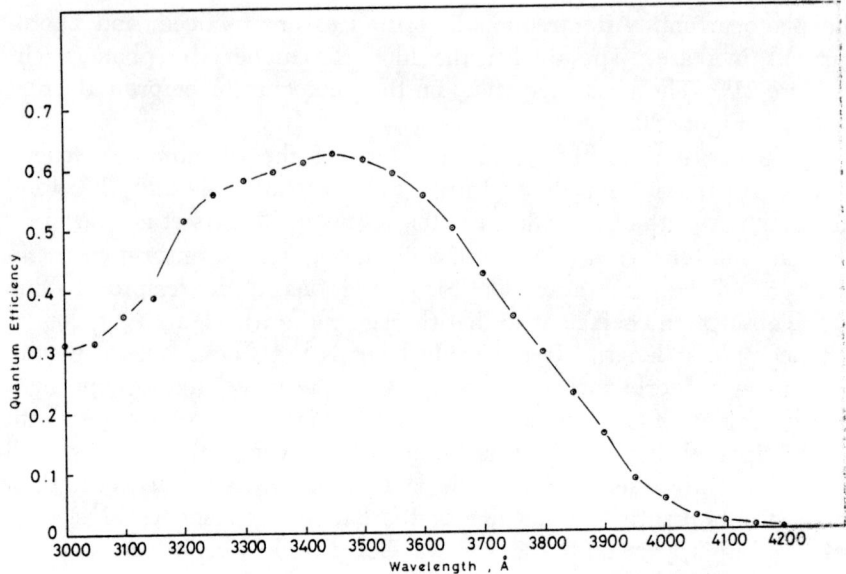

Figure 24. Quantum efficiencies–wavelength relations of Fe_2TiO_5

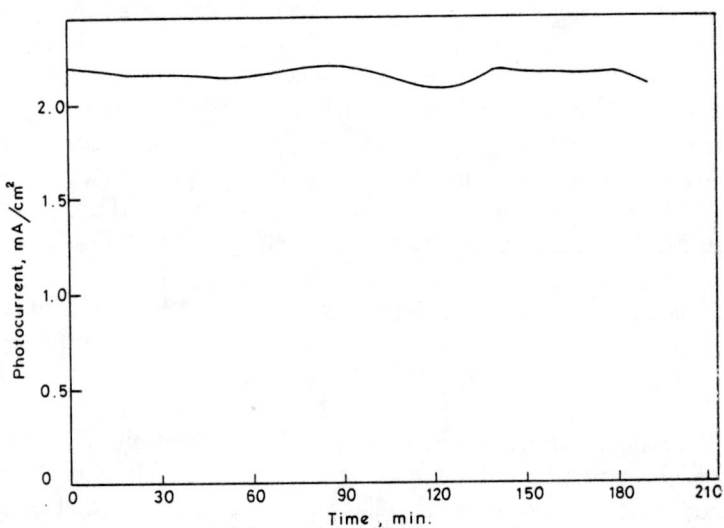

Figure 25. Dependence of the photocurrent on time

Prevention of Anodic Dissolution of Photon Absorber by TiO_2 Coatings (90)

If a semiconductor which has a low energy gap, but dissolves, is protected from dissolution, there may be a wider choice of possibilities. Here CdS ($E_g = 2.5$ V) was coated with a TiO_2 film to prevent dissolution.

Experimental. A film of TiO_2 was produced on a CdS single crystal by chemical vapor deposition. An ohmic contact was made, and the specimen was mounted in a teflon electrode holder so that only the face covered by TiO_2 was exposed to the solution. Experiments were carried out in $1N$ NaOH solution under potentiostatic conditions.

Results. The photocurrent was observed when the electrode was polarized and the spectral response of the photocurrent was measured at 1.0 V(NHE). Figure 26 shows the quantum efficiencies as a function of wavelength for TiO_2 CVD film (on Ti metal), CdS single crystal, and TiO_2-coated CdS single crystal. Although the photocurrent decreased 50% during the first eight hours, the rate of decrease was only 20% of that of the CdS single crystal used alone, i.e., improvement was considerable.

Discussion. Using the result of Figure 27 the spectral response under solar energy of TiO_2–CdS was calculated (Figure 27). The photocurrents under solar illumination are given by the area under these curves.

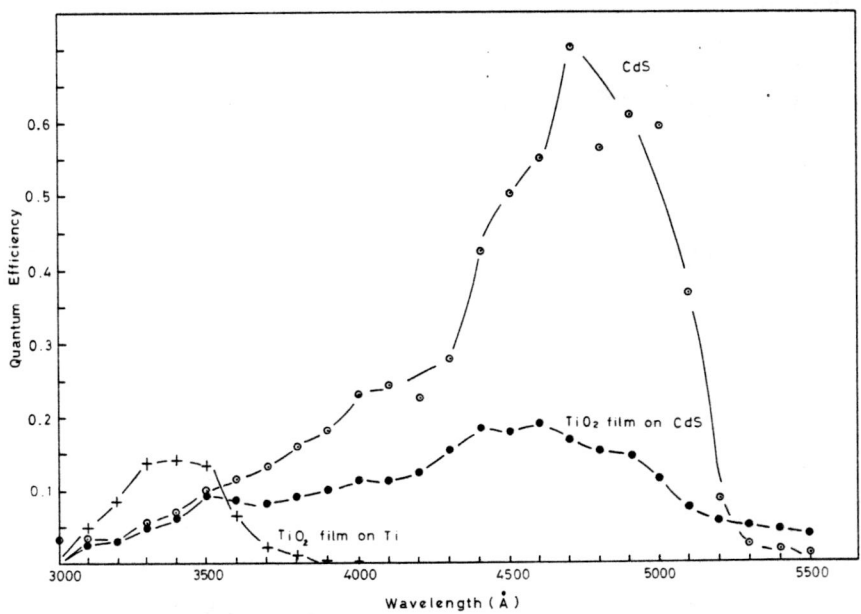

Figure 26. Quantum efficiencies–wavelength relationships of TiO_2 CVD film (on Ti), CdS single crystal, and CdS single crystal coated with TiO_2 CVD film

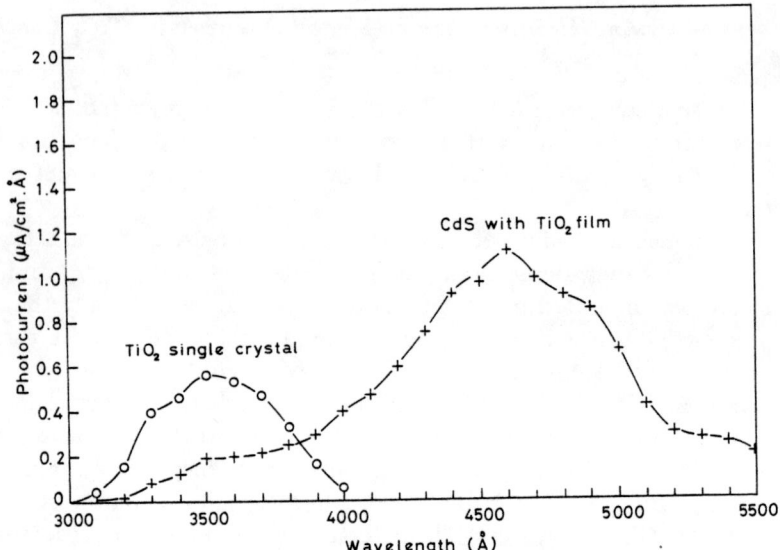

Figure 27. Spectral response of photocurrents of CdS single crystal coated with TiO_2 CVD film and TiO_2 single crystal under solar energy

The TiO_2-coated CdS electrode converts four times more solar energy to electricity than a TiO_2 single crystal although the quantum efficiency of the CdS–TiO_2 electrode is lower than that of TiO_2 single crystal. The photocurrent decrease with time is caused by imperfections in TiO_2 film. Correspondingly, Nakato et al. have reported (*91*) the prevention of anodic dissolution by Au coating.

Acknowledgment

J. O'M. Bockris would like to thank the Petroleum Research and Program Development Funds of the American Chemical Society for financial assistance in attending the New York Bicentennial Meeting of the Society.

Literature Cited

1. Brattain, W. H., Garrett, C. G. B., *Bell System Tech. J.* (1955) **34**, 129.
2. Green, M., in "Modern Aspects of Electrochemistry," Vol. 2, Chapter 5, J. O'M. Bockris, Ed., Butterworths, London, 1959.
3. Gerischer, H., *Z. Phys. Chem. (N.F.)* (1961) **27**, 48.
4. Gerischer, H., in "The Surface Chemistry of Metals and Semiconductors," H. Gatos, Ed., Wiley and Sons, New York, 1960.
5. Gerischer, H., Mattes, I., *Z. Phys. Chem. (N.F.)* (1966) **49**, 112.
6. Gerischer, H., *Surf. Sci.* (1969) **18**, 97–122.
7. Pleskov, Yu. V., Eletsky, V. V., *Electrochim. Acta* (1967) **12**, 707.
8. Memming, R., Schwandt, G., *Angew. Chem.* (1967) **6**, 851.
9. Memming, R., Schwandt, G., *Electrochim. Acta* (1968) **13**, 1299.

10. Dewald, J. F., in "Semiconductors," B. N. Hannay, Ed., Reinhold, New York, 1959.
11. Dewald, J. F., in "The Surface Chemistry of Metals and Semiconductors," H. C. Gatos, Ed., Wiley and Sons, New York, 1960.
12. Efimov, E. A., Erusalimchik, I. G., Sokolova, G. P., *Russ. J. Phys. Chem.* (1962) **36**, 524.
13. Izidinov, S. O., Borisova, T. I., Veselovskii, V. I., *Dokl. Acad. Nauk. SSSR* (1962) **145**, 598.
14. Izidinov, S. O., Borisova, T. I., Veselovskii, V. I., *Russ. J. Phys. Chem.* (1962) **36**, 659.
15. Memming, R., Schwandt, G., *Surf. Sci.* (1966) **4**, 109.
16. Red'ko, F. F., Izidinov, S. O., *Elektrokhimiya* (1966) **2**, 1128.
17. Red'ko, F. F., Izidinov, S. O., *Elektrokhimiya* (1966) **2**, 1282.
18. Jayadevaiah, T. S., *Appl. Phys. Lett.* (1974) **25**, 399.
19. Gerischer, H., *Ber. Bunsenges. Phys. Chem.* (1965) **69**, 578.
20. Gerischer, H., *J. Electrochem. Soc.* (1966) **113**, 1174.
21. Yamamoto, A., Yano, S., *J. Electrochem. Soc.* (1975) **122**, 260.
22. Gerischer, H., Mindt, W., *Electrochim. Acta* (1968) **13**, 1329.
23. Williams, R., *J. Chem. Phys.* (1960) **32**, 1505.
24. Vanden Berghe, R. A. L., Gomes, W. P., Cardon, F., *Z. Phys. Chem. (N.F.)* (1974) **92**, 91.
25. Schöppel, H. R., Gerischer, H., *Ber. Bunsenges. Phys. Chem.* (1971) **75**, 1237.
26. Hauffe, K., Reinhold, K., *Ber. Bunsenges. Phys. Chem.* (1972) **76**, 616.
27. Hardee, K. L., Bard, A. J., *J. Electrochem. Soc.* (1976) **123**, 1024.
28. Gomes, W. P., *Surf. Sci.* (1970) **19**, 172.
29. Taki, K., Kishi, T., Nagai, T., *Denki Kagaku* (1972) **40**, 216.
30. Sapritskii, V. I., Bardina, N. G., *Elektrokhimiya* (1972) **8**, 655.
31. Beckmann, K. H., Memming, K., *J. Electrochem. Soc.* (1969) **116**, 368.
32. Tyagai, V. A., *Fiz. Tverd. Tela* (1964) **6**, 1602.
33. Gerischer, H., *J. Electroanal. Chem.* (1975) **58**, 263.
34. Gerischer, H., *Ber. Bunsenges. Phys. Chem.* (1976) **80**, 327.
35. Williams, R., *J. Electrochem. Soc.* (1967) **114**, 1173.
36. Hodes, G., Cahen, D., Manessen, J., *Nature* (1976) **260**, 312.
37. Gleria, M., Memming, R., *J. Electroanal. Chem.* (1975) **65**, 163.
38. Boddy, P. J., *J. Electrochem. Soc.* (1968) **115**, 199.
39. Fujishima, A., Honda, K., Kikuchi, S., *Kogyo Kagaku Zasshi* (1919) **72**, 108.
40. Fujishima, A., Honda, K., *Nature* (1972) **238**, 37.
41. Shub, D. M., Remnev, A. A., Veselovskii, V. I., *Elektrokhimiya* (1973) **9**, 676.
42. Gissler, W., Lensi, P. L., Pizzini, S., *J. Appl. Electrochem.* (1976) **6**, 9.
43. Lohman, F., *Ber. Bunsenges. Phys. Chem.* (1966) **70**, 87.
44. Morrison, S. R., Freund, T., *Electrochim. Acta* (1968) **13**, 1343.
45. Vanden Berghe, K. A. L., Cardon, F., Gomes, W. P., *Surf. Sci.* (1973) **39**, 368.
46. Cunningham, J., Corkery, S., *J. Phys. Chem.* (1975) **79**, 933.
47. Mavroides, J. G., Kafalas, J. A., Kolesar, D. F., *Appl. Phys. Lett.* (1976) **28**, 241.
48. Watanabe, T., Fujishima, A., Honda, K., *Bull. Chem. Soc. Jpn.* (1970) **49**, 355.
49. Boddy, P. J., Kahng, D., Chen, Y. S., *Electrochim. Acta* (1968) **13**, 1311.
50. Möllers, F., Memming, R., *Ber. Bunsenges. Phys. Chem.* (1972) **76**, 469.
51. Kim, H., Latinen, H. A., *J. Electrochem. Soc.* (1975) **122**, 53.
52. Rouse, T. O., Weininger, J. L., *J. Electrochem. Soc.* (1966) **113**, 184.
53. Yohe, D., Riga, A., Greef, R., Yeager, E., *Electrochim. Acta* (1968) **13**, 1351.

54. Krüger, G., *Electrochim. Acta* (1968) **13**, 1389.
55. Mavroides, J. G., Tchernev, D. I., Kafalas, J. A., Kolesar, D. F., *Mater. Res. Bull.* (1975) **10**, 1023.
56. Fujishima, A., Kohayakawa, K., Honda, K., *J. Electrochem. Soc.* (1975) **122**, 1487.
57. Yoneyama, H., Sakamoto, H., Tamura, H., *Electrochim. Acta* (1975) **20**, 341.
58. Wrighton, M. S., Ginley, D. S., Wolczanski, P. T., Ellis, A. B., Morse, D. C., Linz, A., *Proc. Natl. Acad. Sci. U.S.A.* (1975) **72**, 1518.
59. Ohnishi, T., Nakato, Y., Tsubomura, H., *Ber. Bunsenges. Phys. Chem.* (1975) **79**, 523.
60. Nozik, A. J., *Nature* (1975) **257**, 383.
61. Keeney, J., Weinstein, D. H., Haas, G. M., *Nature* (1975) **253**, 719.
62. Myamlin, V. A., Pleskov, Yu. V., "Electrochemistry of Semiconductors" (Eng. Ed.), Plenum, New York, 1967.
63. Gerischer, H., "Physical Chemistry: An Advanced Treatise," Vol. IX A, Ch. 5, H. Eyring, Ed., Academic, New York, 1970.
64. Archer, M. D., *J. Appl. Electrochem.* (1975) **5**, 17.
65. Paleocrassas, S. N., *Sol. Energy* (1974) **16**, 45.
66. Bockris, J. O'M., Khan, S. U. M., Uosaki, K., *J. Res. Inst. Catal. Hokkaido Univ.* (1976) **24**.
67. Matthews, D. B., Bockris, J. O'M., "Modern Aspects of Electrochemistry," Vol. 6, J. O'M. Bockris, B. E. Conway, Eds., Ch. 4, Plenum, 1971.
68. Schroeder, R., Lippincott, E. R., *J. Phys. Chem.* (1959) **61**, 921.
69. Gobeli, G. W., Allen, F. G., "Semiconductors and Semimetals," Vol. 2, Ch. 11, R. K. Willardson, A. C. Beer, Eds., Academic, 1966.
70. Azaroff, L. V., Brophy, J. J., "Electronic Processes in Materials," Ch. 7, McGraw-Hill, 1963.
71. Kingston, R. H., Neustädter, S. F., *J. Appl. Phys.* (1955) **26**, 718.
72. Ryvkin, S. M., "Photoelectric Effects in Semiconductors" (Eng. Ed.), Ch. XIII, Consultants Bureau, 1964.
73. Bockris, J. O'M., Habib, M. A., *J. Electroanal. Chem.* (1976) **68**, 367.
74. Bockris, J. O'M., Habib, M. A., *Z. Phys. Chem.* (1975) **98**, 43.
75. Fueki, K., Feng, D. F., Kevan, L., *J. Phys. Chem.* (1970) **74**, 1976.
76. Bockris, J. O'M., Reddy, A. K. N., "Modern Electrochemistry," Vol. 2, Ch. 7, Plenum, 1970.
77. Ketelaar, J. A. A., "Chemical Constitution," pp. 28, 29, Elseviter, 1958.
78. Branscomb, L. M., *Phys. Rev.* (1966) **148**, 11.
79. Kerr, J. A., *Chem. Rev.* (1966) **66**, 465.
80. NRC of U.S.A., "International Critical Table," Vol. VII, 1929.
81. Gurney, R. W., *Proc. Roy. Soc. (London)* (1932) **134A**, 137.
82. Seraphin, B. O., Bennett, H. E., "Semiconductor and Semimetals," Vol. 3, Ch. 12, R. K. Willardson, A. C. Beer, Eds., Academic, 1967.
83. Pankove, J. I., "Optical Processes in Semiconductors," Prentice-Hall, 1971.
84. Fujishima, A., Honda, K., Kikuchi, S., *Kogyo Kagaku Zassi* (1969) **72**, 108.
85. Shub, D. M., Remnev, A. A., Veselovskii, V. I., *Elektrokhimiya* (1973) **9**, 1043.
86. Möllers, F., Tolle, H. J., Memming, R., *J. Electrochem. Soc.* (1974) **121**, 1160.
87. Bockris, J. O'M., Reddy, A. K. N., "Modern Electrochemistry," Vol. 2, Ch. 9, Plenum, 1970.
88. Watanabe, T., Fujishima, A., Honda, K., *Chem. Lett.* (1974) 897.
89. Benard, D. J., Handler, P., *Surf. Sci.* (1973) **40**, 141.
90. Bockris, J. O'M., Uosaki, K., *Energy* (1976) **1**, No. 1.
91. Nakato, Y., Ohnishi, T., Tsubomura, H., *Chem. Lett.* (1975) 883.

RECEIVED July 27, 1976.

4

Conversion of Visible Light to Electrical Energy: Stable Cadmium Selenide Photoelectrodes in Aqueous Electrolytes

MARK S. WRIGHTON, ARTHUR B. ELLIS, and STEVEN W. KAISER

Department of Chemistry, Massachusetts Institute of Technology, Cambridge, Mass. 02139

> *Stabilization of n-type CdSe to photoanodic dissolution is reported. The stabilization is accomplished by the competitive oxidation of S^{2-} or S_n^{2-} at the CdSe photoanode in an electrochemical cell. Such stabilized cells are shown to sustain the conversion of low energy (≥ 1.7 eV) visible light to electricity with good efficiency and no deterioration of the CdSe photoelectrode or of the electrolyte. The electrolyte undergoes no net chemical change because the oxidation occurring at the photoelectrode is reversed at the cathode. Conversion of monochromatic light at 633 nm to electricity is shown to be up to $\sim 9\%$ efficient with output potentials of ~ 0.4 V. Conversion of solar energy to electricity is estimated to be $\sim 2\%$ efficient.*

Photoelectrochemical cells such as that schemed in Figure 1 may prove to be useful devices for converting light to electrical energy or to fuels in the form of electrolytic products. It has been known for over a century (1) that irradiation of an electrode in a cell can result in current flow in the external circuit. Light-induced current flow results in photoelectrolysis with oxidation at one electrode (anode) and reduction at the other electrode (cathode). In principle, the current flow can be utilized directly as electricity by merely introducing (in series) an electrical load into the external circuit. Additionally, the electrolytic products may represent fuel(s) which can be utilized with existing technology. Obviously, for example, the photoelectrolysis of H_2O according to either Equation 1 or Equation 2 represents a light-to-chemical energy conversion

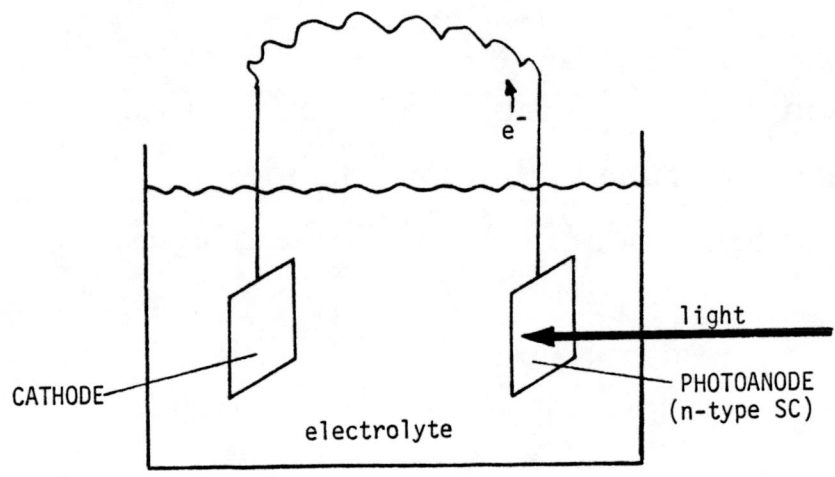

TYPICAL PHOTOELECTROCHEMICAL CELL

Figure 1. Typical photoelectrochemical cell

$$H_2O \rightarrow H_2 + \tfrac{1}{2}O_2 \qquad (1)$$

$$2H_2O \rightarrow H_2 + H_2O_2 \qquad (2)$$

of some potential interest since the fuels H_2 and H_2O_2 have, or could have, considerable utility.

While metal electrodes often yield photocurrents when irradiated, semiconductor photoelectrodes generally give the highest photocurrents. In many cases every photon absorbed by the semiconductor electrode yields an electron in the external circuit. Semiconductor photoelectrodes have at least one basic property which makes them attractive as the photoreceptor in the cells: a mechanism for the inhibition of recombination of photogenerated electron-hole pairs. Figure 2 shows the general model associated with the semiconductor–electrolyte interfacial region, showing that the bands in the semiconductor can be bent near the surface in such a way as to inhibit recombination of electron-hole pairs. Moreover, the bands may be bent in a direction appropriate for the observation of a substantial anodic photocurrent for n-type semiconductors and a cathodic photocurrent for p-type semiconductors. These band bending considerations have been elaborated previously (2). The separation of the photogenerated electron-hole pair allows net redox chemistry to compete very effectively with recombination.

The importance of electron-hole separation immediately after photogeneration can be appreciated by considering an early proposition for the photoassisted conversion of H_2O to H_2 and O_2. It was claimed that

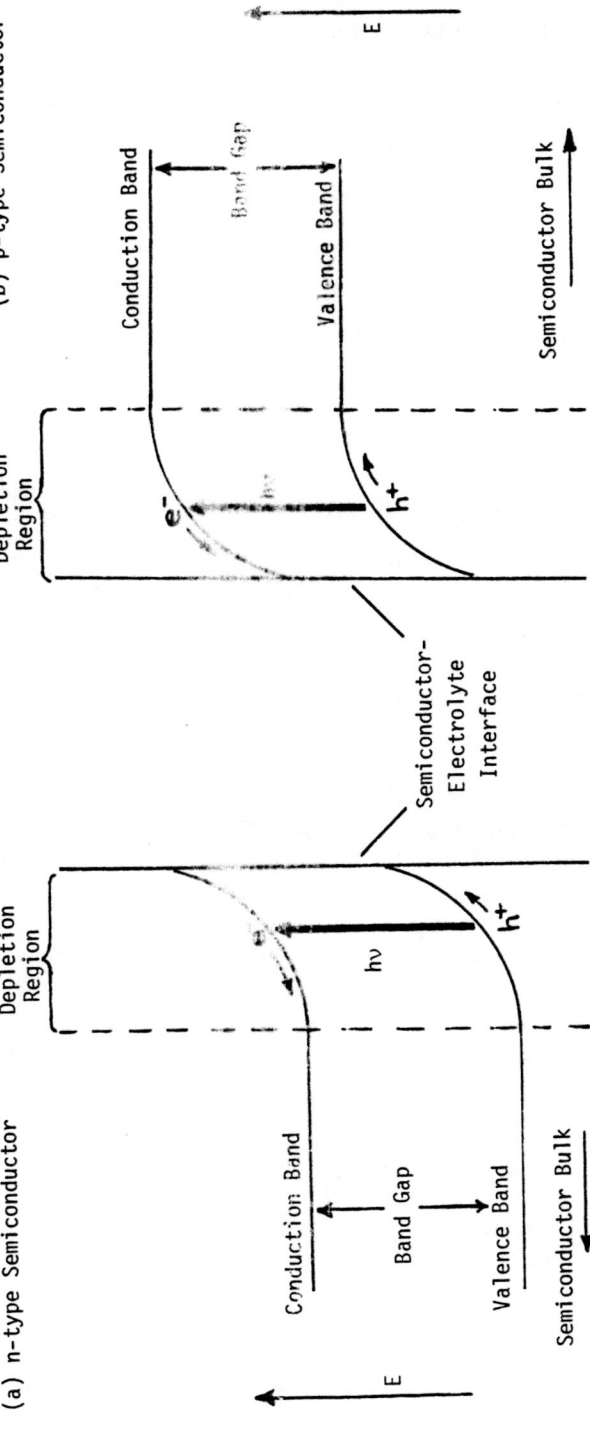

Figure 2. (a) Semiconductor–electrolyte interfacial region showing band bending favorable for observation of a photocurrent for an n-type semiconductor. (b) Same as (a) except for a p-type semiconductor.

one could couple the oxidation and reduction of H_2O to the photo-induced oxidation and reduction of aquo metal ions such as Fe^{2+} and Fe^{3+} (3). That is, irradiation of Fe^{2+} and Fe^{3+} proceeds, at least initially, according to Equations 3 (4–11) and 4 (12, 13, 14, 15), respectively. These sum to give Equation 5, yet the conversion of H_2O to H_2 and O_2

$$Fe^{2+} (aq) + H^+ \xrightarrow[\text{water}]{h\nu} Fe^{3+} (aq) + \tfrac{1}{2} H_2 \qquad (3)$$

$$Fe^{3+} (aq) + OH^- \xrightarrow[\text{water}]{h\nu} Fe^{2+} (aq) + \tfrac{1}{4} O_2 + \tfrac{1}{2} H_2O \qquad (4)$$

$$\tfrac{1}{2} H_2O \xrightarrow[\substack{Fe^{2+}/Fe^{3+} \\ \text{water}}]{2h\nu} \tfrac{1}{2} H_2 + \tfrac{1}{4} O_2 \qquad (5)$$

cannot be sustained using the Fe^{2+}/Fe^{3+} photoassistance agent. The underlying reason lies in the fact that half a molecule of H_2 is a hydrogen atom. To form a gaseous H_2 molecule two hydrogen atoms must combine, but the hydrogen atom can back react with the photogenerated Fe^{3+} which accumulates with time. The problem is that H_2 is not irreversibly, promptly generated with one photon. The high energy hydrogen atom or protonated atom can simply back react to yield no net chemistry. Inhibiting the back reaction of the high energy intermediate is analogous to inhibiting electron-hole recombination in the irradiated semiconductor.

It is widely believed that the degree of band bending in the semiconductor is equal to the difference in the Fermi levels in the semiconductor and the electrolyte (2). Consequently, the band bending may or may not be large enough to prevent electron-hole recombination, depending on the redox active components in the electrolyte and the properties of the semiconductor itself. Moreover, the maximum theoretical open-circuit photopotential is equal to the amount of band bending, and thus the efficiency of the utilization of the light depends on the band bending for a given semiconductor. To affect the band bending favorably an applied bias can be supplied by a power supply in series in the external circuit. It is usually assumed that the entire potential drop occurs in the depletion region of the semiconductor and not in the electrolyte as for metal electrodes (2). If the objective is to produce fuels by photoelectrolysis, the applied potential used must be lower than the thermodynamic reversible electrolysis potential. If the applied potential exceeds the thermodynamic potential, the role of the light, at best, is to serve as a mechanism to reduce overvoltage encountered in conventional electrolysis. The maximum theoretical storage efficiency, η_{max}, by photoelectrolysis is given by Equation 6, where E_p is the reversible electrolysis potential of the reaction being driven (e.g., 1.23 V for Equation 1); E_{BG} is the band gap energy of the semiconductor photoelectrode; and V_{appl} is the potential provided by the power supply in the circuit. To obtain maximum efficiency, then, one must attempt (a) to match the band gap

$$\eta_{max} = \frac{E_p - V_{appl}}{E_{BG}} \qquad (6)$$

of the semiconductor with the electrolysis potential of the reaction, and (b) to seek redox components and semiconductor materials such that the external power supply is not needed.

If the objective is to produce electrical power from light, there must be no external supply. Otherwise, the light-to-electrical energy conversion efficiency would be negative. In more explicit terms, if the photoelectrochemical cell is to produce both a fuel and electricity, the band bending must exceed E_p, and the maximum fraction of energy output as electricity will be the difference between the band bending and E_p. The band bending requirement is for a nonilluminated electrode in equilibrium with the half-cell redox reaction which occurs at the photoelectrode. It is conceivable that conversion of light to electrical energy can be sustained without the production of a fuel; in this case one seeks a chemical redox system like that indicated in Equations 7 and 8; i.e., the electrolyte contains both A and B, and their distribution does not change.

$$A \longrightarrow B \quad \text{(cathode)} \qquad (7)$$

$$B \longrightarrow A \quad \text{(anode)} \qquad (8)$$

In this case the theoretical maximum optical-to-electrical energy conversion efficiency is the extent of band bending divided by the band gap energy.

Besides controlling band bending in the semiconductor, the redox components in the electrolyte can also play a key role in whether the electron transfer processes to and from the semiconductor are fast enough to compete with electron-hole recombination. The fastest rates of electron transfer can be expected when the appropriate semiconductor band overlaps the position of the redox levels in the electrolyte.

While much effort has been applied in providing this understanding of semiconductor photoelectrodes (2, 16, 17, 18, 19), some difficulties are encountered in exploiting photoelectrochemical cells. At least one major problem is that the semiconductor itself is often the electrochemically reactive species, and as such it undergoes irreversible photoelectrolysis.

Until recently, in fact, there existed no n-type semiconductor electrode which could survive irradiation in an aqueous electrolyte without decomposition. A typical situation is encountered with n-type CdS. Irradiation of this material results in dissolution according to Equation 9 (20, 21, 22).

$$CdS \xrightarrow{h\nu} Cd^{2+} (aq) + S (s) + 2e^- \quad (9)$$

The result is that current flows in the external circuit, but the chemistry occurring at the photoanode results in the decomposition of the electrode with zero valent sulfur precipitating on the surface and Cd^{2+} ions going into solution. Initial experiments carried out by Fujishima and Honda (23, 24, 25, 26) on n-type TiO_3 indicated that TiO_2 is an inert photoelectrode, and subsequently others (27–36) have been stimulated to characterize more fully the TiO_2 photoelectrode. All findings are consistent with the conclusion that TiO_2 itself is not susceptible to photoanodic dissolution. It has now been shown that $SrTiO_3$ (37, 38, 39), SnO_2 (38), $KTaO_3$ (40), and $KTa_{0.77}Nb_{0.23}O_3$ (40), WO_3 (41), and Fe_2O_3 (42) are all stable photoelectrodes in aqueous electrolytes.

Stability of the metal oxide n-type semiconductors allows their use as the photoreceptor in photoelectrochemical cells for the photo-assisted electrolysis of H_2O. Indeed all of those listed above as stable have been shown to assist the conversion of H_2O to H_2 and O_2 in cells as shown in Figure 3. Since the reversible electrolysis potential of H_2O is 1.23 V, the ability to sustain the electrolysis at potentials below this applied potential requires the input of energy by some other mechanism. Irradiation of the stable semiconductor electrodes does allow the electrolysis to proceed at applied potentials substantially lower than 1.23 V, and consequently the light can be converted to chemical energy in the form of the electrolytic products H_2 and O_2. If energy from H_2 and O_2 is recoverable at 56.7 kcal/mol H_2, $SrTiO_3$-based cells have $\sim 25\%$ efficiency for the conversion of 330 nm light (37). Since the band gap in $SrTiO_3$ is 3.2 eV (43, 44), the maximum efficiency is 1.23/3.2 or $\sim 38\%$. Thus the closeness of the measured efficiency to this theoretical efficiency implies that the quantum yield is high and that the applied potential required is small, as is the case (37). The high absorptivity of the $SrTiO_3$ is also a useful property in that it allows the photons to be completely absorbed within the depletion region, setting the stage for high observed quantum efficiency.

A major hurdle in improving $SrTiO_3$ efficiency is to lower the band gap without sacrificing current–voltage properties or stability. The low energy visible response of stable Fe_2O_3, for example, is significantly offset by the large V_{appl} necessary to drive the photoelectrolysis of H_2O. In this article we wish to summarize our initial results on one

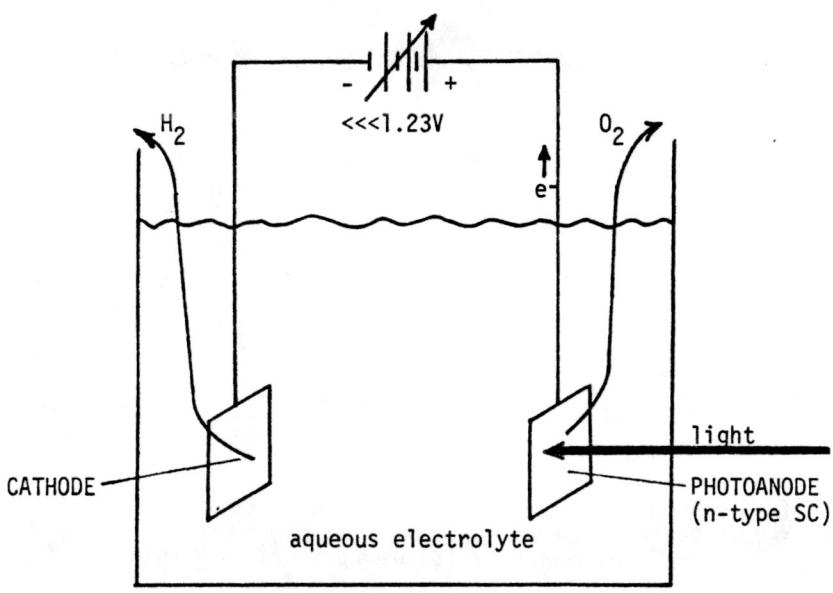

Figure 3. Photoelectrochemical cell used to photoelectrolyze H_2O to H_2 and O_2. Photoelectrodes of n-type TiO_2, SnO_2, $SrTiO_3$, $KTaO_3$, or $KTa_{0.77}Nb_{0.23}O_3$ have been shown to be effective.

basic approach to developing useful photoelectrochemical cells having low energy visible response. The approach is to use redox active electrolytes which can be used competitively to quench photoanodic dissolution by scavenging photogenerated holes before electrode dissolution can occur. Success will be illustrated with the stabilization of n-type CdSe ($E_{BG} = 1.7$ eV) (45) by using polysulfide electrolytes. The full details have been elaborated elsewhere (46, 47).

Results and Discussion

Stabilization of CdSe. Like n-type CdS, n-type CdSe photoelectrodes undergo rapid photoanodic dissolution upon irradiation in an aqueous electrolyte (48, 49). We have found (46, 47) that the photoanodic dissolution of CdSe or CdS can be quenched by adding polysulfide to the aqueous electrolyte. Oxidation of the added polysulfide occurs at the expense of the oxidation of the selenide of the CdSe as schemed in Figure 4. We choose to define a stable photoelectrode here as one which undergoes no weight loss as a consequence of the photocurrent. Table I summarizes typical data which support the claim that CdSe is "stabilized" by the addition of polysulfide to an aqueous electrolyte. In several instances the number of oxidizing equivalents was enough to consume several times the weight of the CdSe crystals.

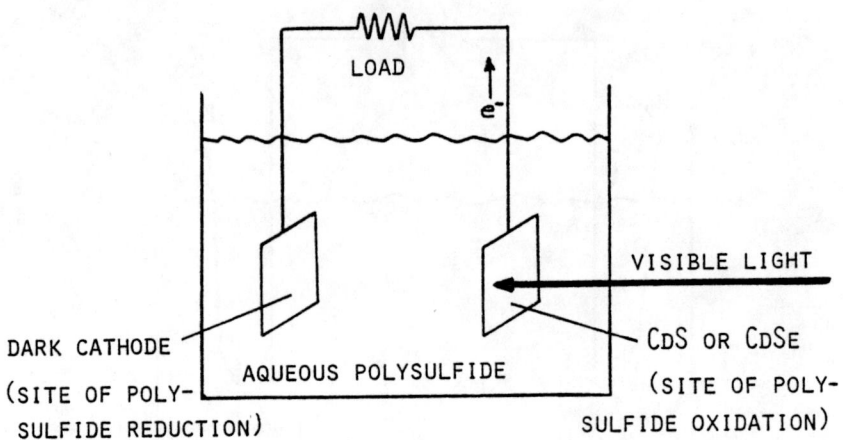

Figure 4. Schematic of n-type CdSe-based photoelectrochemical cell. Photoanodic dissolution of CdSe does not occur in the aqueous polysulfide electrolyte; rather the polysulfide is oxidized at the photoelectrode.

The kind of stability indicated in Table I for CdSe is remarkable. It has been shown in several instances that certain species can be oxidized competitively with the electrode decomposition, but these data for the CdSe are the first to show that the decomposition apparently can be totally quenched. Similar results were obtained for CdS (46, 47). Substances such as $Fe(CN)_6^{4-}$, I^-, and quinones can be oxidized at CdS or

Table I. Stability of n-Type CdSe Photoelectrode in Aqueous Polysulfide Electrolytes[a]

Exp. No.[b]	Crystal	Face	Crystal (mol × 10⁴)		Electrons Generated (mol × 10⁴)	Av i (mA[c])	Time (h)
			Before	After			
1	1	0001	9.39	9.41	4.20	0.64	17.6
2	2	000$\bar{1}$	8.61	8.61	3.31	0.41	21.6
3	3	—	8.78	8.75	23.9	4.2	15.4

[a] Photoelectrochemical cell with CdSe photoanode; see Figure 4.
[b] Exp. 1: Electrolyte is 1.0M NaOH, 1.0M Na_2S, 1.0M S continuously purged with Ar. Irradiation is at 633 nm (2.8 mW) at an applied potential of −0.35 V (negative lead to CdSe) with a Pt gauze cathode. Photoelectrode etched to expose the 5 × 5 mm 0001 face. The photoelectrode is 1 m mthick and the resistivity is $\sim$ 2 Ωcm. Exp. 2: Same as Exp. 1 except 000$\bar{1}$ surface is exposed. Exp. 3: Electrolyte is 1.25M NaOH, 0.2M Na_2S, and the CdSe has not been etched. Irradiation is with wavelengths longer than 420 nm from a 200 W super-pressure Hg lamp. A Pt wire cathode and an applied potential of +2.0 V (positive lead to CdSe) were used. The exposed surface of the 1 mm thick CdSe crystal was 5 × 5 mm and the resistivity was 14 Ωcm.
[c] Multiply by 4.0 cm^{-2} to obtain mA/cm^2.

CdSe photoelectrodes (48, 49, 50, 51), but even at high concentrations these have not given good electrode stability.

It is interesting to speculate on the reason for the remarkable stability of the CdSe and CdS in the polysulfide system. The result is very interesting since it is very evident that other additives are capable of being oxidized to some extent at the photoelectrode. The complete stability of CdSe or CdS in polysulfide electrolytes must be a consequence of the very fast rates of polysulfide oxidation compared with $Se^{2-} \xrightarrow{lattice}$ Se^0 or $S^{2-} \xrightarrow{lattice} S^0$ oxidation. Other additives are only competitively oxidized.

Having established the kind of stability of CdSe indicated in Table I, it is appropriate to consider a criterion for electrode stability which is somewhat more subtle: photocurrent stability. A feeling for photocurrent stability can be gained from examining the drop-off in photocurrent with time in 1.0M NaOH compared with the steady photocurrent obtained using an electrolyte consisting of 1M NaOH, 1M Na_2S, and 1M S (Figure 5). We have found that, even at very high light intensities, photocurrents in the CdSe-based cell are very stable. At the very least, the polysulfide electrolyte provides a mechanism for photocurrent stability for a period which is many orders of magnitude longer than for the 1M NaOH solution. The current profile for Exp. No. 2 of Table I is representative: the photocurrent was initially 0.405 mA and rose to 0.425 mA where it levelled for 5.9 hr. Over the next 7.3 hr the current fell to 0.40 and then held constant for the remaining 8.3 hr. Thus the observed photocurrent is very stable, at least for the 1.0M NaOH, 1.0M Na_2S, 1.0M S electrolyte.

We can discuss, in qualitative terms, the stability of the CdSe as a function of electrolyte composition. First, we have generally found that an electrolyte consisting of 1.0M NaOH, 1.0M Na_2S, and 1.0M S provides a very stable system. Significantly diminishing the amount of S leads to considerable changes in the current–voltage properties (vide infra), and, for etched CdSe electrodes, there seems to be some deterioration of the photocurrent with time, especially at high light intensities. For non-etched samples we have found that solutions containing only 1.0M NaOH and 0.2M Na_2S give satisfactory stability even at very high light intensities. At this point, we simply do not have a good enough measure of electrode stability to do a quantitative study at intermediate concentrations of added sulfide where there is some oxidation of added sulfide and some oxidation of CdSe.

At the very least, CdSe has been stabilized to such an extent that it is now possible to study current–voltage properties, quantum efficiency, wavelength response, etc. without the fear of irreversible decomposition

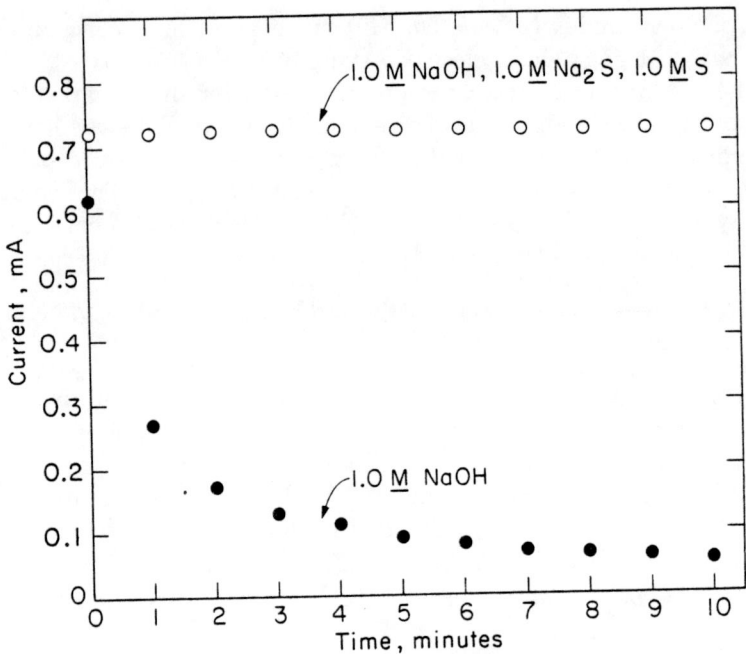

Figure 5. *Photocurrent as a function of time in a CdSe-based cell using 1M NaOH as an electrolyte (●) or using an electrolyte consisting of 1M NaOH, 1M Na_2S, and 1M S (○). The irradiation source is a beam-expanded He–Ne laser with output at 633 nm. The 0001 face of the crystal is exposed to the electrolyte.*

of the electrode. Interestingly, such measurements at high light intensities have really never been possible because of the decomposition problem.

Wavelength Response. Generally, the onset of photocurrent with variation in excitation wavelength will occur near the position of the band gap transition energy. The wavelength response of the CdSe in 1.0M NaOH, 1.0M Na_2S is shown in Figure 6. As seen in the figure, the onset of photocurrent and the onset of the band gap transition coincide at ~750 nm, consistent with the reported E_{BG} of CdSe (45). With regard to potential solar energy conversions, we note that CdSe absorbs ca. 50% of the incident solar insolation (52) at the earth's surface.

Aside from the fact that the onset of the photocurrent is near the band gap energy, the high energy visible response is quite good. In fact, as the excitation energy is increased, there seems to be a gentle increase in response. One possible explanation for the increased response is that there is a greater percentage of the incident light absorbed within the depletion region at the shorter wavelengths. The absolute quantum efficiency for electron flow will be discussed below.

Open Circuit Photopotential. The open circuit photopotential depends on the light intensity, and over a significant range of intensities the open circuit photopotential should increase linearly with the log of the intensity (2). Naturally, at some point the photopotential must reach a value where higher intensities have no effect. In 1.0M NaOH, 1.0M Na_2S, and 1.0M S the open circuit photopotential depends on light intensity as shown in Figure 7. These results compare favorably with those reported (2) for CdSe-based cells in other electrolytes.

According to our model, the limiting photopotential is equal to the band bending which can be no greater than the value of E_{BG}. Measurement of open circuit photopotentials of the order of 0.5 E_{BG} indicates that the band bending is very substantial and that we can expect theoretical energy conversion efficiencies of ~ 50% for irradiation at E_{BG}.

Current–Voltage Properties. Using a standard three-electrode cell configuration and a potentiostat, we have examined the current–voltage

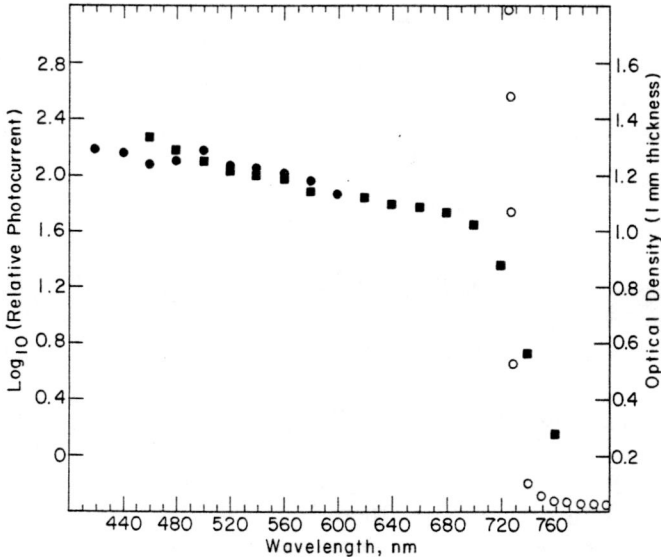

Figure 6. Wavelength response curve for a CdSe-based photoelectrochemical cell with a 1M NaOH, 1M Na_2S electrolyte. The filled circles (●) and squares (■) are relative photocurrents as a function of incident irradiation wavelength after correction for variation in intensity with wavelength. The filled circles are values obtained using the excitation optics of an Aminco-Bowman emission spectrophotometer as the irradiation source, and the filled squares are values using a 600-W tungsten source monochromatized using a Bausch & Lomb high intensity monochromator. The open circles (○) are the optical density for a 1-mm thick polished CdSe crystal.

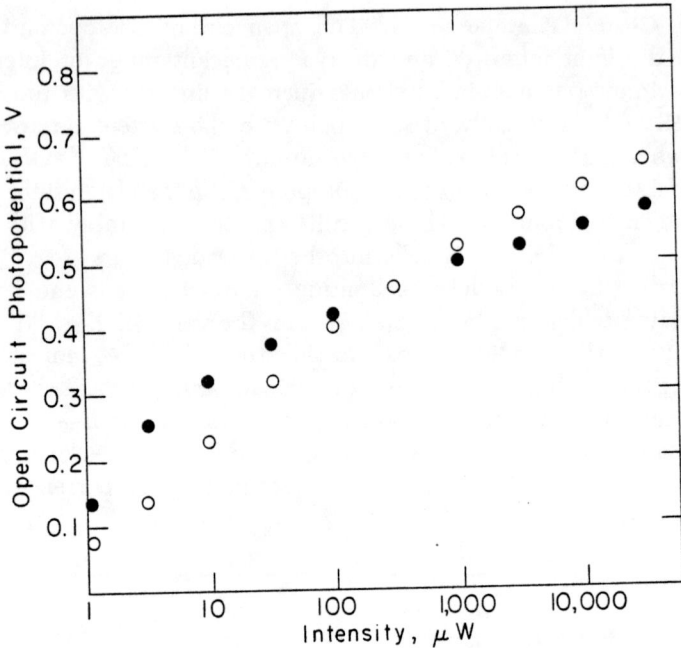

Figure 7. Plot of open circuit photopotential against light intensity (514.5 nm) for CdSe-based cell with a 1M NaOH, 1M Na_2S, 1M S electrolyte. Open circles (○) represent the 0001 face and filled circles (●) represent the 0001 face of the crystal exposed to the electrolyte.

properties of the CdSe-based cells. The reference electrode was a saturated calomel electrode (SCE), and the potentials of both the Pt electrode and the CdSe potential were monitored; the Pt electrode potential did not vary. Current–voltage curves are a very sensitive function of surface treatment (polishing, etching, etc.). All data referred to here are for etched single crystals. As expected, the current–voltage properties depend on light intensity and electrolyte composition as shown in Figures 8, 9, and 10. First, in the dark there is only a small anodic current, consistent with the fact that no minority charge carriers are available to participate in the charge transfer. Irradiation, however, produces holes which give rise to an anodic photocurrent as expected for an n-type semiconductor. The onset of the anodic photocurrents occurs at more negative potentials relative to the SCE as the light intensity is increased. This shift in anodic photocurrent onset can be seen clearly in Figure 8 where the current–voltage properties for CdSe are shown at three different light intensities. The shift in anodic photocurrent onset is consistent with the open-circuit photopotential plots given in Figure 7. From Figure 7 we see a change of ∼ 0.15 V in potential with an order of magnitude change

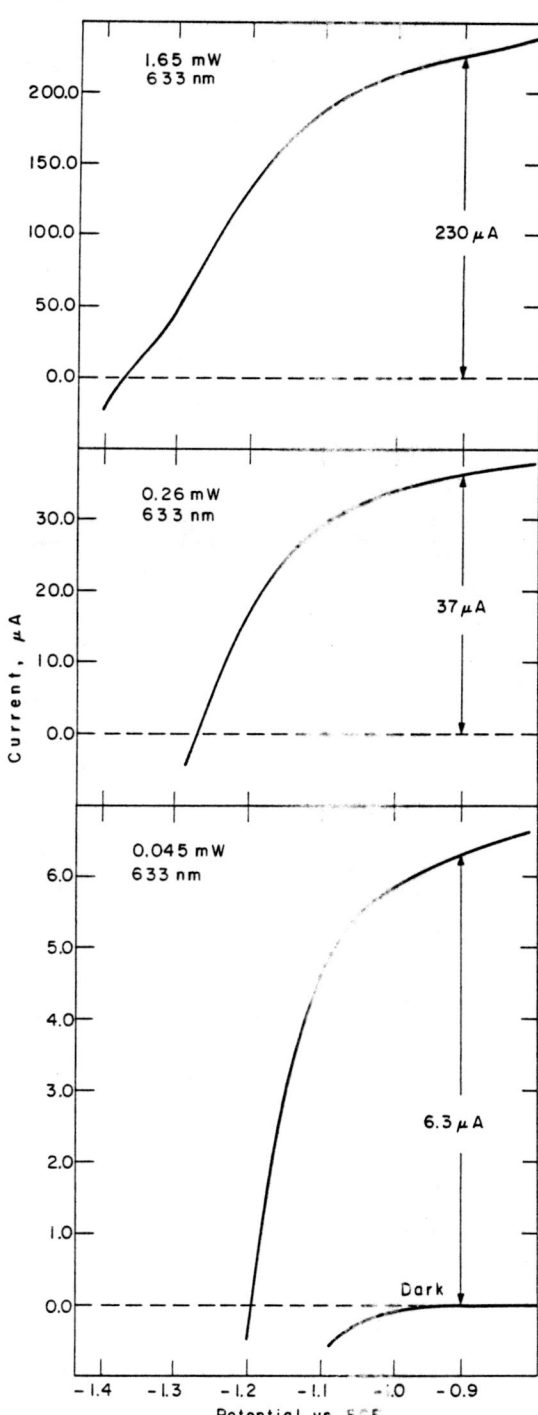

Figure 8. Current–voltage properties for an irradiated (633 nm) n-type CdSe (0.25 cm² surface area) electrode with 0001 face exposed to the 1.0M NaOH, 1.0M Na₂S, 1.0M S electrolyte. The Pt dark electrode potential is constant at −0.71 V vs. SCE for any bias. The sweep rate for all curves is 0.2 V per minute. Note the dependence of the curves on irradiation power.

in light intensity. The shift in the anodic photocurrent onset is a comparable amount with the 10-fold change in light intensity.

Figure 8 shows that the limiting photocurrent is approximately directly proportional to the light intensity. Moreover, the shapes of the curves are fairly similar for the range of intensities studied. At higher light intensities, however, the photocurrent is not directly proportional to light intensity. The saturation effect sets in at different intensities depending on the electrolyte (vide infra) and somewhat on the particular electrode and its surface treatment. Saturation effects of this sort are often encountered and indicate that the hole capture by S_n^{2-} is only competitive with electron-hole recombination, and the competition is apparently influenced by light intensity.

We have investigated the current–voltage properties of CdSe as a function of irradiation wavelength for wavelengths near the band gap (Figure 9). The actual intensity striking the electrode has been held constant. However, as seen in Figure 9, the photocurrent not only increases as the irradiation wavelength is shortened, but the current–voltage curves differ in the same way that they differ with changes in

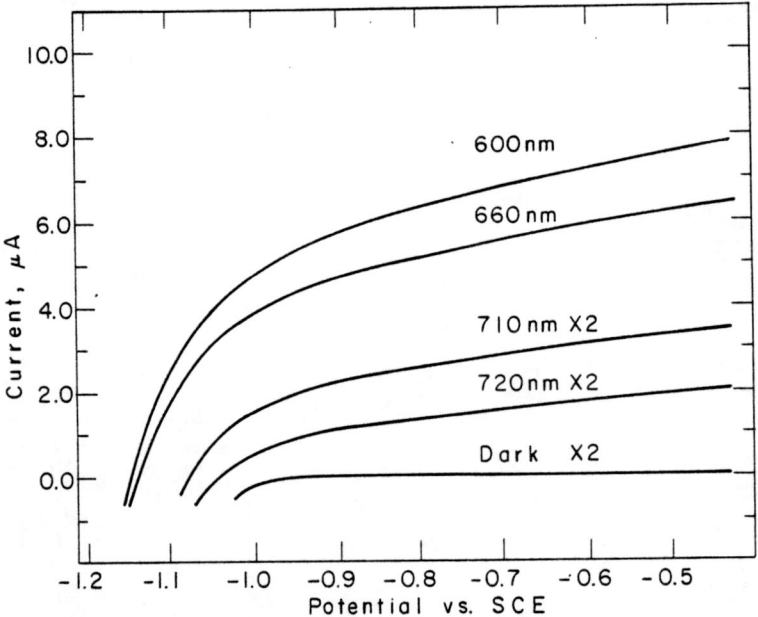

Figure 9. Current–voltage properties of CdSe as a function of incident irradiation wavelength. The 0001 face, 0.25 cm² surface area, is exposed to the 1.0M NaOH, 1.0M Na₂S, 1.0M S electrolyte and illuminated at a constant intensity of 7.2×10^{-10} ein/sec at all four wavelengths. The Pt dark cathode potential was constant at -0.72 V vs. SCE, and the sweep rate was 0.2 V per minute.

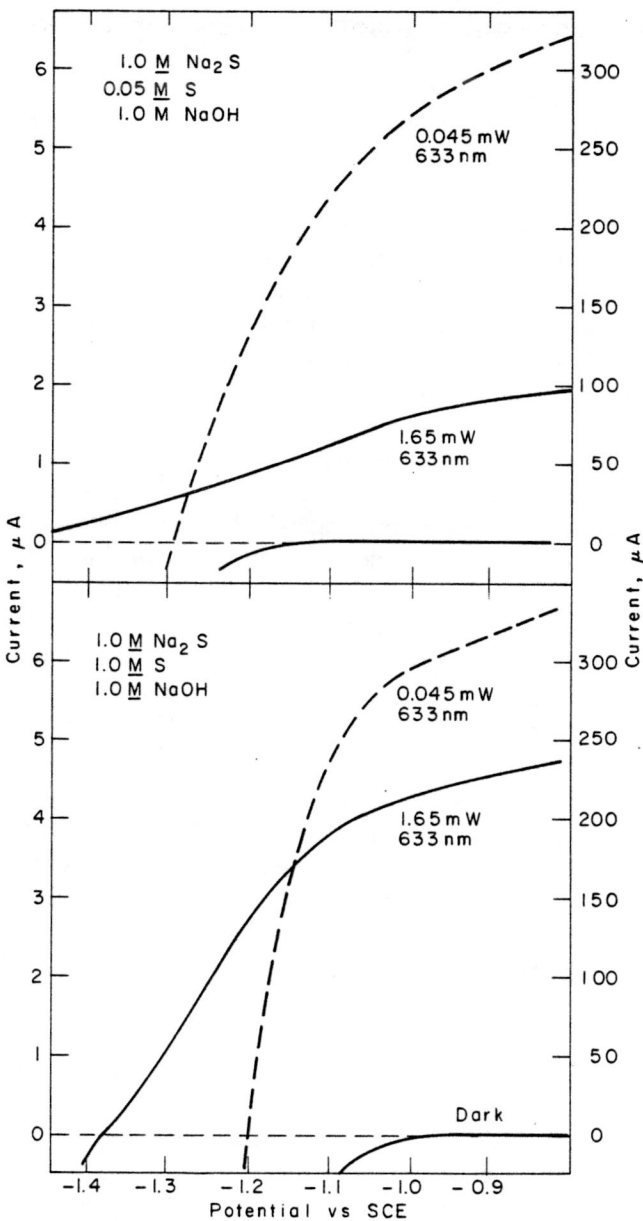

Figure 10. Dependence of CdSe current–voltage properties on intensity and electrolyte composition. In each case the sweep rate was 0.2 V per minute, and the 0.25 cm² 0001 surface is exposed. In the 0.05M S electrolyte the Pt electrode is at −0.78 V vs. SCE, and in the 1.0M S electrolyte the Pt electrode is at −0.71 V vs. SCE.

light intensity at a constant wavelength. We therefore ascribe the changes in current–voltage behavior with variation in wavelength to differences in the amount of light absorbed in the depletion region. At the points where absorptivity of CdSe is relatively small, the amount of light absorbed in the depletion region is small, but at wavelengths substantially shorter than band gap, a very sizable fraction of the incident irradiation is absorbed in the depletion region.

Some extremes in the current–voltage properties of CdSe with changes in electrolyte composition are given in Figure 10. The curves show that the smaller S concentration leads to deleterious effects and that the effect is most pronounced at high light intensity. The photocurrent seems to saturate at a lower light intensity and does not increase as steeply with increasing anodic bias as for the electrolyte consisting of 1.0M NaOH, 1.0M Na$_2$S, and 1.0M S.

Comparison of 000$\bar{1}$ and 0001 Faces. It is possible to expose either the 0001 or 000$\bar{1}$ face of CdSe (53). For the 000$\bar{1}$ face one would have principally Se exposed. Determining whether there is a difference between these two is a very simple experiment in principle, but we have found that the differences are sufficiently small (or our reproducibility so poor) that we cannot, at this time, report differences in either photopotential or current–voltage properties for the 0001 and the 000$\bar{1}$ faces. Generally, for example, we have been able to reproduce current–voltage curves for a given crystal of CdSe such that the anodic photocurrent onsets are within ∼ 100 mV of each other. Recall that the curves depend significantly on etching procedure, etc.

Power Conversion Efficiency. The sustained conversion of light energy to electrical energy in the present system is possible, since the current–voltage curves show that the photocurrent will flow against a negative applied potential. Typically, an anodic bias will assist the flow of current such that oxidations occur at the CdSe. The significance of an anodic photocurrent flowing with a cathodic bias is that the power supply is an electrical load, and the power output of the cell is just current multiplied by voltage. A plot of photocurrent vs. applied potential from a power supply in series in the external circuit is shown in Figure 11. The curve shows very clearly that an anodic photocurrent will flow even at very negative applied potentials. However, the maximum value of current times voltage occurs at some intermediate negative applied potential. The data from Figure 11 reveal that 9.2% of the input optical power at 633 nm is recoverable as electrical power. We have also demonstrated equivalent sustained power conversion efficiencies by replacing the power supply with a resistor in series in the external circuit.

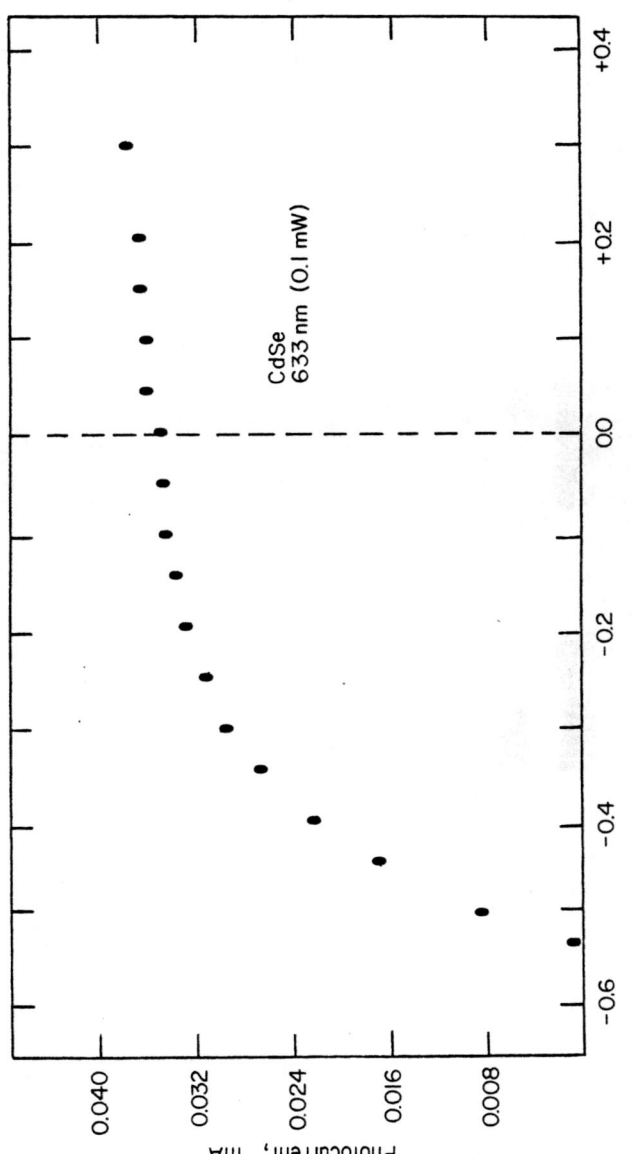

Figure 11. Photocurrent against applied potential from a power supply in series in the external circuit. The 0.25 cm² 0001 face of the CdSe was exposed to the 1.0M NaOH, 1.0M Na₂S, 1.0M S electrolyte. Maximum power conversion efficiency (9.2%) occurs at −0.35 V applied.

Sustained conversion of light to electrical energy demands a totally non-deteriorating system. We have demonstrated that the photoelectrode is stable and have assumed that the Pt cathode is inert. The electrolyte must be capable of being oxidized at CdSe and reduced at Pt with no net chemical change. Aqueous polysulfides meet this criterion in our hands as demonstrated by passing electric current through a 1.0M NaOH, 1.0M Na$_2$S, 1M S electrolyte (2.0 ml) using two Pt electrodes. Current was passed at ~ 0.2 V and ~ 2.0 mA/cm^2 for a very long period with no obvious deterioration of the system. Two other practical considerations should be emphasized. Polysulfides are sensitive to O$_2$, and electrolytes used have been continuously purged with Ar. Secondly, aqueous polysulfides (but not Na$_2$S) absorb blue light quite strongly. The 1.0M NaOH, 1M Na$_2$S, 1M S electrolyte is orange and has an optical density of ~ 1.0 in a 1 mm path length at ~ 490 nm. Thus, very short path lengths are required for complete visible spectral response in this electrolyte.

The 9.2% sustained efficiency shown in Figure 11 is one of the best values we have obtained. Note that this value is for a very low light intensity. Increasing the light intensity causes saturation of the photoeffect, but reasonable sustained efficiencies are found, Table II. Even higher absolute power outputs from the CdSe-based cell are possible, but, of course, the efficiency is less.

The inefficiency in the power conversion rests with the fact that we do not see a unit quantum yield for electron flow at a negative applied

Table II. Power Conversion Efficiency Using CdSe-Based Photoelectrochemical Cells

Exp. No.[a]	Crystal No.	Face Exposed	Irrdn, λ (nm) [Power (mW)][b]	Max Power Out (mW)	η_{max} (%)	Potential @η_{max} (V)	Current @η_{max} (mA)[c]
4	1	0001	632.8 [0.10]	0.0092	9.2	−0.35	0.0263
			[2.8]	0.168	6.0	−0.35	0.480
5	2	000$\bar{1}$	632.8 [0.10]	0.0053	5.3	−0.35	0.0151
			[2.8]	0.117	4.2	−0.35	0.333
6	4	not etched	632.8 [2.2]	0.0082	0.4	−0.20	0.041
7	5	0001	514.5 [0.025]	0.0012	4.8	−0.35	0.0034
			[7.30]	0.176	2.4	−0.55	0.320

[a] All experiments were performed in an electrolyte consisting of 1.0M NaOH, 1.0M Na$_2$S, 1.0M S. The circuit is schemed in Figure 4. See also notes in Table I.
[b] Multiply by 4.0 cm^{-2} to obtain mW/cm^2.
[c] Multiply by 4.0 cm^{-2} to obtain mA/cm^2.

Table III. Quantum Efficiency for Electron Flow for CdSe-Based Photoelectrochemical Cells

Exp. No.	Crystal No.	Face Exposed	Irrdn, λ (nm) [Intensity (ein/sec)][a]	V_{appl}	$\Phi \pm 15\%$
4[a]	1	0001	632.8 [5.29 × 10⁻¹⁰]	−0.35	0.52
				+0.40	0.77
			632.8 [148 × 10⁻¹⁰]	−0.35	0.33
				+0.40	0.55
8[b]	6	0001	632.8 [10.6 × 10⁻¹⁰]	+1.00	0.53
			454.5 [6.76 × 10⁻¹⁰]	+1.00	0.90
9[b]	7	000$\bar{1}$	454.5 [6.46 × 10⁻¹⁰]	+1.00	0.67
			632.8 [10.6 × 10⁻¹⁰]	+1.00	0.34
5[a]	2	000$\bar{1}$	632.8 [5.29 × 10⁻¹⁰]	−0.35	0.30
				+0.40	0.50
			632.8 [148 × 10⁻¹⁰]	−0.35	0.23
				+0.40	0.38
7[a]	5	0001	514.5 [1.07 × 10⁻¹⁰]	−0.35	0.33
				0.00	0.40
			514.5 [3.4 × 10⁻¹⁰]	−0.55	0.11
				0.00	0.17

[a] 1.0M NaOH, 1.0M Na₂S, 1.0M S electrolyte using same circuit as in Figure 4.
[b] 1.0M NaOH, 1.0M Na₂S electrolyte using same circuit as in Figure 4.
[c] Exposed electrode area is 0.25 cm².

potential equal to the band gap energy. This is because of (a) saturation effects at high light intensities and (b) a relatively (compared to band gap energy) small degree of band bending. However, the observed quantum yields for electron flow (not corrected for reflective losses) are rather high (but certainly not unity) at the applied potential for the maximum energy conversion and reach even higher values at positive, applied potentials (Table III). The data in Table II and Figure 11 support the claim that the CdSe-based cell is one of the more efficient reported photoelectrochemical devices for the conversion of optical energy. Conversion of sunlight to electrical energy with an efficiency of at least 2% could be expected using the single-crystal CdSe-based cells.

Summary and Perspective

The stabilization of CdSe to photoanodic dissolution by polysulfide electrolytes has been demonstrated (46, 47). Optical to electrical energy conversion efficiencies of ~ 9% have been obtained with no deterioration of the electrolyte or photoelectrode, and the maximum power output of the CdSe-based photoelectrochemical cell occurs at a potential of a few tenths of a volt. The stability has allowed the first measurements of current–voltage properties of CdSe even at high light intensities without problems associated with photoanodic dissolution.

The crucial result here is the stabilization. This shows that it is possible, by competitive redox processes, to quench completely photoelectrolysis of semiconductors. The essence of the results outlined here has been repeated successfully in other laboratories (54, 55). In addition to CdS, the n-type Bi_2S_3 ($E_{BG} = 1.4$ eV) (56, 57) can be stabilized (55) using the polysulfide electrolyte. Recently, in our own laboratory we have stabilized CdTe ($E_{BG} = 1.4$ eV), GaP ($E_{BG} = 2.24$ eV), GaAs ($E_{BG} = 1.35$ eV), and InP ($E_{BG} = 1.25$ eV) using the polysulfide electrolyte or other chalcogenide-containing electrolytes (58, 59). These several examples of stabilization are promising, but there are disadvantages to the competitive electron transfer approach. The crucial disadvantage is that one clearly restricts the range of chemical reactions that can be driven photoelectrochemically. If the objective is to convert light to electricity, this may be no drawback.

Other approaches to the stabilization of small band gap semiconductors exist, and they are not without disadvantages. One approach is to coat the small band gap material with a thin, inert metal film. It has been claimed, for example, that Au-coated n-type GaP is stable to photoanodic dissolution, and one can oxidize H_2O at such a photoelectrode (60). This approach is no different from a Schottky-barrier photocell, and a difficulty encountered here will be in fabrication of the solid–solid junction. Another tactic has been to attempt to coat an unstable electrode with another semiconductor that is stable. For example, coating CdS with a thin film of TiO_2 has been suggested and tried (61, 62). First, there is no evidence that the technique works at all, and should it be successful one again faces the difficulties associated with solid–solid junctions. One final "coating" technique involves dye sensitization of large band gap materials. Gerischer (2) states that dye sensitization has been known for some time (63, 64, 65), but that the efficiency is limited by the fact that only monolayers of dye can be used. Another factor, with the dye sensitization, is the stability of the dye itself.

In summary, there are some promising avenues of research in photoassisted redox processes at electrodes. The results outlined here are the first of a set showing that it is possible to have interfacial redox processes which occur so fast that electrode decomposition cannot compete. The question now is why do some reductants work while others do not? Study of factors influencing the rate of interfacial electron transfer should yield the answer. Since interfacial electron transfer rates govern efficiency in all cases, these studies should prove useful in all approaches to the ultimate utilization of photoelectrochemical cells.

Acknowledgment

We thank the National Aeronautics and Space Administration for support of this research. We acknowledge the considerable contributions of Peter T. Wolczanski in preparation of electrode crystals.

Literature Cited

1. Becquerel, E., *C. R. Acad. Sci.* (1839) **9**, 561.
2. Gerischer, H., *J. Electroanal. Chem.* (1975) **58**, 263.
3. Heidt, L. J., McMillan, A. F., *Science* (1953) **117**, 75.
4. Jortner, J., Stein, G., *J. Phys. Chem.* (1962) **66**, 1258, 1264.
5. Weiss, J. J., *Nature* (1935) **136**, 794.
6. Potterill, R. H., Walker, O. J., Weiss, J. J., *Proc. R. Soc. London* (1936) **A156**, 561.
7. Farkas, A., Farkas, L., *Trans. Faraday Soc.* (1936) **34**, 1113, 1126.
8. Weiss, J. J., *Trans. Faraday Soc.* (1941) **37**, 463.
9. Rigg, T., Weiss, J. J., *J. Chem. Phys.* (1952) **20**, 1194.
10. Hayon, E., Weiss, J. J., *J. Chem. Soc.* (1960) 3866.
11. Heidt, L. J., Mullin, M. G., Martin, W. B., Jr., Johnson Beatty, A. M., *J. Phys. Chem.* (1962) **66**, 336.
12. Collinson, E., Dainton, F. S., Holmes, B., *Nature* (1950) **165**, 267.
13. Baxendale, J. A., Magee, J., *Trans. Faraday Soc.* (1955) **51**, 205.
14. Adamson, M. G., Baulch, D. L., Dainton, F. S., *Trans. Faraday Soc.* (1962) **58**, 1388.
15. Buston, G. V., Wilford, S. P., Williams, R. J., *J. Chem. Soc.* (1962) 4957.
16. Gerischer, H., in "Physical Chemistry: An Advanced Treatise," Vol. 9A, H. Eyring, D. Henderson, and W. Jost, Eds., Academic, New York, 1970, Ch. 5.
17. Gerischer, H., *Adv. Electrochem. Electrochem. Eng.* (1961) **1**, 139.
18. Myamlin, V. A., Pleskov, Yu. V., "Electrochemistry of Semiconductors," Plenum, New York, 1967.
19. Freund, T., Gomes, W. P., *Catal. Rev.* (1969) **3**, 1.
20. Bohenkamp, K., Engell, H. J., *Z. Elektrochem.* (1957) **61**, 1184.
21. Harten, J., *Z. Naturforsch.* (1961) **16a**, 1401.
22. Williams, R., *J. Chem. Phys.* (1960) **32**, 1505.
23. Fujishima, A., Honda, K., *Bull. Chem. Soc. Jpn.* (1971) **44**, 1148.
24. Fujishima, A., Honda, K., *Nature* (1972) **238**, 37.
25. Fujishima, A., Kohayakawa, K., Honda, K., *J. Electrochem. Soc.* (1975) **122**, 1437.
26. Fujishima, A., Kohayakawa, K., Honda, K., *Bull. Chem. Soc. Jpn.* (1975) **45**, 1041.
27. Wrighton, M. S., Ginley, D. S., Wolczanski, P. T., Ellis, A. B., Morse, D. L., Linz, A., *Proc. Natl. Acad. Sci. U.S.A.* (1975) **72**, 1518.
28. Keeney, J., Weinstein, D. H., Haas, G. M., *Nature* (1975) **253**, 719.
29. Frank, S. N., Bard, A. J., *J. Am. Chem. Soc.* (1975) **97**, 7427.
30. Hardee, K. L., Bard, A. J., *J. Electrochem. Soc.* (1975) **122**, 739.
31. Ohnishi, T., Nakato, Y., Tsubomura, H., *Ber. Bunsenges Phys. Chem.* (1975) **79**, 523.
32. Nozik, A. J., *Nature* (1975) **257**, 383.
33. Gissler, W., Lensi, P. L., Pizzini, S., *J. Appl. Electrochem.* (1976) **6**, 9.
34. Harris, L. A., Wilson, R. H., *J. Electrochem. Soc.* (1976) **123**, 1010.
35. Mavroides, J. G., Tchernev, D. I., Kafalas, J. A., Kolesar, D. F., *Mater. Res. Bull.* (1975) **10**, 1023.
36. Carey, J. H., Oliver, B. G., *Nature* (1976) **259**, 554.

37. Wrighton, M. S., Ellis, A. B., Wolczanski, P. T., Morse, D. L., Abrahamson, H. B., Ginley, D. S., *J. Amer. Chem. Soc.* (1976) **98**, 44, 2774.
38. Mavroides, J. G., Kaalas, J. A., Kolesar, D. F., *Appl. Phys. Lett.* (1976) **28**, 241.
39. Watanabe, T., Fujishima, A., Honda, K., *Bull. Chem. Soc. Jpn* (1976) **49**, 355.
40. Ellis, A. B., Kaiser, S. W., Wrighton, M. S., *J. Phys. Chem.* (1976) **80**, 1325.
41. Hodes, G., Cohen, D., Manassen, J., *Nature* (1976) **260**, 312.
42. Hardee, K. L., Bard, A. J., *J. Electrochem. Soc.* (1976) **123**, 1024.
43. Noland, T. A., *Phys. Rev.* (1954) **94**, 724.
44. Cardona, M., *Phys. Rev.* (1965) **140**, A651.
45. Wysoki, J. J., Rappaport, P., *J. Appl. Phys.* (1960) **31**, 571.
46. Ellis, A. B., Kaiser, S. W., Wrighton, M. S., *J. Am. Chem. Soc.* (1976) **98**, 1635.
47. Ellis, A. B., Kaiser, S. W., Wrighton, M. S., *J. Am. Chem. Soc.* (1976) **98**, 6855.
48. Tyagai, V. A., Kolbasov, G. Ya., *Surface Science* (1971) **28**, 423.
49. Vanden Berghe, R. A. L., Gomes, W. P., Cardon, F., *Z. Phys. Chem. N. F.* (1974) **92**, 91.
50. Fujishima, A., Sugiyama, E., Honda, K., *Bull. Chem. Soc. Jpn.* (1971) **44**, 304.
51. Sitabkhan, F., *Ber. Bunsenges. Phys. Chem.* (1972) **76**, 383.
52. Approximately 50% of the solar energy output is at wavelengths shorter than 700nm; Archer, M. D., *J. Appl. Electrochem.* (1975) **5**, 17.
53. Warekois, E. P., Lavine, M. C., Mariano, A. N., Gatos, H. C., *J. Appl. Phys.* (1962) **33**, 690.
54. Hodes, G., Manassen, J., Cahen, D., *Nature* (1976) **261**, 403.
55. Miller, B., Heller, A., *Nature* (1976) **262**, 680.
56. Black, J. F., Conwell, E. M., Seigle, L., Spencer, C. W., *J. Phys. Chem. Solids* (1957) **2**, 240.
57. Gildart, L., Kline, J. M., Mattox, D. M., *J. Phys. Chem. Solids* (1961) **18**, 286.
58. Ellis, A. B., Kaiser, S. W., Wrighton, M. S., *J. Am. Chem. Soc.* (1976) **98**, 6418.
59. Ellis, A. B., Kaiser, S. W., Bolts, J. M., Wrighton, M. S., *J. Am. Chem. Soc.* (1977) **99**, in press.
60. Nakato, Y., Ohishi, T., Tsubomura, H., *Chem. Lett.* (1975) 883.
61. Kohl, P. A., Frank, S. N., Bard, A. J., submitted for publication, and private communication.
62. Bockris, J. O'M., Uosaki, K., Abstract No. 14, Inorganic Division, Centennial ACS Meeting, New York, 1976.
63. Gerischer, H., Tributsch, H., *Ber. Bunsenges. Phys. Chem.* (1968) **72**, 437.
64. *Ibid.* (1969) **73**, 251.
65. Gerischer, H., Michel-Beyerle, M. E., Rebentrost, F., Tributsch, H., *Electrochim. Acta* (1968) **13**, 1509.

RECEIVED July 27, 1976.

5

Solar Energy Conversion through Photosynthesis?

RODERICK K. CLAYTON

Division of Biological Sciences, Cornell University, Ithaca, N. Y. 14853

> *Ultimately the use of sunlight offers the clearest way to meet our needs for energy without running afoul of serious environmental problems. Direct solar heating is approaching technical and economic feasibility on a large scale. Generation of fuel from organic wastes can be economically profitable now, especially when combined with the waste-assisted growth of algae. Growth of energy-efficient plants such as sugarcane and harvesting of wild plants such as water hyacinth as sources of fuel are attractive possibilities. Schemes to modify normal photosynthesis or to use extracted parts of photosynthetic tissues to generate hydrogen or electricity from sunlight are visionary, but they are in their infancy and should be developed along with all other reasonable approaches.*

We estimate that in the United States we have enough fossil fuel, in one form or another, to meet our energy requirements for more than a thousand years at the present rate of consumption, between 10^{19} and 10^{20} J (or about 10^{17} Btu) annually. Rough estimates of energy reserves, modified from data listed by Hammond (1, 2), are shown in Table I. However, the use of all our fossil fuel reserves will require new solutions to economic and environmental problems. In the long run, only solar energy is relatively free from thermal and chemical pollution. Even the increased absorption of solar radiation because of its collection as an energy source could be offset by compensating reflectors, so that the absorption of heat by the planet as a whole remains unaltered. Currently the U.S. Energy Research and Development Administration (ERDA) projects that 6% of our energy needs will be met with solar systems by the year 2000, and 25% by 2020 (3).

Table I. Rough Estimates of Energy Reserves in the United States at the Present Rate of Consumption

Energy Source	Years
Presently developed oil and gas	< 10
Estimated total oil and gas	10^2
Fission (conventional)	$< 10^2$
Coal and oil shale	$> 10^3$
Geothermal	$> 10^3$
Fusion	10^6–10^9
Solar	$> 10^9$

Let us list energy sources that can be made quantitatively important in the United States, and that are directly or indirectly of contemporary solar origin. We shall then review these briefly and consider in more detail those that involve photosynthesis.

1. Wind-powered turbine.
2. Ocean thermal gradients; heat engine.
3. Direct solar heating; heat engine or chemical reaction cycle producing hydrogen from water.
4. Photosynthetic systems. (a) Variations of agriculture; growth of plants including algae, and conversion to fuels. (b) Photosynthetic hydrogen production.
5. Photoelectric devices, possibly using materials derived from photosynthetic tissues.

In many of these systems hydrogen is likely to play a central role (4) both as a clean fuel and as a means of storing and transporting energy. As a fuel, the only chemical pollutant might be oxides of nitrogen if atmospheric nitrogen is exposed to a high combustion temperature. As a medium of exchange hydrogen can be converted to electricity with about 40% efficiency, and electricity to hydrogen at about 80% efficiency. It has been estimated (5) that the transport of energy as hydrogen is cheaper than the transmission of electricity if the distance is greater than about 250 miles.

Wind

Wind-powered electric generators can be good decentralized sources of power in rural areas, although the optimum size for efficiency is estimated at several megawatts. The cost of such a power plant is estimated (3) to be two to four times greater than the present cost, about $400 per kilowatt, of a nuclear power plant.

An extensive array of major wind-powered plants, absorbing much of the surface wind over a large region, could have an effect on the local weather that merits serious study.

Ocean Thermal Gradients

A temperature difference of 20°C exists vertically through a depth of 1000 m in tropical waters, and a similar difference is found at the edge of the Gulf Stream near the coast of Florida. This difference around 300°K could provide ~ 6% efficiency for a heat engine with a Carnot cycle, and an efficiency greater than 2% for the generation of electricity or hydrogen could be expected. There would be formidable engineering problems (2) concerned with thermal conductivity between the sea water and the working fluid, perhaps ammonia, when the differential in temperature is so small. The corrosive action of sea water must also be considered in estimating cost and lifetime of such a power plant.

Abstraction of heat from the ocean would be compensated by the fact that the optical absorbance of cold water is greater than that of warmer water. Thus there would be a net increase in the flow of heat from the sun to the earth's surface, where the energy is eventually released, but the local temperature of the ocean would remain fairly constant.

With a vertical gradient the upwelling of deeper water would carry organic nutrients that might be used to feed phytoplankton at the surface. Also, the system could be arranged to provide fresh water, a commodity that may soon become critically scarce in Florida.

Direct Solar Heating

The total flux at the earth's surface from the sun at its zenith is about 1 kW/m^2 (6). The annual mean flux in the United States is about 200 W/m^2. The consumption of power in the United States amounts to about 0.2 W/m^2. Thus a solar power plant operating at 5% overall efficiency would require about 1/50 of the country's surface, or the average size of one state.

Rooftop solar heating of water in individual houses already finds widespread use, especially in tropical regions. Even in the northeastern United States one can readily provide 20% of the total domestic power requirement in this way, at a cost competitive with electric heating derived from the costlier fuels (about twice the cost of energy from a nuclear or a coal burning plant).

The practicality of direct solar heating depends on the "greenhouse effect." The glass of a greenhouse transmits nearly all the solar radiation. When absorbed by the earth, most of this radiation is converted to heat, corresponding to quanta in the infrared. The re-radiated heat at these wavelengths is almost entirely reflected or absorbed by the glass of the greenhouse, and is thereby trapped inside except for the part that is radiated from the outer surface of the glass. An ideal material for trapping heat in this way would transmit all wavelengths of the solar spectrum below about 2000 nm and reflect all greater wavelengths. This can be achieved with a multilayer interference filter and can be approximated by depositing a thin film of silicon on glass. However, the use of such high-performance "greenhouse" materials entails unsolved problems of cost and long term stability.

This type of solar heat trap can reach temperatures approaching 300°C with no focusing of the sun. Focusing the sun into a linear image with about tenfold concentration of intensity can give about 500°C (7), and an approximate point focus can give temperatures greater than 1000°C. Of course the focusing systems (arrays of mirrors that track the sun's movement) entail greatly increased costs. Also, the unfocused system can function with a cloudy sky. Finally, the performance of the "greenhouse" material becomes poorer as the stored temperature increases, because the spectrum of thermal radiation is shifted to shorter wavelengths and overlaps more with the spectrum of sunlight. Degradation of the material should also become a greater problem at the higher temperatures.

The trapped solar heat can be transferred to a large reservoir by means of a circulating fluid or a "heat pipe" which relies on gaseous convection. The reservoir can then drive an engine to produce electricity, hopefully with overall efficiency (from solar radiation to electricity) as great as 20%. Hydrogen could then be produced electrolytically with comparable efficiency.

Alternatively, the heat could be used to drive a cycle of chemical reactions that produce hydrogen and oxygen from water. Many such cycles have been conceived; some have been brought to a semblance of technical feasibility, and none has been established on a large scale. A single example proposed by Abraham and Schreiner (8) follows.

$$LiNO_2 + I_2 + H_2O \xrightarrow{300°K} LiNO_3 + 2HI$$

$$2HI \xrightarrow{700°K} I_2 + H_2$$

$$LiNO_3 \xrightarrow{750°K} LiNO_2 + \tfrac{1}{2}O_2$$

Net: $\quad H_2O \longrightarrow \tfrac{1}{2}O_2 + H_2$

This reaction sequence might be driven at temperatures attained with a line focus of the sun.

A primitive but effective type of solar heat collector is the solar pond (6). The pond should be about 1 m deep and 50 m or more in diameter, to minimize heat loss at the edges into the earth. Heat transferred to earth through the bottom is largely retained and can be retrieved at night. The bottom of the pond is black to maximize absorption, and the surface is covered to retard evaporation and consequent cooling. An ingenious approach (9) is to use brine covered by a shallow layer of fresh water. The surface layer remains cool but does not mix with the denser brine below. Daytime temperatures at the bottom could reach about 100°C.

In conclusion, direct solar heating appears to be especially favorable as a method for large-scale solar energy conversion, with much of the technology already proven.

Variations of Agriculture

The overall efficiency with which solar energy is stored as organic matter (mostly carbohydrate) in crops and forests is in the range of 0.1–0.4% for "C_3" plants [those that use the Calvin–Benson cycle (10) for carbon assimilation], and about twice that for tropical grasses and other "C_4" plants, using the Hatch–Slack (11) pathway for fixing CO_2. In the latter category sugarcane can show an annual yield of 50 tons dry weight per acre and overall efficiency about 2.5% for storing the energy of sunlight in organic matter, about half of which is sucrose. The sugar can be fermented to ethyl alcohol at an estimated cost less than $1 per gallon (12). The same could be done with other ready sources of carbohydrate, some in the category of waste, such as spoiled grain in our major wheat farming areas.

Plant materials as well as organic wastes can also be converted to fuels by other means (2). Pyrolysis, effected by anaerobic heating to about 500°C at atmospheric pressure, produces a mixture of crude oil, low-grade combustible gas, and "char." The gas and char can be burned to maintain the high temperature and some of the gas recycled to preserve an anaerobic atmosphere. A more efficient process, but technically more difficult, is hydrogenation, conducted at a lower temperature (about 300°C) but at $\sim$ 200 atm. Carbon monoxide and steam are introduced to provide a reducing environment; the organic matter is converted almost entirely to crude oil. A third alternative is anaerobic bacterial fermentation. This avoids the use of high pressures or temperatures and gives principally a very desirable fuel, methane. A drawback of bacterial fermentation is that about 40% of the organic matter is left as a residual sludge that must be disposed of or treated by one of the other methods.

If all the readily collected waste were processed in one of these ways, we could provide 3% of our present oil use, or 6% of the output of present power stations. These schemes as applied to waste are commercially advantageous now, when the cost of alternative disposal (landfill) has been discounted. When these schemes are applied to products of agriculture, the conversion to fuel must compete with use as sources of food, fiber, and chemicals. However, with some plants the conversion to fuel is clearly advantageous. One such plant is the water hyacinth, which grows so prolifically in shallow tropical waters that it is a nuisance to navigation and at the same time is easy to harvest. Its wholesale removal might, however, threaten the extinction of some animals, such as the manatee.

Algae and photosynthetic bacteria comprise a special category in "energy farming" (13). In ponds, with their growth assisted by organic wastes and with adequate sunlight, algae can yield in a few days what a cane field can yield in an entire crop, with about 4% of the solar energy stored as organic matter, mainly lipids and protein. (Unlike the algae, the photosynthetic bacteria cannot grow without a source of reduced compounds, as they cannot use water as a source of hydrogen for the fixation of carbon dioxide. A great variety of organic compounds can serve this purpose.) A one-acre "waste plus algae" pond, with the harvest converted to fuel and thence to electricity, might give about 15 kw at about one cent per kilowatt hour.

Cattle feed lots provide a large concentrated source of waste that can be coupled to an algae farm. Some of the algae might be fed to the cattle. Anaerobic fermentation of the algae would yield methane, sludge, and some by-products (CO_2, N, and P) that could be recycled to the algae.

The amount of easily collected waste is not enough to provide the greater part of our needs for fuel and energy by these means, but the greatest possible development of the foregoing systems is clearly advantageous and economically sound. Much of the land and money required for such projects would be needed for waste disposal in any case.

Photosynthetic Hydrogen Production

A far more visionary variation of agriculture is found in schemes, presently being explored, to redirect the photosynthetic process to yield hydrogen instead of carbohydrate.

Photosynthesis in green plants and algae involves two distinct photochemical systems (14). In one of these, called system 2, a quantum of light energy absorbed by chlorophyll induces a photochemical oxidation–reduction reaction which generates a strong oxidant and a weak reductant. The oxidant is strong enough to remove electrons from water, releasing oxygen from the water. The reductant feeds electrons, through a chain of electron transporting molecules, to the second photochemical system (system 1). In system 1 a photochemical process, again driven by the energy of light absorbed by chlorophyll, generates a strong reductant and a weak oxidant. The weak oxidant is neutralized by the electrons flowing from system 2. The strong reductant produced by system 1 is stored initially in the form of reduced ferredoxin, an iron-containing protein that can in turn reduce other substances. The normal end result is the reduction of CO_2 to carbohydrate. The plan of schemes for producing hydrogen through photosynthesis is to alter the normal utilization of reduced ferredoxin. Instead of flowing to the enzymes that catalyze CO_2 fixation, the electrons from reduced ferredoxin cause reduction of H^+ ions to H_2:

$$2H^+ + 2e^- \rightarrow H_2$$

This reaction is catalyzed by either of two enzymes that occur in many algae and bacteria: hydrogenase and nitrogenase (15).

Hydrogenase is found in green algae and in some bacteria. In the algae its natural function may be to rid the organism of excess reductants that can arise under certain conditions, thereby restoring a desirable balance between oxidized and reduced states of the chain of electron carriers (16).

Nitrogenase, found in blue-green algae and in both photosynthetic and some non-photosynthetic bacteria, has the primary function of nitrogen fixation, reducing N_2 to ammonia. However, this enzyme can also mediate the reduction of H^+ to H_2. Its activity requires adenosine triphosphate and is inhibited by N_2; hydrogenase is not subject to these complications.

Both hydrogenase and nitrogenase are inhibited by oxygen, which introduces a serious problem since the photosynthesis of green plants generates oxygen along with reducing power. In fact, the hydrogenase activity of algae was discovered in cultures that had been kept in the dark under anaerobic conditions for several hours (17). Upon illumination, the hydrogenase actvity declined as the algae began to produce oxygen. No one has found conditions under which intact algae produce hydrogen and oxygen concomitantly at rates exceeding a miniscule fraction of the normal rate of photosynthesis (18).

We are thus led to consider schemes, proposed by Krampitz and others (15, 19), in which the plant cells are broken and their photochemical components are dissected and rearranged to produce hydrogen while shielding the hydrogenase or nitrogenase from oxygen. There are several reasons why the oxygen-evolving system must be kept separate from the reductants and catalysts that produce hydrogen. First, hydrogenase and nitrogenase are sensitive to oxygen. Second, the reductants must not be allowed to react wastefully with oxygen before they can drive the conversion of H^+ to H_2. This applies to natural reduced products and also artificial ones that might usefully be interposed to link the reducing side of system 1 to the evolution of hydrogen. Third, the oxygen-evolving system must be shielded from the strong reductants which could otherwise prevent the operation of system 2. Finally, the oxygen and hydrogen must be separated if the hydrogen is to be stored and used. The alternative, burning the evolved hydrogen immediately with the evolved oxygen, would be hazardous on a large scale and would restrict the options for use of the hydrogen.

Three examples can be cited of photosynthetic systems in which the evolution of hydrogen is coupled with the release of oxygen from water.

(a) Chloroplasts from spinach leaves are mixed with bacterial hydrogenase, and the inhibitory action of oxygen is minimized with an oxygen-absorbing system composed of glucose plus glucose oxidase (*20*). Hydrogen is then evolved as a product of photosynthesis.

(b) The filamentous alga *Anabena cylindrica* has cells which perform photosynthesis, and other cells (heterocysts) that are specialized for nitrogen fixation. Respiration in the heterocysts keeps the concentration of oxygen low in those cells. Reduced products can diffuse from the photosynthetic cells into the heterocysts, where they promote hydrogen evolution using the nitrogenase as catalyst (*21, 22*).

(c) The water fern *Azolla* lives symbiotically with nitrogen-fixing blue-green algae. When provided with nitrate and shielded from nitrogen, the system grows, and hydrogen evolves (*23*).

In none of these systems is the rate of hydrogen evolution more than a small fraction, ca. 0.2–3%, of "normal" photosynthesis.

Some current efforts and proposals to improve this approach (*15, 19, 24*) would begin with the isolation of the essential components (systems 1 and 2 and hydrogenase) from their native tissues. The components can be kept separate by means of semipermeable membranes, by adsorption on solid particles, or by microencapsulation. Functional communication between the components is maintained by diffusible electron carriers in aqueous media. Some of the necessary but fragile enzyme systems might be stabilized by cross-linking their proteins internally through glutaraldehyde fixation, or by adsorption or microencapsulation. Obviously these approaches raise difficult problems, especially with regard to stability, but one can at least envision a complete functional system within the bounds of present technology.

We have yet to consider questions of efficiency and rate in photosynthetic systems that produce hydrogen. Of all the solar energy that falls on a leaf or a dense culture of algae, only the part with wavelengths below 680 nm is absorbed by chlorophyll and other pigments. Of the radiation that is absorbed, the quanta of shorter wavelengths have relatively more energy, but this is quickly degraded to the level of the long wave absorption band of chlorophyll near 680 nm. This degradation, with the excess energy dissipated as heat, happens before the light energy can be used for photochemistry. Because of the combined effects of non-absorption at wavelengths beyond 680 nm and degradation of energy absorbed at shorter wavelengths, about 40% of the incident solar energy becomes available for photochemistry in the form of 680 nm quanta. At 680 nm each quantum has an energy of 1.8 eV. When systems 1 and 2 cooperate to remove an electron from water and promote it to the level of hydrogen, an energy of 1.2 eV is stored (*14, 25*). This requires an input of two quanta, one in each system, or a total input of 3.6 eV. The maximum efficiency theoretically attainable in this way, for the storage of solar energy as hydrogen with the concomitant release of oxygen from water, is therefore (1.2/3.6) × 40% or 13%. In practice one would do well to realize 5%, and if the hydrogen were burned to produce electricity, an overall efficiency of 2% might be attained. This is close to the efficiency of growing sugarcane, but electricity is a far more desirable end product than sugar. We would need to commit 1/10 of the area of the country, on land or in the neighboring ocean, to satisfy our total energy requirement from sunlight at an overall efficiency of 2%.

Another problem has to do with the rate at which a photosynthetic system can keep pace with incoming light energy. The architecture of the photosynthetic tissue is such (*14*) that each photochemical reaction center, whether of system 2 or system 1, is served by about 200 molecules of "light harvesting" or "antenna" chlorophyll. The antenna chlorophyll molecules do not participate directly in the photochemistry; they absorb light and deliver the energy to the reaction centers. If every quantum absorbed by the antenna in full sunlight were used photochemically at the reaction centers, electrons would be flowing through the complete chain, from water to ferredoxin, at a rate of approximately 2000 per sec. In fact the electron transport chain can transport no more than about 200 electrons per sec, so in full sunlight only 1/10 of the incoming quanta can be used. The prospect of increasing the rate at which the electron transport system can operate is limited, but there is a better way to solve this problem (*16*). If each reaction center were served by an antenna of only 20 chlorophyll molecules rather than 200, quanta would be delivered to the reaction centers at a rate of 200/sec in full sunlight, rather than 2000/sec. The electron transport machinery could then keep pace. The only qualification is that the system be optically dense enough to absorb most of the incident quanta at wavelengths below 680 nm. It should not be difficult to delete most of the antenna chlorophylls from an in vitro preparation, or to start with mutant plants or algae that have an abnormally low ratio of antenna chlorophylls to reaction centers.

The vision of generating hydrogen by photosynthesis with reasonable efficiency is most attractive. There are formidable problems to be overcome, but efforts to solve them have just begun.

Photoelectric Devices

The direct conversion of solar energy to electricity has great potential advantages of high efficiency, minimal machinery, and flexible use of the product. In special applications such as the provision of power to instruments on spacecraft, we already use photovoltaic silicon cells.

Contemporary solar batteries are semiconductive devices that operate in the manner suggested by Figure 1. Light promotes an electron in the material from a ground state (valence band) to an excited state, leaving an electron vacancy or hole (+) in the valence band. The excited electron loses some energy and enters a continuum of excited states, the conduction band. The cell is divided by a junction into two regions of different composition; in one region the electron has a high mobility and in the other the mobility of the hole is high. The difference in composition is designed to shift the energies of the valence and con-

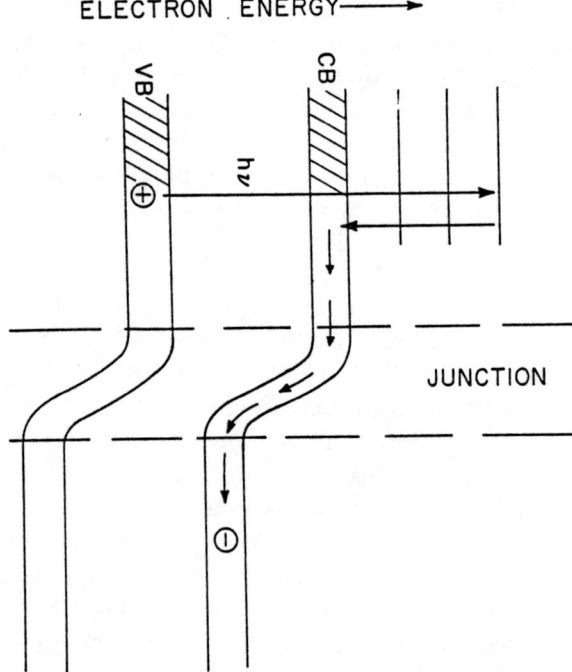

Figure 1. Diagram of the function of a solar battery. VB, valence band; CB, conduction band.

duction bands as shown in the figure. This affords a high probability that the electron in the conduction band will migrate across the junction, to a conduction band of lower energy, rather than give up all its extra energy and recombine with the hole. The separated positive and negative charges can then be collected through electrodes. The energy gap between conduction and valence bands is typically about 1 eV, and a drop across the junction of about 0.4 eV assures efficient separation and stabilization. The residual 0.6 eV is the energy stored by the separated electron and hole; in other words, the electromotive force of the solar battery is 0.6 V.

There are no allowed states for the absorption of quanta between the valence band and the conduction band, so quanta of energy less than the gap between these bands (about 1 eV) are not absorbed. For a silicon cell, in which the two regions are silicon with different kinds of trace substances added, the non-absorption of solar flux amounts to ca. 23% of the total reaching the earth's surface (26). Quanta of energy greater than this are absorbed, but the excess energy is lost as the excited electron settles into the conduction band. In silicon this loss is 33% of the total solar flux. Finally, the drop in energy across the junction necessary to prevent recombination, plus some irreducible "frictional" losses (internal resistance), cost another 24% or more. The combined loss is then 23 + 33 + 24 or 80% of the incident solar energy. The maximum efficiency of a silicon cell is therefore about 20%; efficiencies of 10–13% are attained in the silicon cells used in aerospace technology.

These limitations on efficiency apply to any device (not necessarily a semiconductor) that absorbs sunlight, passes through a succession of excited states, and enters a state in which some of the energy is retained, as shown in Figure 2. We have already applied these principles in estimating the maximum efficiency of photosynthetic hydrogen production.

The energy of the lowest excited state relative to the ground state can be raised or lowered by a suitable choice of absorbing material. This will shift the partitioning of losses between "quanta not absorbed" and "relaxation-dissipation," but the sum of these losses is not altered greatly. Even so, it is interesting to note that the optimum choice for the lowest excited state energy falls at 1.4 eV, or 890 nm wavelength. This is exactly the long-wave absorption limit of bacteriochlorophyll in most photosynthetic bacteria, the most primitive of contemporary photosynthetic creatures.

The crucial problem with photovoltaic solar energy conversion is economic. The extremely pure crystals of silicon needed for a solar cell are costly to produce. To be competitive with conventional energy systems the cost would have to be reduced 50- to 100-fold. Cheaper methods, such as the deposition of silicon crystals in a chemical reaction, are

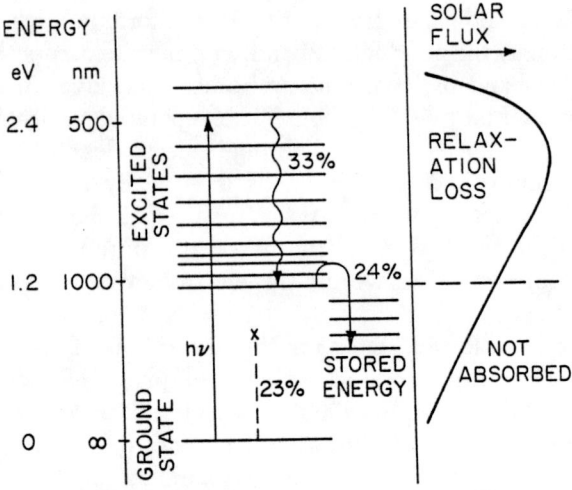

Figure 2. Energy level diagram for a system that absorbs light and stores some of the energy. The ordinate represents energy, shown here as electron volts and also as wavelength (in nm). Loss figures for a silicon solar cell are given, for quanta not absorbed (23%), thermal relaxation from higher excited states (33%), and stabilization of the excitation energy (24%). The sum of these losses, 80%, is characteristic of devices with long-wave absorption limit in the red or near infrared. A crude representation of the spectrum of sunlight reaching the earth's surface is shown at the right.

being explored but have not been realized. Solar cells using cadmium sulfide are much cheaper, but their efficiency is less (about 6%), and cadmium is extremely toxic. Also, with all known photovoltaic cells there are unsolved problems of long term stability when exposed to natural conditions (for perspective, a fossil fuel driven power plant is expected to last about 25 years).

The high cost per unit area of silicon cells can be alleviated by concentrating the sunlight onto a smaller area. This would surely aggravate the problem of stability.

Although photovoltaic cells made of silicon or cadmium sulfide are regarded as the most promising, a great variety of other photoelectric devices have been known for decades and have been studied with a view to solar energy conversion (6). Prominent among these is the photogalvanic cell using thionine (a purple dye) and ferrous ions in aqueous solution.

In a photochemical reaction the thionine is reduced by Fe^{2+}, forming the colorless leucothionine and Fe^{3+}. The reverse reaction proceeds in the dark:

$$\text{thionine} + 2\text{H}^+ + 2\text{Fe}^{2+} \underset{\text{dark}}{\overset{\text{light}}{\rightleftharpoons}} \text{leucothionine} + 2\text{Fe}^{3+}$$

With a suitable arrangement of these components and electrodes, some of the current of electrons from leucothionine to Fe^{3+} in the back reaction can be made to flow through an external circuit.

Of course, it has not escaped the attention of scientists that chlorophyll and similar substances might be used to make a photoelectric device, in an imitation of the natural photochemical oxidoreduction of photosynthesis (27). One possibility would be to use the photosynthetic reaction centers themselves in constructing a photovoltaic cell. Photochemical reaction centers are readily isolated from photosynthetic bacteria (28). They are pigment–protein complexes; a specific protein bears four molecules of bacteriochlorophyll (the bacterial analogue of green plant chlorophyll) and two of bacteriopheophytin, which is the same as bacteriochlorophyll except that the central magnesium atom has been replaced by two hydrogen atoms (29). The reaction center particle also contains ubiquinone and iron. Absorption of light by the pigments of the reaction center causes a single electron to be transferred from two of the four bacteriochlorophylls to the ubiquinone, giving an oxidized bacteriochlorophyll dimer, $(B)_2^{+\cdot}$, and reduced ubiquinone, $UQ^{-\cdot}$ (30). The function of the iron is not clear; it may help to maintain an appropriate structural arrangement. The earliest photochemical event appears to be the displacement of an electron from the "active" bacteriochlorophyll dimer to one of the two bacteriopheophytin molecules (31, 32, 33, 34). In ~ 0.2 nsec the electron moves on from bacteriopheophytin to ubiquinone. The products $(B)_2^{+\cdot}$ and $UQ^{-\cdot}$ are stable against recombination for more than 50 msec at room temperature. If extra ubiquinone is present, as it is in the intact photosynthetic bacteria, the "primary" ubiquinone transfers its electron rapidly to the secondary pool of ubiquinone (35). The products are then stable against recombination for more than a second. The separation of oxidized bacteriochlorophyll and reduced ubiquinone gives a potential of about 0.6 V. The quantum efficiency of the photochemical electron transfer is close to 100% (36).

A solar cell might therefore be built by first depositing reaction centers onto a piece of glass bearing a semitransparent film of metal. A dried film of reaction centers remains fully capable of performing its photochemistry efficiently. One would need to find the proper metal or to interpose a suitable mediator, to ensure that electrons can flow readily from the metal electrode to the oxidized bacteriochlorophyll, $(B)_2^{+\cdot}$, made photochemically in the reaction centers. The film of reaction centers might then be coated with a layer of ubiquinone to draw electrons away from the "primary" reduced ubiquinone and endow the system with electric polarity. A second electrode in contact with the added ubiquinone would complete the "sandwich."

This model has not been built or tested. We can expect a problem of electric conductivity between the electrode surfaces and the layers of reaction centers and ubiquinone respectively. The conductivity needed for a successful solar cell of this kind can be computed. Applying the principles shown in Figure 2, we predict a maximum overall efficiency of 18% if the cell produces 0.6 V. At high noon in June the cell would then have to handle a power density of 18 mW/cm^2 and a current density of 30 mA. In a completed circuit, a 0.6-V battery will deliver a current of 30 ma if the total resistance is 20 ohm. For efficient power transfer to an external load, no more than half of this resistance should be internal. Thus if the internal resistance of the solar cell can be held to 10 ohm for every sq cm of surface, half of the total power can be delivered usefully to an external load. This would give an overall operating efficiency of 9% for the cell.

We can also estimate that a bacterial reaction center particle exposed to full sunlight absorbs ca. 50 quanta/sec. Efficient use of these quanta requires that the turnover time for the entire process (photochemistry plus the secondary electron transfer steps) be 1/50 sec or 20 msec. The time needed for photochemical electron transfer from bacteriochlorophyll to secondary ubiquinone is far less than that, so contact with the electrodes must be effected within 20 msec. This calculation is equivalent to the foregoing calculation that the resistance must be < 10 ohm/cm^2. It also shows that the requisite rate of turnover is fast compared with a potentially wasteful back-reaction between oxidized bacteriochlorophyll and reduced ubiquinone. Thus if the necessary conductivity is achieved, the back-reaction will not be a problem.

As with photosynthetic hydrogen evolution, the development of a photosynthetic solar battery as a major source of energy is in its infancy. Those who would attempt to make a solar cell from purified chlorophyll and other components might consider adding pheophytin to the mixture, in view of our present picture of the early steps in the photochemistry of bacterial reaction centers. In developing photosynthetic solar cells we will surely encounter great problems, especially in the realm of stability. The most promising photoelectric device at present seems to be the silicon cell.

Conclusion

Sunlight provides our best long-term hope for an inexhaustible supply of pollution-free energy. The most practical schemes for capturing solar energy appear at this time to be directly heating and harnessing the wind. Silicon photovoltaic cells could contend with these if a means were discovered for reducing their cost sharply.

The conversion of organic waste, plants, and algae into fuel has immediate practicality, but on a limited scale relative to our total needs.

In comparison with these approaches, the schemes for using photosynthetic tissues in special ways seem highly visionary. However, this article has been written from the point of view of a specialist in photosynthesis, and a specialist can easily identify and foresee the many difficulties that beset work in his own field. We should not capitulate to the proponents of other approaches merely because we are ignorant of details and magnitudes of the difficulties that confront them. Rather, we should press forward on all fronts until one or more working systems are established and proven.

Literature Cited

1. Hammond, A. L., *Science* (1972) **177,** 875.
2. Hammond, A. L., Metz, W. D., Maugh, T. H., II, "Energy and the Future," Amer. Assoc. Adv. Sci., Washington, D. C., 1973.
3. Hammond, A. L., *Science* (1975) **189,** 538.
4. Gregory, D. P., *Sci. Am.* (1973) **228,** 13.
5. Gregory, D. P., *Public Utilities Fortnightly* (1972) **89,** 21.
6. Daniels, F., "Direct Use of the Sun's Energy," Yale Univ., 1964. Reprinted by Ballantine, New York, 1974.
7. Meinel, A. B., Meinel, M. P., *Phys. Today* (1972) **25,** 44.
8. Abraham, B. M., Schreiner, F., *Science* (1973) **180,** 959.
9. Tabor, H., *Sol. Energy* (1963) **7,** 189.
10. Bassham, J. A., Calvin, M., "The Path of Carbon in Photosynthesis," Pentice-Hall, Englewood Cliffs, N. J., 1957.
11. Hatch, M. D., Slack, C. R., *Ann. Rev. Plant Physiol.* (1970) **21,** 141.
12. Calvin, M., *Science* (1974) **184,** 375.
13. Oswald, W. J., Golueke, C. G., *in* "Single Cell Protein," R. I. Mateles and S. R. Tannenbaum, Eds., M.I.T., Cambridge, Mass., 1968.
14. Clayton, R. K., "Light and Living Matter," Vol. II, Ch. 1, McGraw-Hill, New York, 1971.
15. Stuart, T. J., "Proceedings of the Workshop on Bio-Solar Conversion" (Report to NSF/RANN), pp. 45–55, Bethesda, Md., Sept. 1973.
16. Kok, B., "Proceedings of the Workshop on Bio-Solar Conversion" (Report to NSF/RANN) pp. 22–30, Bethesda, Md., Sept. 1973.
17. Gaffron, H., Rubin, J., *J. Gen. Physiol.* (1942) **26,** 219.
18. Bishop, N. I., "Proceedings of the Workshop on Bio-Solar Conversion" (Report to NSF/RANN), pp. 9–13, Bethesda, Md., Sept. 1973.
19. Krampitz, L. O., "An Inquiry into Biological Energy Conversion" (Report to NSF/RANN), pp. 22–23, Gatlinburg, Tenn., Oct. 1972.
20. Benemann, J. R., Berenson, J. A., Kaplan, N. O., Kamen, M. D., *Proc. Natl. Acad. Sci. U. S.* (1973) **70,** 2317.
21. Benemann, J. R., Weare, N. M., *Science* (1974) **184,** 174.
22. Jones, L. W., Bishop, N. I., *Plant Physiol.* (1976) **57,** 659.
23. Newton, J. W., *Science* (1976) **191,** 559.
24. Lien, S., San Pietro, A., "An Inquiry into Biophotolysis of Water to Produce Hydrogen" (Report to NSF/RANN), 1975.
25. Clayton, R. K., "Photosynthesis," Modules in Biology No. 13, Addison Wesley, Reading, Mass., 1974.

26. "Solar Cells: Outlook for Improved Efficiency," Report of Ad Hoc Panel on Solar Cell Efficiency, Space Sciences Board, National Research Council, 1972.
27. Wang, J. H., *Proc. Natl. Acad. Sci., U. S.* (1969) **62**, 653.
28. Clayton, R. K., Wang, R. T., *Methods Enzymol.* (1971) **23**, 696.
29. Straley, S. C., Parson, W. W., Mauzerall, D. C., Clayton, R. K., *Biochim. Biophys. Acta* (1973) **305**, 597.
30. Clayton, R. K., *Ann. Rev. Biophys. Bioeng.* (1973) **2**, 131.
31. Parson, W. W., Clayton, R. K., Cogdell, R. J., *Biochim. Biophys. Acta* (1975) **387**, 265.
32. Kaufmann, K. J., Dutton, P. L., Netzel, T. L., Leigh, J. S., Rentzepis, P. M., *Science* (1975) **188**, 1301.
33. Rockley, M. G., Windsor, M. M., Cogdell, R. J., Parson, W. W., *Proc. Natl. Acad. Sci., U. S.* (1975) **72**, 2251.
34. Fajer, J., Brune, D. C., Davis, M. S., Forman, A., Spaulding, L. D., *Proc. Natl. Acad. Sci., U. S.* (1975) **72**, 4956.
35. Clayton, R. K., Yau, H. F., *Biophys. J.* (1972) **12**, 867.
36. Wraight, C. A., Clayton, R. K., *Biochim. Biophys. Acta* (1973) **333**, 246.

RECEIVED July 27, 1976.

6

Photovoltaic Solar Cells

SIGURD WAGNER

Bell Laboratories, Holmdel, N. J. 07733

> *Solar cells convert incident light to electrical power. They are semiconductor diodes with two key functions: separation of electrical charge in energy, and in space. Absorption of light quanta by the semiconductor separates electron-hole pairs by the band gap energy; the output voltage is proportional to this energy. The electric field associated with the semiconductor junction separates electrons and holes in space, leading to an external current. The voltage-current product, or output power, thus depends on light absorption, charge transport, and type of junction. In this paper we consider cell characteristics, power conversion efficiencies, alternative cell structures, and approaches to the development of inexpensive cells.*

Photovoltaic solar cells convert incident light to electricity. Solar power, the product of photon flux and photon energy, is turned into electrical power, the product of electrical current and output voltage. Solar cells are conceptually simple and rugged devices. Therefore, widespread use of photovoltaic converters is very attractive. However, solar electricity is about one hundred times more expensive than conventional power. To a large extent, it will be the task of chemists to find improved materials and processes to make photovoltaic power cost competitive.

Semiconductor Diodes

Semiconductors combine two characteristic properties which make them suitable for photovoltaic cells (*1, 2, 3*). First, numerous semiconductors exhibit the proper absorption characteristics for solar radiation. Second, a space charge can be introduced in a small region of a semiconductor, while most of it remains neutral and conducting. This is the space charge of a junction. A simple way to picture the construction of a *pn* junction is shown in Figure 1a. We start with two pieces of the semiconductor (e.g., silicon), one of which is n-type, the other, p-type. In n-Si, typically 10^{17} cm^{-3} (2 ppma), donor impurities with five valence electrons (e.g., phosphorus) have been introduced to make the highly resistive pure Si an electric conductor with the current being carried by free electrons. The piece of n-Si is neutral, however, because the free electrons are exactly compensated by the fixed (nonmobile) ionized donor impurities. In p-Si, doped with acceptor impurities like boron, current is carried by free holes. However, there is still a finite density of electrons in p-Si, and of holes in n-Si. This density is determined by the concentration of the respective majority carrier, n or p, through the equilibrium constant

$$pn = N_c N_v \exp(-E_g/kT) = n_i^2$$

where N_c and N_v are the effective densities of states in the conduction and valence bands, E_g is the band gap, and k is Boltzmann's constant. In Si, pn at 300°K is 2.1×10^{20} cm^{-6}. Here n_i is the carrier concentration for undoped (intrinsic) Si. Imagine joining the n and the p halves. When they make contact, electrons diffuse from n into p and holes from p into n because of their respective concentration drop across the interface. Once majority carriers have diffused into the other side and thus have become minority carriers, they recombine with the local majority carrier

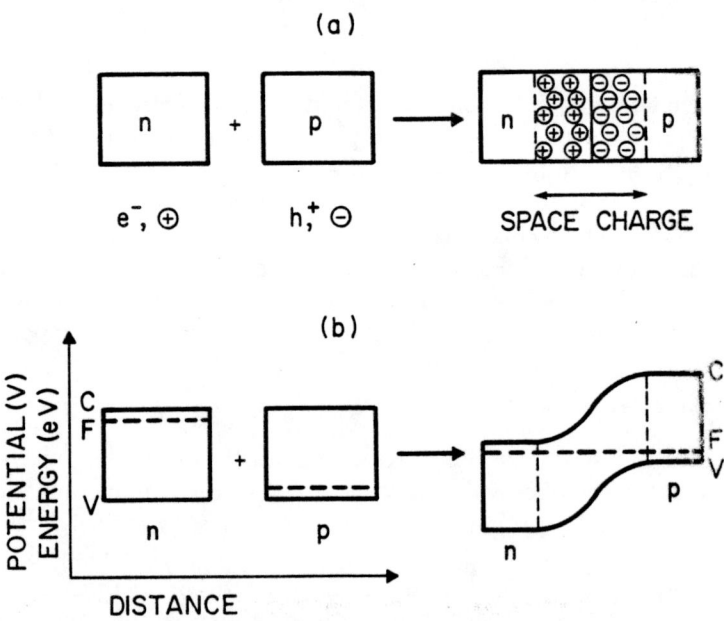

Figure 1. *Schematic construction of a* pn *homodiode*

to prevent the product pn from exceeding the equilibrium value. The excess charge is taken up by the fixed ionized impurities which now become uncompensated. Net charge is thereby introduced to the originally neutral p and n halves. This charge, the space charge denoted in Figure 1a, introduces a field according to

$$\frac{dE}{dx} = \frac{\rho}{\epsilon_0 \epsilon}$$

where x is the one-dimensional coordinate for distance, E the electric field, ρ the charge density, ϵ the relative dielectric constant, and ϵ_0 the permittivity of free space.

The field increases until it prevents net diffusion resulting from the difference in carrier concentration. In other words, the electrochemical potential of a given carrier, electron or hole, is now uniform throughout the p–n diode. This electrochemical equilibrium condition is usually expressed with the Fermi level, F in Figure 1b. The Fermi level denotes the electrochemical potential of electrons. It is high (close to the conduction band C) in n-Si and low (close to valence band V) in p-Si. The reference level is usually the center of the band gap E_i, with the value of the Fermi level given by

$$E_n = E_i + kT \ln \frac{n}{n_i}$$

Equalization of the Fermi level in the two halves of a diode requires the introduction of an electrical potential difference as shown on the right in Figure 1b. This potential difference, the diffusion voltage V_D, is given by the initial difference between the Fermi levels in the n and p regions, $E_n - E_p$:

$$V_D = E_n - E_p = \frac{kT}{q} \ln \frac{n_n}{n_p}$$

(n_n and n_p denote electron concentration on the n-side and on the p-side, respectively, and q is the magnitude of the electronic charge.) The most outstanding characteristic of a diode is that it passes current easily in one direction but not in the other when an external voltage is applied. In the forward, or "easy" direction, electrons flow from n to p, and holes from p to n. In the reverse, or "difficult" direction, electrons flow from p to n, and holes from n to p. This effect results from the availability of a large density of electrons for transport into p-Si, and of holes into n-Si (forward), and of the nonavailability of electrons for current transport from p into n, and of holes from n into p (reverse). Theoretical treatment of the most simple case shows that the reverse current is a constant I_o independent of applied voltage and that the forward current I increases approximately exponentially with applied voltage V,

$$I = I_o \left[\exp\left(\frac{q}{kT}\frac{V}{A}\right) - 1 \right]$$

where A is a parameter which depends on the detailed mechanism of current flow.

When junctions are made between n- and p-type regions of one semiconductor as in the preceding example of nSi/pSi, pn homodiodes are formed (Figure 2). The semiconductor space charge associated with diodes can be introduced in two other ways which prove of increasing importance in solar cell research and development. Heterodiodes [pGaAs/nAlAs (4), pInP/nCdS (5), pCu$_2$S/nCdS (6), etc.] are prepared from two different semiconductors of opposite conductivity type. In Schottky barrier diodes, which are produced by depositing a metal film on a semiconductor [pSi/Cr (7), nGaAs/Pt (8)], the space charge is built up only in the semiconductor. Heterodiodes with one highly

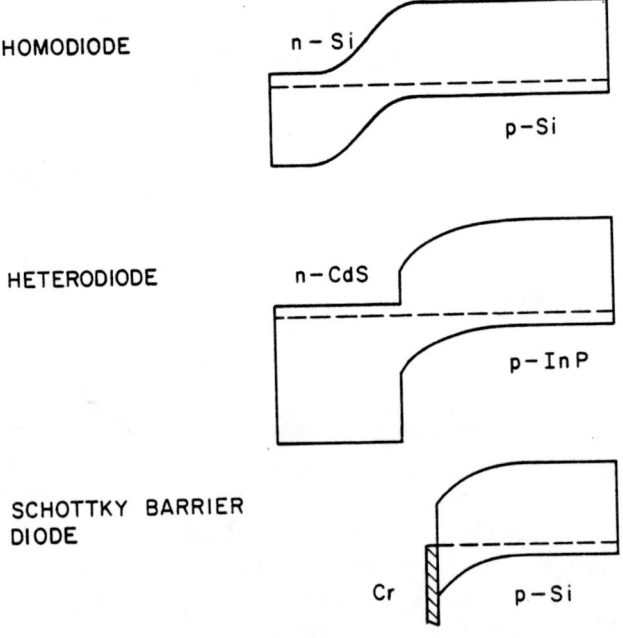

Journal of Crystal Growth

Figure 2. Band diagrams of a homodiode, a heterodiode, and a Schottky barrier diode (63)

conducting ("degenerate") partner can also be viewed as Schottky barrier diodes. This subgroup includes diodes made of a semiconductor and a conducting transparent glass (pSi/nIn_2O_3) (9, 10). Promising results have been obtained recently with metal oxide–semiconductor cells. These are modified Schottky carrier diodes which contain a very thin (10–30Å) oxide layer between the semiconductor ($nGaAs$) and the metal film (Au) (11). Thin oxide layers have also been detected in specimens of silicon-based cells that had originally been conceived as Schottky barrier diodes (7).

Diodes Operating as Solar Cells

When a solar cell is illuminated, a reverse current I_L, which is large compared with I_o, is generated. Figure 3a shows how light quanta arriving at the cell's surface penetrate the diode, are absorbed, and generate electron hole pair in either the n or the p region.

These additional charge carriers increase the pn product above the equilibrium value of n_i^2. The carrier population tends to return to equilibrium, and can do this in two ways. The minority carrier can lose its energy and disappear by immediately recombining with a majority carrier, i.e., by reversing the process of its generation, or the minority carrier can diffuse to the junction and drift in the field of the p–n junction to the side where it is the majority carrier. This is desirable for a solar cell because excess negative and positive charge is not annihilated by recombination within the diode but only after flowing through an external circuit where it can do work. The two extreme modes of operating a cell are shown in Figures 3b and 3c. Under short-circuit conditions (Figure 3b) the external circuit does not offer any resistance to current flow. All the photocurrent then flows through the external circuit. This short-circuit current I_{sc} is the maximum current one can obtain from a solar cell. Under open-circuit conditions with infinite external resistance (Figure 3c) the reverse photocurrent flows initially, but the carriers cannot recombine through the external circuit. They accumulate in their respective halves of the diode, electrons in the n portion and holes in the p portion, and partially compensate its space charge. This effect is identical to that of the external application of a forward bias, i.e., a forward current begins to flow. Steady state is reached when the reverse photocurrent is compensated by that forward current. The corresponding steady-state forward voltage is called the open-circuit voltage, V_{oc}, and is the maximum voltage attainable.

Note that no power ($I \times V$) is drawn from the cell in either the short-circuit (I_{sc}, $V = 0$) or the open-circuit (V_{oc}, $I = 0$) condition. Power is delivered only when the external load resistor R_{ex} is finite (Figure 3d). The voltage V_{op} is smaller than V_{oc} because most of the

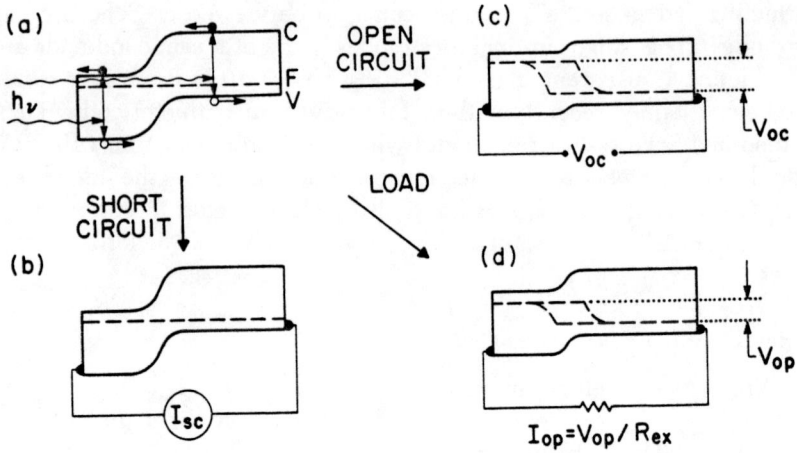

Figure 3. Band diagram of a homodiode solar cell: (a) showing creation of electron-hole pairs by absorption of light quanta; (b) short circuit condition; (c) open circuit condition; and (d) under finite external load (63)

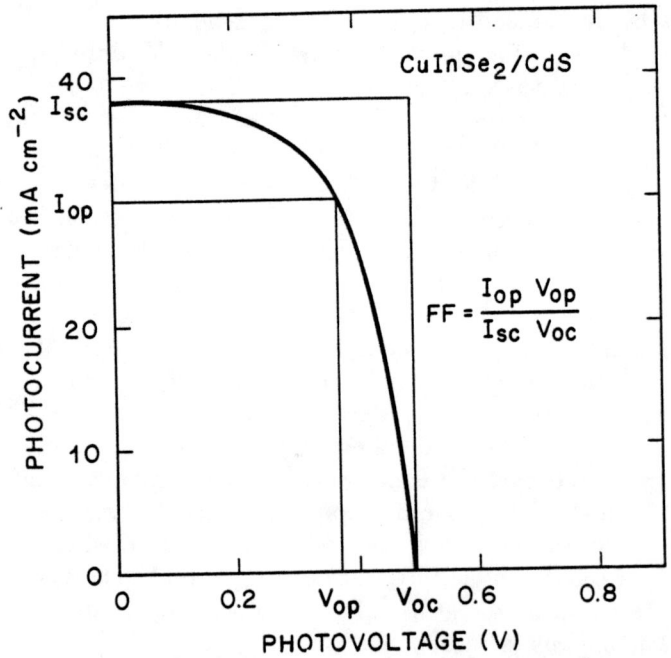

Figure 4. Photocurrent–photovoltage curve of a pCuInSe$_2$/ nCdS heterodiode under 92 mW cm^{-2} solar illumination. Definition of the fill factor FF (51).

photocurrent does not build a forward voltage; it is drawn off through the external circuit. The current I_{op} is smaller than I_{sc} because some of the reverse photocurrent does not flow through the external circuit; as forward current it returns internally to the region where it was generated. However, the product $I_{op} \times V_{op}$ is finite; thus, power is delivered externally.

The photocurrent–photovoltage characteristic of a solar cell is displayed in Figure 4. Aside from I_{sc}, V_{oc}, I_{op}, and V_{op}, the fill factor, or curve factor, FF is an important cell characteristic. While the cell delivers power at any point of the I–V curve except at $V=0$ (I_{sc}) and $I=0$ (V_{oc}), there is one point at which the power is maximum. In experimental work this point is determined by inscribing the I–V rectangle with maximum area (i.e., maximum power) to the I–V curve. In an operating cell the external load resistor must be matched with the solar cell so that it operates at the maximum power point. This point is shown in Figure 4. The fill factor is given by $FF = I_{op}V_{op}/I_{sc}V_{oc}$. It is the ratio of maximum deliverable power to the power represented by the $I_{sc}V_{oc}$ product.

In studying cell behavior it is often useful to resort to equivalent circuits to represent ideal and nonideal contributions to I_{op} and V_{op}. An ideal solar cell comprises only the reverse photocurrent generator and the forward-biased diode with the current–voltage characteristics given by the equation:

$$I = -I_L + I_0 \left[\exp\left(\frac{q}{kT}\frac{V}{A}\right) - 1 \right]$$

A typical nonideal cell contains two further elements, the shunt resistor R_{sh}, and the series resistor R_s. These reduce both output current and output voltage. Their effect is included in the equation for a nonideal cell:

$$I = -I_L + I_0 \left[\exp\left(\frac{q}{kT}\frac{V+|I|R_s}{A}\right) - 1 \right] + \frac{V+|I|R_s}{R_{sh}}$$

Photovoltage

Under open-circuit conditions $I = 0$, the open circuit voltage can be expressed as a function of the photocurrent I_L:

$$V_{oc} = A\frac{kT}{q}\ln\left(\frac{I_L}{I_0}+1\right) \cong A\frac{kT}{q}\ln\left(\frac{I_L}{I_0}\right)$$

For a given I_L, V_{oc} depends primarily on I_o, the saturation current, and on the factor A which is a function of the mechanism of carrier transport in the forward biased diode. Electrons flowing from the n to the p side can recombine with holes either within the space charge ("recombination current," A = 2) or after diffusing into the p-type bulk ("diffusion current," A = 1). For the case of diffusion current in an ideal homodiode, I_0 is given by

$$I_o = q n_i^2 \left[\frac{1}{N_A} \left(\frac{D_{np}}{\tau_{np}} \right)^{\frac{1}{2}} + \frac{1}{N_D} \left(\frac{D_{pn}}{\tau_{pn}} \right)^{\frac{1}{2}} \right]$$

where N_A and N_D are the concentrations of ionized acceptors and donors in the p and n regions respectively, D_{np} and D_{pn} are the diffusion coefficients of electrons in the p region and of holes in the n region, and τ_{np} and τ_{pn} are the corresponding lifetimes. The diffusion coefficients D are related to the mobilities by the Einstein relationship:

$$D = \frac{kT}{q} \mu$$

Large dopant concentrations N_A and N_D, and long minority carrier lifetimes τ_{np} and τ_{pn} entrain low values of I_o and high V_{oc}. I_o and V_{oc} so calculated are ideal limits. Recombination within the space charge frequently dominates forward current transport in solar cells. It is brought about by crystallographic defects and by certain impurities. I_0 is approximately proportional to their concentration, and can be many orders of magnitude larger than the ideal value. The highest V_{oc} reached lie between about $\frac{1}{2}$ and $\frac{3}{4}$ of the band gap E_g of the absorbing semiconductor. In Si ($E_g = 1.11$ eV), a $V_{oc} = 0.637$ V has been reported (12); in GaAs ($E_g = 1.43$ eV), $V_{oc} = 1.13$ V (13) in concentrated ($\times 896$) sunlight. The efficiencies of numerous existing cells is less than the theoretically achievable values primarily because of lower-than-ideal V_{oc}.

Optimum Band Gap

Figure 5a displays the solar power spectrum outside the earth's atmosphere (Air Mass 0) and at sea level with the sun 30° above the horizon (AM2) (14, 15). The spectra are measured on surfaces perpendicular to the incoming ray. The term AM denotes the air mass value, a measure of the length of the optical path through the earth's atmosphere.

The smallest air mass value which can be reached at sea level is 1. Values between 1 and 0 can be attained at altitudes above sea level. Back-reflection and scattering in the atmosphere and absorption reduce the total intensity and introduce characteristic bands into the spectrum.

Rayleigh scattering losses on air molecules are most pronounced at short wavelength. The absorption bands displayed in the spectrum are caused mainly by water vapor. There is also Mie scattering from large particles such as water droplets, dust, and other aerosols. This type of scattering accounts for the variability of the solar spectrum for a given air mass.

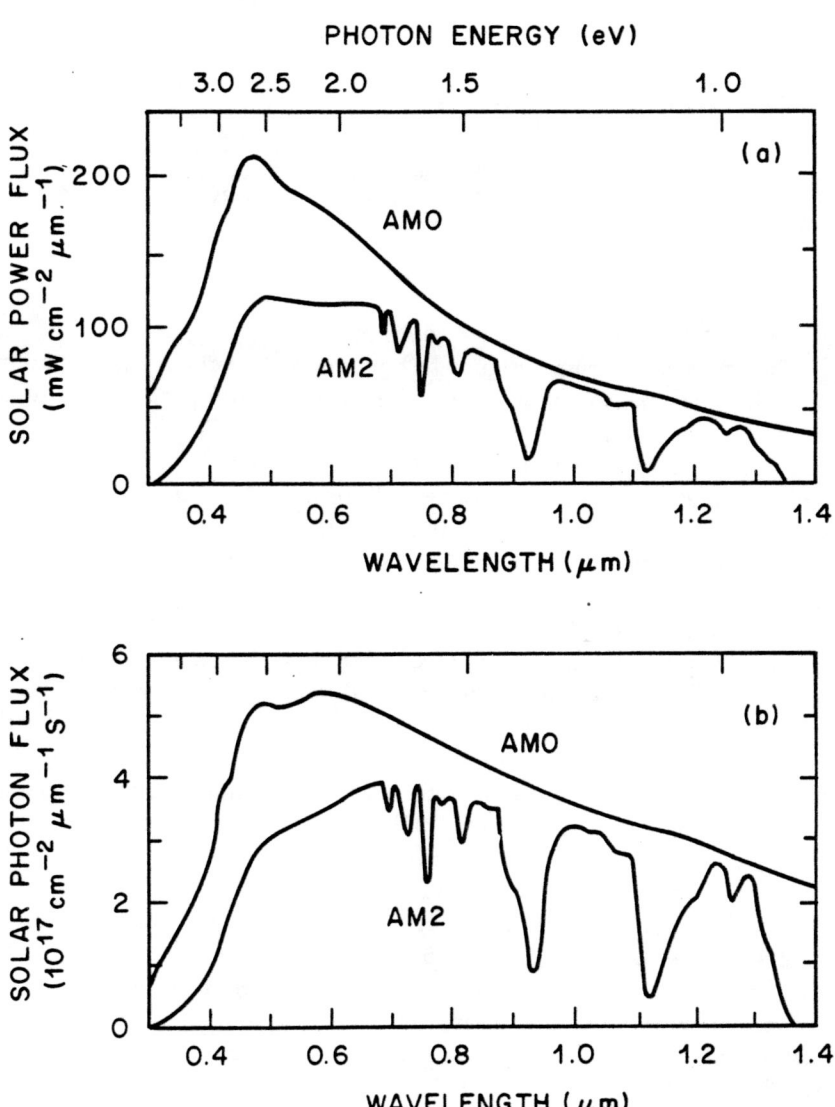

Figure 5. Solar power (a) and solar photon flux (b) spectra for above-atmosphere (AM0) and typical terrestrial (AM2) conditions (15)

Table I. Total Energy of Direct Solar Radiation Incident on a Surface Perpendicular to the Sun's Rays[a] (15)

Air Mass	Energy (kWm^{-2})
0	1.40
1.0	0.865
1.5	0.750
2.0	0.640
4.0	0.372
6.0	0.240
8.0	0.152

[a] Air Mass is a measure of the length of the optical path through the atmosphere, AM $= 1/\cos\theta$, where θ is the angle between the sun and the zenith. $1 > \text{AM} > 0$ obtains for locations above sea level, with a maximum energy flux (the solar constant) at AM0 above the earth's atmosphere.

Science

The total power available at vertical incidence is listed in Table I. The value for AM0 is called the solar constant.

The power spectrum (Figure 5a) provides the primary information about the solar energy available for conversion. More pertinent to solar cell work is the spectrum of the photon flux density (Figure 5b), because the basic process in a cell is the photon-to-electron/hole-pair conversion. Note that under typical terrestrial conditions (AM2) the photon flux is weighted toward higher wavelengths than the power flux.

With the aid of this photon flux spectrum we can estimate the optimum band gap value for semiconductors used in solar cells (16, 17). Assume that all photons which fall on the cell are converted to external current as long as their energy $h\nu$ is larger than the energy of the band gap E_g. Reducing the semiconductor band gap E_g raises the output current, because more photons are absorbed. The fraction of photons with energies above a certain value is shown, for AM2 illumination, by the curve $F(h\nu > E_g)$ in Figure 6a. Obviously, for maximum output current one will use semiconductors with small band gaps. However, the output power depends on the product of current I_{op} and voltage V_{op}. V_{oc} and in turn V_{op} is proportional to the band gap energy. Because of the ensuing trade-off between current and voltage, the output power has a maximum value which lies between 1.0 and 1.5 eV, as indicated by the $F \times E_g$ curve in Figure 6b. In a more detailed consideration which includes the mechanism of current flow in the diode, the typical maximum efficiency curves marked η are obtained. The power efficiency η is defined as the ratio of extracted electrical power to incident solar power P_{in}. Note that the efficiency drops with increasing temperature, and drastically so for semiconductors with low band gaps (18). Applications involving high operating temperatures, e.g., with solar concentrators, will favor relatively large band-gap materials. An alternative approach toward estimating the

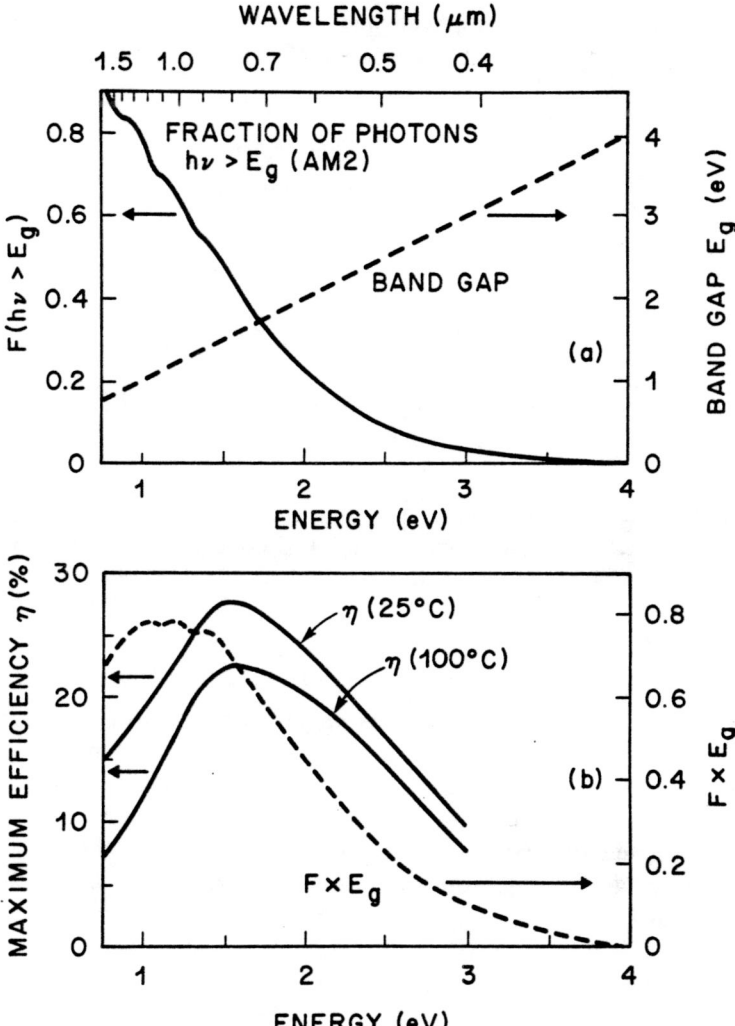

Figure 6. (a) Fraction of solar photons (AM2) with energy higher than the band gap E_g, $F(h\nu > E_g)$, as a function of energy. The band gap (dashed line) illustrates the contribution of output voltage to the current-voltage product. (b) The product $F \times E_g$ showing a maximum between 1.0 and 1.5 eV. Typical theoretical solar efficiencies η at 25° and 100°C from a detailed calculation.

optimum band gap considers the thermodynamic equilibrium between the sun and the solar cell; the result is an ultimate (ideal) efficiency, $\eta = E_g \times F(h\nu > E_g)/P_{in}$, of 44% at $E_g = 1.1$ eV (19). A number of reported semiconductors with band gaps in the vicinity of the optimum range are listed in Table II.

Table II. Properties of Semiconductors Used in Solar Cells

Material	Band Gap at 300K (eV) Direct or Indirect	Structure and Lattice Parameters (Å)	Use (Absorber/ Window)
Si	1.11, i	diamond, $a = 5.431$	A
AlAs	2.16, i	zinc blende, $a = 5.661$	W
GaP	2.25, i	zinc blende, $a = 5.451$	W
GaAs	1.43, d	zinc blende, $a = 5.654$	A
InP	1.34, d	zinc blende, $a = 5.869$	A
CdS	2.42, d	wurtzite, $a = 4.137, c = 6.716$	W
CdSe	1.7, d	wurtzite, $a = 4.29, c = 7.03$	A
CdTe	1.44, d	zinc blende, $a = 6.488$	A
CuInSe$_2$	1.01, d	chalcopyrite, $a = 5.78, c = 11.60$	A
Cu$_2$S	1.2, i	orthorhombic (chalcocite), $a = 11.86, b = 27.32, c = 13.49$	A
Cu$_2$Se	1.2, d(?)	fcc (fluorite), $a = 5.75$	A
Cu$_2$Te	1.1, d(?)	hexagonal, $a = 12.5, b = 21.7$	A
In$_2$O$_3$	2.62, i	b.c.c., $a = 10.11$	W

Direct and Indirect Band Gap

Apart from the energy of the forbidden gap, the nature of the absorption process, direct or indirect, is an important consideration in the selection of a semiconductor (Figure 7). In indirect gap materials the lowest conduction band minimum lies at a momentum different from the valence band maximum. Only a photon $h\nu$ is required for excitation of an electron from the valence to the conduction band in the direct-gap case. With an indirect gap the transition takes place only when assisted by a phonon $h\psi$, the quantum of lattice vibration. Although the typical energy of a phonon is small (~ 0.05 meV), its momentum, $\sim h/a$, is large in comparison with that of a photon, $\sim h/\lambda$ (where a is the crystal lattice parameter, λ the wavelength of the absorbed light). The absorption of the photon can be accompanied by either absorption (as depicted in Figure 7) or emission of phonons. In either case, the need for phonon assistance greatly reduces the transition probability and therefore the absorption coefficient α for the incident light. Attenuating the light to 1/e of its initial intensity requires a path length (the absorption length $1/\alpha$) which is greater for indirect gap than for direct gap materials. This difference is illustrated in Figure 8 with the absorption curves for a typical direct (InP) (20, 21) and a typical indirect (Si) (22, 23) semiconductor. In InP ($E_g = 1.34$ eV, $\lambda_g = hc/E_g = 0.93$ μm) $1/\alpha$ is smaller than 1 μm for any photon energy above E_g. In Si ($E_g = 1.11$ eV, $\lambda_g = 1.12$ μm) $1/\alpha$ is 100 μm at $\lambda = 1$ μm while it approaches 1 μm, the value typical for direct gaps, only at a wavelength of 0.5 μm.

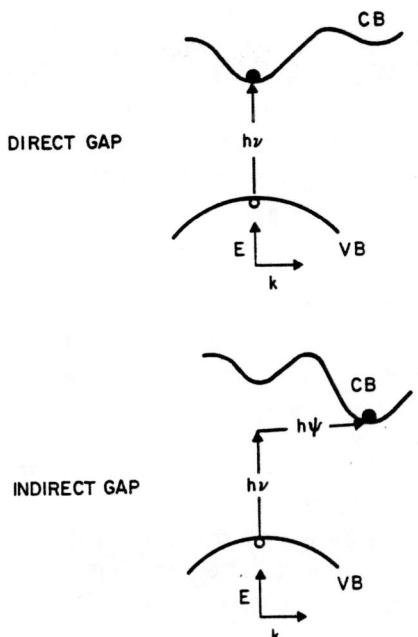

Figure 7. Absorption of a photon in (a) a direct gap semiconductor and (b) an indirect gap semiconductor with phonon assistance

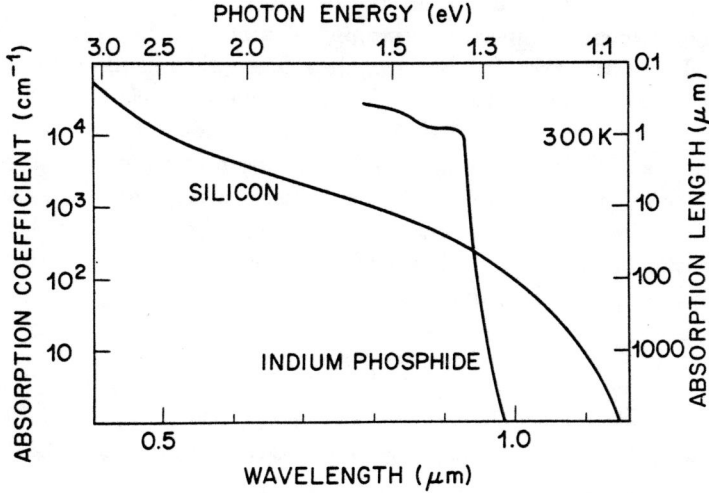

Figure 8. Absorption coefficient (α) and absorption length ($1/\alpha$) for a typical direct-gap semiconductor, InP, and a typical indirect gap semiconductor, Si (63)

Photocurrent

In solar cells made of direct-gap semiconductors the p–n junction can be positioned such that all the incident light is absorbed in its vicinity. As a result, photogenerated minority carriers need not travel farther than a few micrometers to cross the junction space charge. In an indirect-gap material like Si the absorption takes place within a slab ~ 100 μm thick. Minority carriers must travel over lengths of up to 100 μm to reach the junction. It is very expensive to produce silicon of a quality permitting minority carriers to diffuse over this distance instead of recombining with a majority carrier. Solar grade silicon must be of high purity and crystalline perfection. Many impurities, other than those intended as dopants, and imperfections reduce the minority carrier lifetime τ, the time the density of photoexcited carrier decays to $1/e$ of its initial value. During its life the minority carrier diffuses, under its own concentration gradient, toward the junction. If its lifetime is too short, its diffusion length will be too short. It will not reach the junction and will therefore not flow through the external circuit.

Several recombination processes participate in limiting the lifetime of a carrier:

$$\frac{1}{\tau} = \sum_i \frac{1}{\tau_i}$$

Presently it suffices to consider only two of these. Band-to-band recombination is the reverse of the excitation process. If it is the only recombination mechanism operating, the highest lifetime achievable for a given material is obtained. In direct materials band-to-band recombination and also recombination through electronic levels associated with crystalline imperfections or impurities have high probability and lead to a relatively short lifetime of ~ 10 ns. In indirect materials it requires assistance by phonons and thus permits long carrier life, ~ 10 μs. These typical lifetimes are order-of-magnitude values because they depend on the density of majority carriers (with one of whom the minority carrier is to recombine). The carrier diffusion length

$$L = \left(\frac{kT}{q}\mu\tau\right)^{\frac{1}{2}}$$

is proportional to the square root of the carrier lifetime and is therefore typically one to two orders of magnitudes larger for indirect than for direct material. (L is also larger for electrons than for holes since in most semiconductors the electron mobility is 10–100 times greater than that of holes.) On one hand, the longer indirect diffusion length appears to compensate for the equally longer indirect absorption length. On the other hand, the diffusion length in indirect gap materials is more susceptible to impurities and defects which introduce electronic levels near the center of the band gap and promote recombination by alternate emission of holes and electrons. The appropriate lifetime τ_t is inversely proportional to the concentration of these defects N_t.

$$\tau_t = \frac{1}{N_t v \sigma_t}$$

(Here v is the thermal velocity of charge carriers, and σ_t the recombination cross section of the participating defect.) High defect density and associated electric fields make grain boundaries effective sinks for minority carriers. For efficient current collection indirect gap materials must contain single crystals larger than $\sim 100\ \mu\text{m}$. Grains in direct gap materials need not be larger than a few micrometers. For this reason polycrystalline cells in thin film form are an attractive alternative when prepared from direct gap materials.

Free surfaces are to an even greater extent than grain boundaries sites of high defect density N_{st}. Because of the large values of N_{st} encountered, the lifetime for a minority carrier reaching the surface, $\tau_{st} = 1/N_{st} v \sigma_{st}$, is so short that the surface can be a sink for minority carriers almost as effective as a p–n junction. The effect of free surfaces can be reduced in several ways. The surface can be passivated, i.e., provided with a coating which reduces N_{st}. Silicon cells can be covered with a thin film of SiO_2 grown by thermal oxidation. The p–n junction can be moved close to the surface so that few photons are absorbed between the surface and the junction, and only a small fraction of photogenerated carriers is susceptible to surface recombination. Increasing resistance of the thin layer between the p–n junction and the surface imposes a practical limit to this method of reducing the fraction of photocurrent lost by surface recombination. Metal contacts to the front of the solar cell are spaced to allow illumination of the semiconductor. To reach these contacts current in the top layer flows parallel to the p–n junction. When p–n junctions are shallower than $\sim 1\ \mu\text{m}$, resistance loss in the top layer leads to a reduced fill factor and may outweigh the gain in photocurrent. An additional drawback of this approach lies in the difficult technology of preparing very shallow junctions with reproducibly high solar efficiency. Irreproducibility has plagued homodiode cells made of indirect gap (Si) and direct gap [GaAs (*24*), InP (*25*)] materials. In Si this problem has been largely overcome by efficient use of the large fraction of the photocurrent generated below the p–n junction of this indirect gap material, and by introduction of electrical fields in the bulk regions (*26*).

Heterodiodes and Schottky Barrier Diodes

In direct-gap semiconductors high efficiencies have been reached with heterodiodes and Schottky barrier diodes. Ideally the large-bandgap semiconductor in a heterodiode should not absorb solar light. It should act solely as a window through which light penetrates to be absorbed by the small-gap semiconductor. In true heterodiodes the space charge lies at the interface between the two semiconductors of opposite conductivity type. Examples are pGaAs/nAlAs, pInP/nCdS, pCdTe/nCdS (27), pCu$_2$S/nCdS, pCuInSe$_2$/nCdS (28), (by convention, the absorbing small-gap semiconductor is written first). A heteroface cell contains a homodiode with a shallow junction that is passivated with a large-gap semiconductor. The prominent example for this type is the nGaAs/pGaAs/pAl$_x$Ga$_{1-x}$As (13, 29, 30) cell which has reached efficiencies as high as 21%.

Many combinations between semiconductors are potentially available for heterodiode cells. High solar efficiency can be expected for a smaller number of true heterodiodes because of the numerous requirements their components have to meet (31, 32). The small-gap material must be in the optimum range for high efficiency, 1.0–1.5 eV. It must be combined with a large-gap material of opposite conductivity type. CdS ($E_g = 2.42$ eV) is used as a "window" in several heterodiodes. Because of self-compensation of acceptor impurities, it can be made only n-type and thus usually requires p-type partners. Another important condition is that of matching lattice structures and interatomic distances. For instance, GaAs and AlAs both have zinc blende structures with a difference in lattice parameters of only 0.12% at room temperature. (111) planes of zinc blende type InP and (0001) of wurtzite CdS match to within 0.32%. Unsaturated ("dangling") bonds resulting from lattice mismatch lead to a high density of electronic states of energies within the band gap. These states, when located in the junction space charge, can act as recombination centers raising I_o and reducing V_{oc}. They can also trap charge permanently, introduce a sheet of charge in the interface and thereby form electrostatic barriers to the passage of photocurrent. A requirement more precise than that for different conductivity type is that the partners exhibit a large difference in work function ϕ_i, the energy required to move an electron from the Fermi to the vacuum level. The diffusion voltage V_D of a heterodiode consisting of materials A and B is

$$V_D = |\phi_A - \phi_B|$$

and determines the maximum attainable V_{oc}. Another requirement relates to the electron affinity χ_i, i.e., the potential difference between the conduction band and the vacuum level. χ_i of the partners has to be such that no "spike" or "well" is produced, at the interface, in that band edge where the photo-generated minority carriers flow.

It may appear that these conditions, some of which can be relaxed in specific devices, are so numerous and stringent as to be prohibitive. However, heterodiodes have been the main vehicles for the incorporation of direct gap semiconductors in solar cells, both for high efficiency (e.g., GaAs/Al$_x$Ga$_{1-x}$As) and for thin-film cells (e.g., Cu$_2$S/CdS). Therefore, research in this field is intensive.

In a broad sense, Schottky barriers are also heterodiodes since a space charge is established at the interface between two materials. The space charge resides exclusively in the semiconductor since the metal with its high concentration of free electrons cannot support an electric field. In ideal Schottky barrier diodes the band bending is equal to the difference between the work functions ϕ_i of the semiconductor and the metal. Typical metal/semiconductor combinations are Au (11) or Pt (8) (high ϕ) on n-GaAs (low ϕ) and Al (33) or Cr (7) (low ϕ) on p-Si (high ϕ). Because of electronic states in the band gap at the metal/semiconductor interface, the built-in voltage is nearly independent of ϕ_{metal} for barriers or semiconductors with $E_g \lesssim 2$ eV. These electronic states "pin" the band edges at the interface by releasing or picking up charge. The pinning reduces V_D from the ideal value which in turn results in low V_{oc}. Nevertheless, high efficiencies have been obtained with very thin (50–100 Å) metal layers which are virtually transparent to solar light. Such Schottky barrier cells generate high photocurrents because of efficient use of the short wavelength portion to which homodiodes are comparatively insensitive because of surface recombination losses (heterodiodes are insensitive to short wavelength light because of absorption in the window material).

Metal–insulator–semiconductor (MIS) diodes represent an approach toward improvement of V_{oc} over Schottky barrier cells (11, 34). The insulator frequently is a native oxide formed on the surface of semiconductor wafers (Si, GaAs) during storage. When true Schottky barriers are formed, this oxide layer is cleaned off before the metal film is deposited. For the fabrication of MIS diodes, it is left on the semi-conductor. MIS cells exhibit larger V_{oc} than simple Schottky barrier cells. The increased V_{oc} is tentatively ascribed to either a reduction in I_o by the insulator or to an additional voltage drop across the insulator originating at charge trapped in electronic states that reside at the insulator–semiconductor interface. The insulator can also reduce the photocurrent I_L. Thickness and electronic properties of the thin insulator films are critical for optimum tradeoff between increased V_{oc} and reduced I_L.

Table III. Solar Cells[a]

Cell	Single/ Polycryst.	Efficiency (%)	Air Mass	Reference
Silicon:				
pSi/nSi	S	6	—	35
nSi/pSi	S	15	0	37
nSi/pSi	ribbon	10	0	38
nSi/pSi	P	1	—	39
nSi/pSi	P	6	1	40
pSi/nSi	amorphous	2	1	60
pSi/nCdS	S	5	—	41
pSi/nIn$_2$O$_3$	S	6	—	9, 10
pSi/Cr	S	8	1	7
pSi/Al	S	8	1	33
Gallium arsenide:				
pGaAs/nGaAs	S	11	—	24
pGaAs/nAlAs	S	19	1.4	4
pnGaAs/pAl$_x$Ga$_{1-x}$As	S	21	1.4	13
nGaAs/pGaP	S	8	—	42
pGaAs/nGaP	S	7	—	43
nGaAs/Au(MIS)	S	15	—	11
nGaAs/Pt	P	5	—	8
Indium phosphide:				
pInP/nInP	S	7	—	25
pInP/nCdS	S	15	2	44
pInP/nCdS	P	5	2	45
Cadmium telluride:				
pCdTe/nCdS	S	12	2	61
pCdTe/nCdS	P	6	—	27
Semiconductors containing transition metals:				
pCu$_2$S/nSi	S	4	—	46
pCu$_2$S/nCdS	S	~2	—	36
pCu$_2$S/nCdS	P	7	0	47
pCu$_{1.8}$Se/nGaAs	P	5	—	8
pCu$_{1.8}$Se/nInP	P	1	—	48
pCu$_2$Se/nCdSe	P	4	—	49
pCu$_2$Te/nCdTe	S	8	—	50
pCu$_2$Te/nCdTe	P	6	—	50
pCuInS$_2$/nCdS	P	2	—	62
pCuInSe$_2$/nCdS	S	12	1	51
pCuInSe$_2$/nCdS	P	4	—	52

[a] The layer on top of the cell is listed first for homodiodes; in heterodiodes the principal absorbing semiconductor is written first. Published efficiencies are rounded off to integers.

Table III presents a list of published photovoltaic solar cells with data about crystallinity, efficiency, and testing conditions. Cells with maximum reported efficiencies are shown in addition to the first reported silicon (35) and Cu_2S/CdS (36) cells.

Reduction of Cost of Solar Cells

The only solar cells that are commercially available are silicon homodiodes. Their capital cost per unit power delivered is approximately 100 times that of conventional power sources. One reason for its high cost is the small volume of production which does not permit economy of scale. The principal factor in the high price, however, is the need for wafers of highly pure and highly perfect material. Metallurgical-grade Si is prepared by reduction of SiO_2 with C in arc furnaces. For purification it is converted to $SiHCl_3$ which is distilled. Pure polycrystalline silicon is then obtained by pyrolysis. Single crystals are pulled by the Czochralski method from the melt. The single-crystal boules are cut into slices which are polished to remove mechanical damage. The actual cell fabrication involves controlled in-diffusion of a dopant to form a *p–n* junction, application of back contact and front contact grid, and evaporation of an antireflection coating. It is obvious that the present-day manufacture of Si cell produces small active areas while being labor-intensive.

Several strategies are being pursued to reduce cell costs. The most obvious is to raise the output per cell area by increasing cell efficiency, and by using inexpensive collectors, or concentrators, to achieve higher power density. Figure 9 presents a schematic of power conversion losses occurring in a typical Si cell (53). The efficiency can still be improved by raising V_{oc} (reducing the voltage loss), by more efficient collection of photo-generated carriers (reducing recombination loss), by using better matched antireflection coatings, and by reducing series resistance. At best, one might double the efficiency per unit area, probably with a simultaneous increase in manufacturing cost. Inexpensive concentrators combined with small active cell area, i.e., large concentration ratios, are presently under consideration for Si (54) and GaAs (55) cells. Single-crystal GaAs cells when produced individually are expected to be considerably more expensive than Si cells because of the high cost of GaAs wafers (about 10 times that of Si wafers), and particularly because most GaAs-based cells are produced by the very expensive liquid-phase-epitaxy method. However, GaAs heterodiodes have reached the highest efficiency yet obtained, and, compared with Si, the efficiency of GaAs-based cell is less affected by operation at elevated temperature, an important advantage for concentrator applications.

The other approach toward less expensive solar energy is to reduce the cost per cell area while retaining useful efficiency. The two main

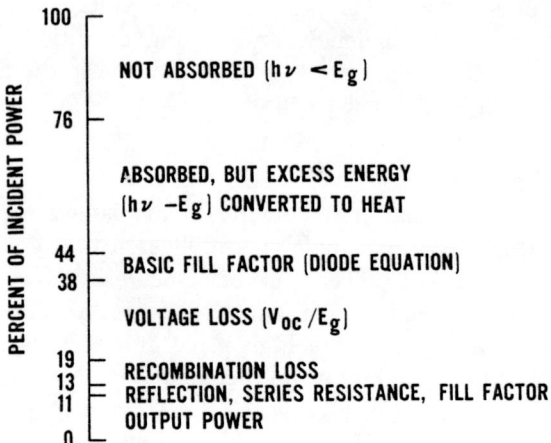

Figure 9. Illustration of the principal contributions to power loss during photovoltaic conversion in a typical silicon cell (53)

avenues are the development of methods for inexpensive growth of single crystals and the fabrication of polycrystalline thin-film cells.

Silicon ribbons 2.5 cm wide and ~ 0.02 cm thick can be grown by the edge-defined film-fed technique at rates of ~ 2 cm/min (38). These ribbons exhibit very high crystallinity; solar cells up to 10% efficient have been produced. The technique is based on the controlled solidification of a ribbon of molten silicon pulled, with a seed crystal, from a slot-like capillary which is immersed in a silicon melt. This process, which is amenable to automation, avoids the usual slicing and polishing step. A similar but less developed technology is the dendritic web growth from silicon melts (56). Here a ribbon is pulled, without a die, by two bounding dendrites whose growth is started with a seed.

Thin-film polycrystalline cells represent another approach to cost reduction. Production of the Cu_2S/CdS cells, the best known example for such cells (6), typically involves preparation of a conducting substrate, e.g., Zn-plated Cu sheet, evaporation of a 20–40 μm thick CdS film, a brief etch of the CdS followed by a 10-sec dip in $CuCl_2$ solution which forms a Cu_2S layer with a few thousand Å thickness, a 2-min activation anneal in air at 250°C; finally, contact grid and anti-reflection coating are produced, and a transparent cover is applied with epoxy resin. This thin-film cell should be amenable to highly automated production. Another advantage is the economical use of semiconductors which may contain comparatively rare elements (Cd, Ga, In) and which even when abundant (Si) are expensive when purified for solar use. Highest reported efficiencies for the Cu_2S/CdS cell are $\sim 7\%$. Higher values are predicted for modified cells which conform better to certain requirements for ideal heterodiodes, viz., lattice match and proper electron affinities; the alloying of CdS with ZnS has been a step in this direction (*47*). The Cu_2S/CdS cell has been plagued by rapid degradation while in operation. It has been proposed to reduce failure caused by reaction of Cu_2S with ambient O_2 and H_2O by careful encapsulation, and to avoid electrochemical decomposition of Cu_2S under the cell's own photovoltage (*57*) by making this semiconductor strictly stoichiometric.

It is nevertheless desirable to develop alternatives. Interesting results have been obtained with polycrystalline Si homodiodes prepared by chemical vapor deposition (*40*). The small crystallite size, and to some extent the comparatively high impurity content, conflict with the necessity for large minority carrier diffusion length and have limited the efficiency to date to about 6%.

The $CuInSe_2$/CdS (*51*) and InP/CdS (*44*) heterodiodes prepared in single crystal form in or laboratory are attractive candidates for thin-film cells. Although the first thin-film $CuInSe_2$/CdS cells with $\eta = 4\%$ have already been produced (*52*), we have focused our attention on the InP/CdS cell because of the greater experience accumulated with InP. Single-crystal InP/CdS cells are very stable to degradation in the atmosphere. We have produced thin-film InP/CdS cells on carbon substrates. The efficiency of current polycrystalline samples is 5%. However, based on V_{oc} and I_{sc} of these early cells, efficiencies of at least 7–8% can be expected (*45*).

Future efforts to produce inexpensive solar cells will go beyond existing materials and processes. For instance, a large number of semiconductors exist whose band gap and conductivity have not been sufficiently characterized to permit a decision even about their potential usefulness in solar cells (*58*). Many of these semiconductors are composed of inexpensive raw materials. Purification to solar-grade semiconductors, currently an expensive step, will have to be carried out by continuous methods. An example is a unit combining reduction of SiO_2 to Si, reaction to SiF_2, distillation of SiF_2, and disproportionation to pure polycrystalline Si and to SiF_4 (*59*).

Conclusion

While much progress has been made in analyzing and improving the performance of solar cells, it is not yet possible to predict materials and processes for inexpensive converters. The present situation calls for an increase in the number of available options and for the development of new production techniques, both with a substantial input from chemists.

Nomenclature

A	diode factor
a, b, c	lattice parameters (cm or Å)
D_{np}	diffusion coefficient of electrons in p-type material (cm² s⁻¹)
D_{pn}	diffusion coefficient of holes in n-type material (cm² s⁻¹)
E_g	band-gap energy (eV)
E_i	Fermi level in an intrinsic semiconductor (eV)
E_n	Fermi level in an n-type semiconductor (eV)
E_p	Fermi level in a p-type semiconductor (eV)
FF	fill factor, or curve factor
h	Planck's constant (6.62×10^{-34} J s)
I	electrical current density (A cm⁻²)
I_L	photocurrent density (A cm⁻²)
I_{op}	operating current density (A cm⁻²)
I_{sc}	short circuit current density (A cm⁻²)
I_0	reverse saturation current density (A cm⁻²)
k	Boltzmann's constant (1.380×10^{-23} J K⁻¹)
L	diffusion length (cm)
N_A	concentration of ionized acceptors (cm⁻³)
N_C	effective density of states in conduction band (cm⁻³)
N_D	concentration of ionized donors (cm⁻³)
N_V	effective density of states in valence band (cm⁻³)
N_{st}	density of surface recombination centers (cm⁻²)
N_t	density of bulk recombination centers (cm⁻³)
n	concentration of electrons (cm⁻³)
n_i	carrier concentration in an intrinsic semiconductor (cm⁻³)
n_n	concentration of electrons in n-type material (cm⁻³)
P_{in}	incident solar power flux (W cm⁻²)
p	concentration of holes (cm⁻³)
p_p	concentration of holes in p-type material (cm⁻³)
q	electronic charge (1.60×10^{-19} C)
R_{ex}	external load resistance (Ω)
R_s	internal series resistance (Ω)
R_{sh}	internal shunt resistance (Ω)
T	temperature (K)
V_D	diffusion voltage (V)
V_{oc}	open-circuit voltage (V)
V_{op}	operating voltage (V)
v	thermal velocity of charge carriers (cm s⁻¹)
x	distance (cm)
α	optical absorption coefficient (cm⁻¹)
E	electric field (V cm⁻¹)

ϵ	relative dielectric constant
ϵ_0	permittivity of free space (8.86×10^{-14} f cm^{-1})
λ	wavelength (cm)
ν	frequency (s^{-1})
ρ	density of charge (C cm^{-3})
σ_{st}	capture cross section of a surface defect (cm^2)
σ_t	capture cross section of a bulk defect (cm^2)
τ	lifetime of a charge carrier (s)
τ_i	lifetime determined by recombination of type i (s)
τ_{np}	lifetime of electrons in p-type material (s)
τ_{pn}	lifetime of holes in n-type material (s)
τ_{st}	lifetime determined by surface recombination (s)
ϕ_i	work function of semiconductor i (eV)
χ_i	electron affinity of semiconductor i (eV)

Literature Cited

1. Sze, S. M., "Physics of Semiconductor Devices," Wiley-Interscience, New York, 1969.
2. Pankove, J. I., "Optical Processes in Semiconductors," Prentice-Hall, Englewood Cliffs, N. J., 1971.
3. Hovel, H. J., "Solar Cells," Academic, New York, 1975.
4. Johnston, W. D., Callahan, W. M., *Appl. Phys. Lett.* (1976) **28**, 150.
5. Wagner, S., Shay, J. L., Bachmann, K. J., Buehler, E., *Appl. Phys. Lett.* (1975) **26**, 229.
6. Rothwarf, A., Böer, K. W., *Prog. Solid State Chem.* (1975) **10**, 71.
7. Anderson, W. A., Delahoy, A. E., Milano, R. A., *J. Appl. Phys.* (1974) **45**, 3913; and unpublished results.
8. Vohl, P., Perkins, D. M., Ellis, S. G., Addiss, R. R., Hui, W., Noel, G., *IEEE Trans. Electron Devices* (1967) **ED-14**, 26.
9. Matsunami, H., Oo, K., Ito, H., Tanaka, T., *Jpn. J. Appl. Phys.* (1975) **14**, 915.
10. Lai, S. W., Franz, S. L., Kent, G., Anderson, R. L., Clifton, J. K., Masi, J. V., See *Proc. IEEE Conf. Rec. Photovoltaic Specialists Conference, 11*, Scottsdale, Arizona, May 6–8, 1975, IEEE, New York, 1975, p 398.
11. Stirn, R. J., Yeh, Y. C. M., *Appl. Phys. Lett.* (1975) **27**, 95.
12. D'Aiello, R. V., Robinson, P. H., Kressel, H., *Tech. Dig. Intern. Electron Devices Mtg.*, Washington, D. C., December 1–3, 1975. IEEE, New York, 1975, p. 335.
13. James, L. W., Moon, R. L., IEEE *Conf. Rec. Photovoltaic Specialists, 11*, Scottsdale, Arizona, May 6–8, 1975, IEEE, New York, 1975, p. 402.
14. Moon, P., *J. Franklin Inst.* (1940) **230**, 583.
15. Gates, D. M., *Science* (1966) **151**, 523.
16. Prince, M. B., *J. Appl. Phys.* (1955) **26**, 534.
17. Loferski, J. J., *J. Appl. Phys.* (1956) **27**, 777.
18. Halsted, R. E., *J. Appl. Phys.* (1957) **28**, 1131.
19. Shockley, W., Queisser, H. J., *J. Appl. Phys.* (1961) **32**, 510.
20. Turner, W. J., Reese, W. E., Pettit, G. D., *Phys. Rev.* (1964) **136A**, 1467.
21. Emlin, R. V., Zverev, L. P., Rut, O. E., *Sov. Phys. Semicond.* (1974) **8**, 796.
22. Dash, W. C., Newman, R., *Phys. Rev.* (1955) **99**, 1151.
23. Philipp, H. R., Taft, E. A., *Phys. Rev. Lett.* (1962) **8**, 13.

24. Gobat, A. R., Lamorte, M. F., McIver, G. W., *IRE Trans. Mil. Electron.* (1962) **6**, 20.
25. Galavanov, V. V., Nasledov, D. M., *Sov. Phys. Solid State* (1967) **8**, 2723.
26. Ellis, B., Moss, S. T., *Solid-State Electron.* (1970) **13**, 1.
27. Bonnet, D., Rabenhorst, H., *IEEE Conf. Rec. Photovoltaic Specialists*, 9, Silver Spring, Maryland, 1972, IEEE, New York, 1972, p. 129.
28. Wagner, S., Shay, J. L., Migliorato, P., Kasper, H. M., *Appl. Phys. Lett.* (1974) **25**, 434.
29. Alfërov, Zh. I., Andreev, V. M., Kagan, M. B., Protasov, I. I., Trofim, V. G., *Sov. Phys. Semicond.* (1971) **4**, 2047.
30. Hovel, H. J., Woodall, J. M., *J. Electrochem. Soc.* (1973) **120**, 1246.
31. Milnes, A. G., Feucht, D. L., "Heterojunctions and Metal-Semiconductor Junctions," Academic, New York, 1972.
32. Sharma, B. L., Purohit, R. K., "Semiconductor Heterojunctions," Pergamon, New York, 1974.
33. Charlson, E. J., Lien, J. C., *J. Appl. Phys.* (1975) **46**, 3982.
34. Fonash, S. J., *J. Appl. Phys.* (1975) **45**, 1286.
35. Chapin, D. M., Fuller, C. S., Pearson, G. L., *J. Appl. Phys.* (1954) **25**, 676.
36. Reynolds, D. C., Leies, G., Antes, L. L., Marburger, R. E., *Phys. Rev.* (1954) **96**, 533.
37. Arndt, R. A., Allison, J. F., Haynos, J. G., Meulenberg, A., *IEEE Conf. Rec. Photovoltaic Specialists*, 11, Scottsdale, Arizona, May 6-8, 1975, IEEE, New York, 1975, 40.
38. Ravi, K. V., Serreze, H. B., Bates, H. E., Morrison, A. D., Jewett, D. J., Ho, J. C. T., *IEEE Conf. Rec. Photovoltaic Specialists*, 11, Scottsdale, Arizona, May 6-8, 1975, IEEE, New York, 1975, p. 280.
39. Heaps, J. D., Tufte, O. N., Nussbaum, A., *IRE Trans. Electron Devices* (1961) **ED-8**, 560.
40. Chu, T. L., *J. Vac. Sci. Technol.* (1975) **12**, 912; and unpublished results.
41. Okimura, H., Kawakami, M., Sahai, Y., *Jpn. J. Appl. Phys.* (1967) **6**, 908.
42. Alfërov, Zh. I., Zimogorova, N. S., Trukan, M. K., Tuchkevich, V. M., *Sov. Phys. Solid State* (1965) **7**, 990.
43. Purohit, R. K., *Phys. Status Solidi* (1967) **24**, K57.
44. Shay, J. L., Wagner, S., Bachmann, K. J., Buehler, E., *J. Appl. Phys.* (1976) **47**, 614; and unpublished results.
45. Bachmann, K. J., Buehler, E., Shay, J. L., Wagner, S., *Appl. Phys. Lett.* (1976) **29**, 121; and unpublished results.
46. Drozdov, V. A., Mel'nikov, M. M., *Sov. Phys. Semicond.* (1973) **7**, 801.
47. Palz, W., Besson, J., Nguyen Duy, T., Vedel, J., *IEEE Conf. Rec. Photovoltaic Specialists*, 10, Palo Alto, California, November 13-15, 1973, IEEE, New York, 1974, p. 69.
48. Fischer, H., Ph.D. Thesis, Technische Universität Braunschweig, Germany, 1970.
49. Komashchenko, V. N., Fedorus, G. A., *Ukr. Fiz. Zh.* (1968) **13**, 688.
50. Cusano, D. A., *Solid State Electron.* (1963) **6**, 217.
51. Shay, J. L., Wagner, S., Kasper, H. M., *Appl. Phys. Lett.* (1975) **27**, 89.
52. Kazmerski, L. L., White, F. R., Morgan, G. K., *Appl. Phys. Lett.* (1976) **29**, 268.
53. Wolf, M., *IEEE Conf. Rec. Photovoltaic Specialists*, 10, Palo Alto, California, November 13-15, 1973, IEEE, New York, 1974, p. 5.
54. Dean, R. H., Napoli, L. S., Liu, S. G., *RCA Rev.* (1975) **36**, 324.
55. James, L. W., *Tech. Dig. Intern. Electron. Device Mtg.*, Wash., D. C., December 1-3, 1975, IEEE, New York, 1975, p. 87.
56. Seidensticker, R. G., Scudder, L., Brandhorst, H. W., Jr., *IEEE Conf. Rec. Photovoltaic Specialists*, 11, Scottsdale, Arizona, May 6-8, 1975, IEEE, New York, 1975, p. 299.

57. Besson, J., Nguyen Duy, T., Gauthier, A., Palz, W., Martin, C., Vedel, J., *IEEE Conf. Rec. Photovoltaic Specialists, 11,* Scottsdale, Arizona, May 6–8, 1975, IEEE, New York, 1975, p. 468.
58. Strehlow, W. H., Cook, E. L., *J. Phys. Chem. Ref. Data* (1973) **2,** 163.
59. Wolf, M., *IEEE Conf. Rec. Photovoltaic Specialists, 11,* Scottsdale, Arizona, May 6–8, 1975, IEEE, New York, 1975, p. 306.
60. Carlson, D. E., Wronski, C. R., *Appl. Phys. Lett.* (1976) **28,** 671.
61. Yamaguchi, K., Matsumoto, H., Nakayama, N., Ikegami, S., *Jpn. J. Appl. Phys.* (1976) **15,** 1575.
62. Kazmerski, L. L., White, F. R., Ayyagari, M. S., Juang, Y. J., Patterson, R. R., *J. Vac. Sci. Technol.,* to be published.
63. Wagner, S., *J. Cryst. Growth* (1975) **31,** 113.

RECEIVED July 27, 1976.

7
Recrystallization of Semiconducting Polycrystalline Ribbons Using the Peltier Effect

S. VOJDANI and R. HASHEMIAN

Materials and Energy Research Center, Arya Mehr University of Technology, P.O. Box 41-2927, Tehran, Iran

> *A new approach to zone refining thin semiconductor ribbons or films necessary for the production of low-cost solar cells is investigated using the Peltier effect. The results indicate that under certain conditions the Peltier current tends to stabilize the freezing interface allowing an increase in the grain size of a thin film.*

Commercial production of silicon solar cells routinely yields highly reliable devices having adequate (~12%) efficiencies. These devices have been designed to operate in the space environment and have proved very suitable for this application. However, for terrestrial applications they have a serious defect—their cost is too high by at least one order of magnitude. The most important factors in the device cost are the expensive production of large single-crystal boules and the wafering of these crystals to give thin slices suitable for use in devices. Two processes have been considered for producing cheap wafers: (a) the growth of single crystals in the form of thin ribbons so that expensive wafering is avoided (1), and (b) the deposition of films on suitable substrates by heteroepitaxial techniques—CVD, sputtering, evaporation (2), and more recently LPE (3). Heteroepitaxial films generally give low efficiency when used in solar cells because of a reduction in open-circuit voltage and minority carrier lifetime associated with the presence of grain boundaries.

However, if the grain size is sufficiently large (4, 5) (e.g., for Si, 100–1000 μm), adequate efficiency is obtained. The production of silicon solar cells by thin or thick film techniques will probably require a processing step that enlarges the crystallite size in the film. Two techniques have already been suggested for this prupose: crystallization of Si films by means of a scanning electron or laser beam (6) and heat treatment of CVD-grown Si films in an inert atmosphere (7). Another alternative will be passage of molten zone (zone refining) across the film under controlled conditions to increase the grain size.

For the purpose of zone refining a thin polycrystalline film, four problems must be considered:

(a) the molten zone must be very narrow to prevent the break-up of the liquid into globules because of surface tension, unless the liquid wets the substrate (e.g., Si wets carbon);

(b) the molten semiconductor must not be contaminated by the substrate;

(c) the molten zone must move across the film with uniform velocity so that solidification can proceed, in a controlled manner; and

(d) the interface between solid and liquid must be planar so that, upon solidification of the molten zone, an improvement in crystallite size is attained.

Achieving acceptable results with thin films demands control of the zoning process, and this is difficult to attain in an inherently small-volume crystallization process. Hence it is interesting to examine the possibility of using the Peltier effect. Since the two solid–melt interfaces also constitute boundaries between phases having different electrical resistivities, the passage of a direct current through the sample causes Peltier heating at one interface and cooling at the other. This could cause the zone to move and has the advantage of localized heat supply and extraction precisely at the interfaces, facilitating control. The process was tried many years ago for bulk crystals (8) but was discarded because a large current was needed for large-area samples to provide adequate interface heating and cooling. This limitation is not important for zoning thin films, and, additionally, there is no need to provide all the heat for zone melting from the direct current; the sample can be placed in a furnace to provide auxiliary heating. The purpose of using Peltier current is to allow the zone width and possibly the interface topology to be stabilized.

The remainder of this chapter presents a theoretical and experimental investigation of the solidification process in the presence of a direct current as a first step towards Peltier zoning.

Theory

All symbols used in this analysis are defined in the "Nomenclature" section. The rate per unit area at which Peltier heat is delivered to (or extracted from) a solid–liquid interface through which a direct current of density J_p is passing is (9)

$$Q_p = \alpha T_m J_p \tag{1}$$

For the theoretical analysis, a thin ribbon of semiconductor is considered as shown in Figure 1. The sample is placed in a cylindrical furnace at

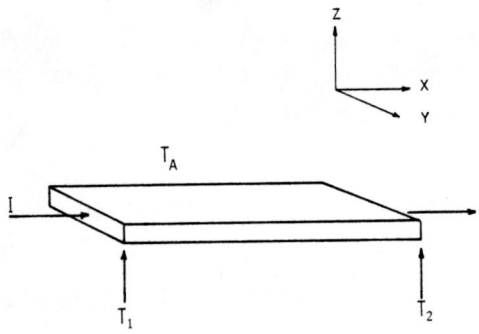

Figure 1. A thin ribbon of semiconductor with the reference axes used for the theoretical modeling

ambient temperature T_A. A direct current I is passed through it, while the ends of the sample are kept at temperatures T_1 and T_2 respectively. The relevant heat balance equations are as follows. In the solid region for unit volume:

$$K_s \frac{\partial^2 T}{\partial x^2} + J_s^2 \rho_s - \frac{P}{A} \epsilon_s \sigma (T_s^4 - T_A^4) = c_{ps} \frac{\partial T_s}{\partial t} \tag{2}$$

In the liquid region for unit volume:

$$K_l \frac{\partial^2 T}{\partial x^2} + J_l \rho_l - \frac{P}{A} \epsilon_l \sigma (T_l^4 - T_A^4) = c_{pl} \frac{\partial T_l}{\partial t} \tag{3}$$

At the interface between solid and liquid:

$$K_s \frac{\partial T}{\partial x} - K_l \frac{\partial T}{\partial x} + \alpha J_p T_m = L \frac{\partial x}{\partial t} \tag{4}$$

In the formulation of Equations 2, 3, and 4 the following assumptions have been made:

(a) The sample consists of solid and liquid regions.

(b) Heat loss from the sample results from radiation from the surfaces and conduction through the ends only.

(c) ρ_s and ρ_l are considered constants.

(d) The sample is symmetrically located in the furnace with radial symmetry, and the ambient temperature T_A is constant.

(e) The ambient temperature is close to the melting point.

(f) The temperature is constant along the y and z axes; only variation along the x axis is considered. In the region of the molten zone there can be no variation in T along the x axis because of the two-phase condition.

(g) At $x = 0$, $T = T_1$; at $x = a$, $T = T_2$.

(h) Steady state conditions exist.

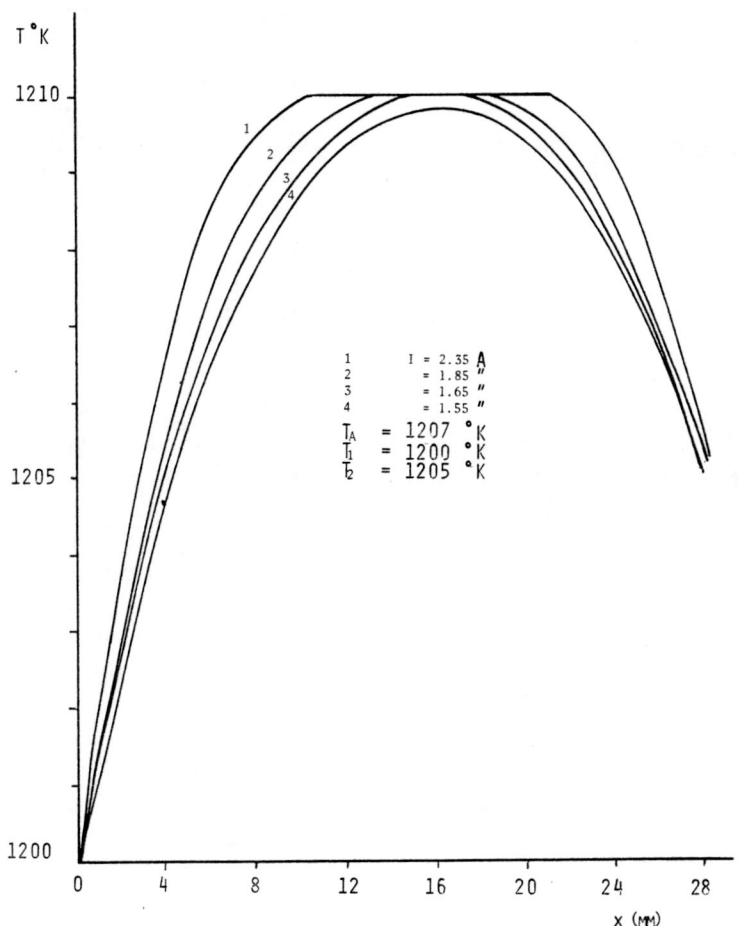

Figure 2. Temperature profile across the sample as a function of current. Peltier heating and cooling are neglected.

Using the above assumptions, Equations 2, 3, and 4 were solved computationally for germanium, since this material was to be used, for convenience, in initial experiments. Representative values used for the parameters in the equations are shown in the Nomenclature section. The solutions allowed the zone width and the completeness of one melting to be related to the experimental conditions used and particularly to the direct current flowing.

Theoretical Modeling. Passage of a current through a sample of semiconductor ribbon containing a molten zone has two effects: it will change the zone width W and the degree of melting within the zone. This latter is described in terms of a parameter γ which will be a function of distance along the x axis; $\gamma = 0$ defines a solid region, $\gamma = 1$ a liquid region, and $0 < \gamma < 1$ a region of partial melt, so that, across the area of the zone at a point x, the fraction $\gamma(x)$ of the area will be molten.

Figure 2 shows the calculated temperature profile along the sample for various currents when Peltier heating and cooling are neglected in the calculations. The changes are caused simply by different levels of Joule heating. The zone width depends on the temperatures T_1 and T_2 and also, as shown in the figure, on the current flowing.

Figure 3 shows the dependence of zone width (W) and γ on the temperatures T_1 and T_2 (assumed equal) for three different currents. At the melting point W is about 3 cm, but reducing T_1 and T_2 to about 23°K below this value reduces W to less than 1 mm. The width is again shown to depend on the current. The degree of melting of the zone is determined by the current only (for given ambient temperature) and not by the temperature at the ends of the sample. The dependence of W and γ on ambient temperature is shown in Figure 4, where both are seen to decrease as T_A decreases.

Thus incomplete melting of the zone with the passage of current has been observed and predicted from the mathematical model. The reason for this phenomenon is associated with the different resistivities of the melt and the solid ($\rho_l : \rho_s = 1:8$ for Ge), which, in the event of incomplete zone melting, causes the current to channel through the melt region. This will increase the Joule heating in this region, thus causing the molten region to grow. A steady state will be reached when Joule heating in the melt is balanced by heat loss from the melt to the solid and also by the usual heat losses by conduction along the x axis and by radiation.

We can now proceed to incorporate the effect of Peltier heating and cooling at the two interfaces. Let us assume that a molten zone is formed, penetrating the specimen uniformly to a depth y_0 as shown in Figure 5a. The molten zone is confined by the dotted line ABCD. Peltier cooling will occur at the interface AB (current going from solid to liquid), and Peltier heating will occur at the interface DC (current going from liquid

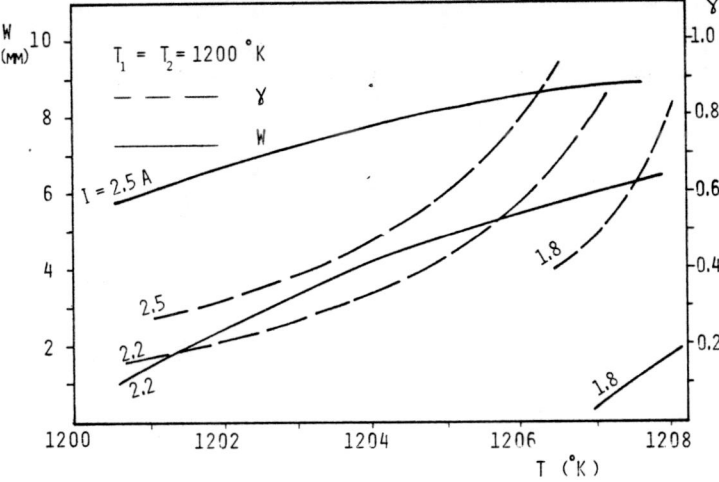

Figure 3. Dependence of W and γ on the ambient temperature T_A for different currents

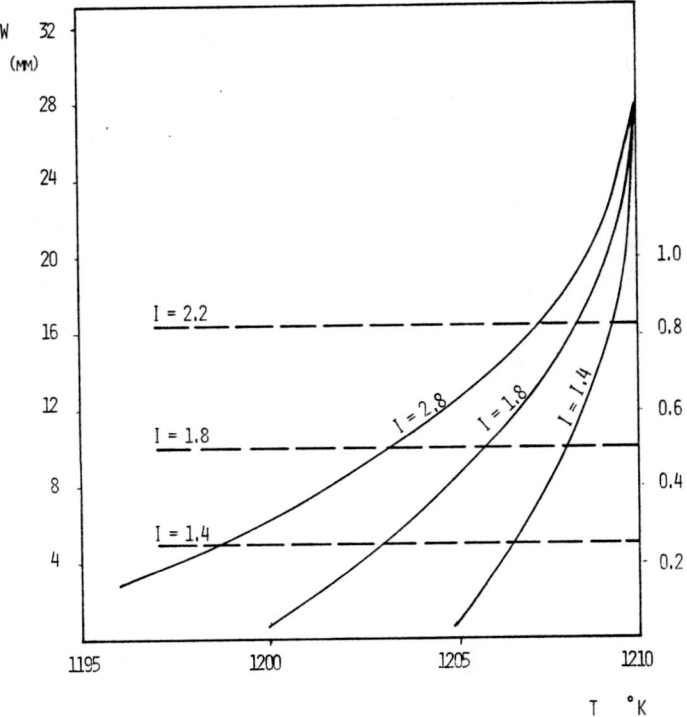

Figure 4. Dependence of W and γ on the end temperatures as a function of different currents: $T_1 = 1200°K$, $T_2 = 1205°K$

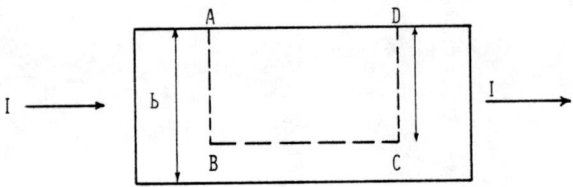

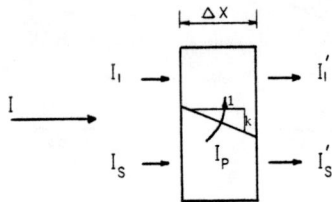

Figure 5. Schematic illustrating the effect of I_p on zone solidification

to solid). As the molten region is at the melting point, Peltier heat absorbed from the interface AB and evolved at the interface DC is expected to change the shape and position of these interfaces. Let us now take a segment of width Δx in the molten zone as shown in Figure 5b. Since the bottom solid–liquid boundary is not parallel to the current flow (the slope $dJ/dx = 0$), and since the two phases have different resistivities, there is a net current I_p across the interface, which causes the Peltier effect. The interface between solid and liquid is taken as a line segment of gradient dJ/dx found computationally. The formulation to find dJ/dx, and consequently the final shapes and positions of interfaces AB and DC upon the passage of current I, are given as follows.

The equations for the currents passing through the segment are:

$$I_1 + I_s = I_1' + I_s' = I \tag{5}$$

$$I_1 + I_p = I_1' \tag{6}$$

$$J_1 = \delta_r J_s \tag{7}$$

$$J_1' = \delta_r J_s' \tag{8}$$

Where δ_r is the ratio of the solid to the liquid resistivity. Rearranging Equations 5, 6, 7, and 8 gives:

$$I_p = \frac{\delta_r b I}{(\delta y + b)^2} \Delta y \tag{9}$$

where $\delta = \delta_r - 1$. From Equation 9 the Peltier cooling at the interface kk' is given as:

$$Q_p = \alpha T_m I_p = \frac{\alpha \delta_r T_m b I}{(\delta y + b)^2} \tag{10}$$

However, the heat generated in the slice caused by the Joule heating is given by:

$$Q_j = \frac{\delta_r \rho_1 I^2}{(\delta y + b) m} \Delta x \tag{11}$$

while originally, when no zone shaping caused by the Peltier effect is considered, we would get

$$Q_{jo} = \frac{\delta_r \rho_1 I^2}{(\delta y + b) \omega} \Delta x \tag{12}$$

as the amount of Joule heating in the segment. For the equilibrium situation the excess Joule heating caused by such zone shaping must be equal to the loss of energy caused by the Peltier effect, that is,

$$Q_p = Q_j - Q_{jo} = \frac{\delta_r \rho_1 I^2}{W} \left[\frac{1}{\delta y + b} - \frac{1}{\delta y_o + b} \right] \Delta x \tag{13}$$

and after substituting for Q_p from Equation 10 we obtain

$$K \frac{dy}{dx} = -\delta y^2 + (\delta y_o - b) + b y_o \tag{14}$$

where K is a constant and

$$K = \frac{\alpha T_m A (\delta y_o + b)}{\delta \rho_1 I} \tag{15}$$

Solution of the differential equation, 14, with the boundary conditions gives the shape of the solid–liquid interface (AB' in Figure 5a). This interface shape for two different currents $I = 2$ A and $I = 2.3$ A has been computationally evaluated; the results are given in Figures 6a and 6b.

Note that the interface curve always starts from point A (*see* Figure 5a). To determine the position of the liquid–solid boundary, i.e., the segment B'D', we consider the cross section shown in Figure 7a at the boundary B'D'. The equations for equilibrium heat flow and the current condition at such an interface is given as:

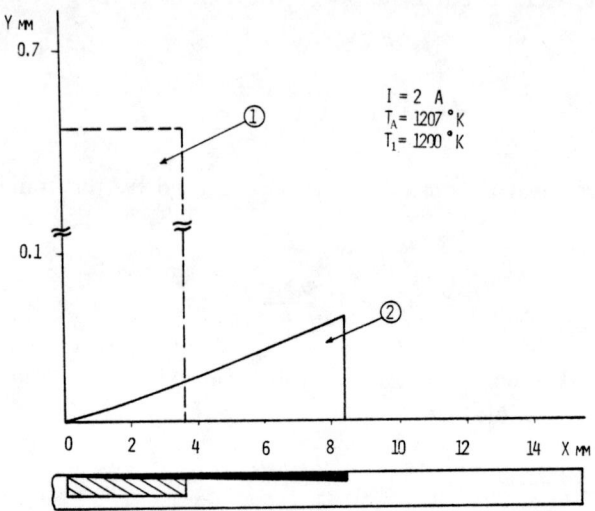

Figure 6a. Molten zone movement: (1) without Peltier current, (2) with Peltier current. $I_p = 2.0$ A.

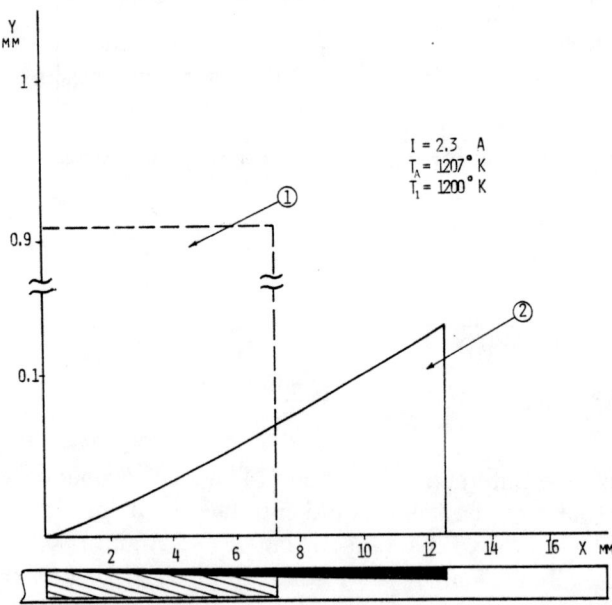

Figure 6b. Molten zone movement: (1) without Peltier current, (2) with Peltier current. $I_p = 2.3$ A.

7. VOJDANI AND HASHEMIAN   *Semiconducting Polycrystalline Ribbons*

$$K_1 y \frac{\partial T_L}{\partial x} + K_s (b-y) \frac{\partial T_L}{\partial x} - K_s b \frac{\partial T_R}{\partial x} = \alpha J_1 y T_m \tag{16}$$

where

$$J_1 = \frac{\delta_r}{c(\delta y + b)} I \tag{17}$$

and T_L and T_R stand for the temperature of the sample at the left side and the right side of B'D', respectively. However, the temperature at the left side of B'D' is constant (T_m), because both melt and solid are present (assumption vi), and therefore, we get $\partial T_L/\partial x = 0$. Thus Equation 16 is reduced to

$$-K_s b \frac{\partial T_R}{\partial x} = \alpha J_1 y T_m \tag{18}$$

or after substituting for J_1 from Equation 17 we obtain

$$\frac{\partial T_R}{\partial x} = -\alpha \frac{\delta_r T_m I}{K_s A} \frac{y}{\delta y + b} \tag{19}$$

On the other hand, having the boundary temperatures T_m and T_2 in the solid region at the right side of B'D' (Figure 7) we can compute the

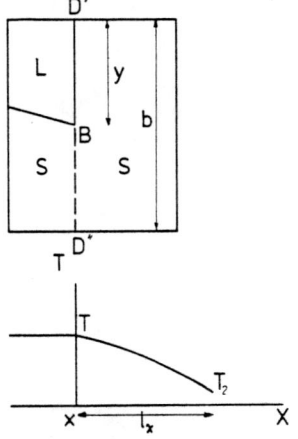

Figure 7. Schematic for determining the shape and position of the interface

temperature profile in this region. More specifically, since the length of this solid region L_s is a function of x, the temperature gradient $\partial T_R/\partial x$ is obtained as a function of x, i.e.,

$$\frac{\partial T_R}{\partial x} = g(x) \tag{20}$$

or after substituting for $\partial T_R/\partial x$ in Equation 19 we obtain

$$g(x) = -\frac{\alpha \delta_r T_m I}{K_s A} \frac{y}{\delta y + b} \qquad (21)$$

or after the appropriate manipulation y is found as a function of x

$$y = f(x) \qquad (22)$$

As a result, solution of Equation 14 with boundary condition given in Equation 22 will provide enough information for computing the shape and the size of the molten zone when both Joule heating and the Peltier phenomenon are effective.

Experimental

For the present experiments Ge and InSb ribbons were prepared from polycrystalline ingots. The dimensions of the ribbon were varied, always keeping the ribbon thickness below one mm. The specimen was held between two carbon blocks attached to a ceramic substrate. Nickel-chromium wires were connected to the carbon blocks to pass the direct current. The assembly was then located in a vacuum chamber, and the sample was heated by a tungsten element wound around a silica tube. Figure 8 shows a typical sample and the schematic experimental arrangement. Thermocouples T_1, T_2, T_A, and T_s continuously monitored the end temperatures, ambient temperature, and the specimen's temperature in the middle. The experimental procedure was to raise the temperature

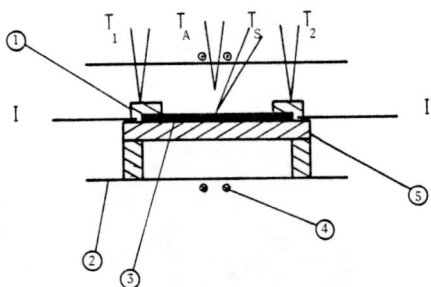

Figure 8a. Schematic showing the sample setup: (1) holding clamp, (2) silica tube, (3) sample, (4) heating coil, and (5) supporting base.

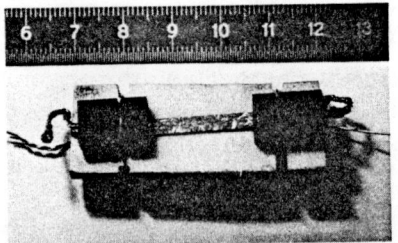

Figure 8b. Actual Ge ribbon in a sample holder.

of the furnace slowly until the specimen was close to T_m, so that the formation of the zone could be observed under various experimental conditions. When the zone was established in the absence of any direct current, it was possible to observe the effect of applying such a current by observing the associated zone movement.

Results and Discussion

Solidification of the Molten Zone in the Presence of Peltier Current. When the temperature of a ribbon was slowly raised to near the melting point with no current passing through, usually the specimen began to melt near the middle, the molten zone extending outwards in a symmetrical manner until the steady state was reached. To assist the formation of the molten zone, we often made a transverse cut of 0.3 mm deep across the middle of the ribbon to provide a region of high resistance. In many cases the temperature of the sample was raised to just below the melting point while a direct current was passed through the sample, so that the molten zone was created partly by Joule heating. Under these conditions the zone became extended more to one side than the other (relative to the point of initiation) during the non-steady state period, the directionality of the effect depending on the direction of the current (Figure 9). The molten zone has moved to the left side of the transverse cut, with no melting on the right side as is evident from the saw marks. When the current was reversed, the molten zone reversed its direction of movement.

Figure 9. InSb sample with a transverse cut in the middle showing surface movement of the molten zone to the left. $I_p = 1.95$ A, $T_1 = T_2 = 361°C$.

There were two important features observed during most of the experiments. First, the molten zone did not extend completely across the section of the specimen. Examples of these samples are shown in Figures 4 and 5. In Figure 10 the molten zone has moved to the right and has not completely penetrated the depth of the ribbon. In Figure 11, when the current is reduced, the melt only partially covers the surface of the ribbon.

The second feature concerns the change in the zone shape with and without the direct current. It had been assumed originally that the applied current would cool one interface and heat the other, causing

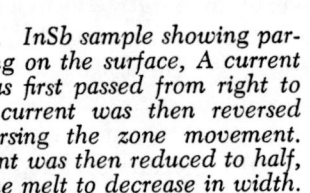

Figure 10. InSb sample showing partial melting across the sample thickness. $I_p = 1.5$ A, $T_1 = T_2 = 398°C$.

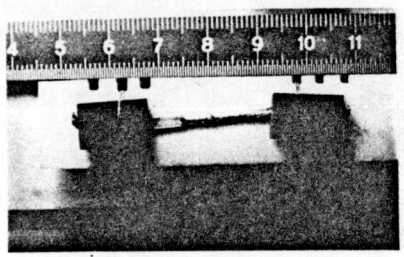

Figure 11. InSb sample showing partial melting on the surface, A current of 1 A was first passed from right to left; the current was then reversed thus reversing the zone movement. The current was then reduced to half, causing the melt to decrease in width.

the zone to move. This was not observed. Instead, only the leading edge moved, extending the width of the zone. Theoretical modelling of the system predicts this behavior because the current causes Joule heating as well as Peltier heating and cooling. When the specimen was cooled, globules formed on the surface. This was caused by the freezing of the surface of the sample while molten material still existed below. This melt was subjected to pressure during the cooling process and forced its way up through weak spots in the frozen surface. Similar behavior is observed when molten germanium is solidified in a crucible.

Conclusion

The results presented in this paper indicate that the Peltier effect itself is not adequate for the process of zoning thin films. There are two reasons for this: (a) it is difficult to obtain a fully melted narrow zone across a ribbon or thin film, mainly because of the effect of Joule heating; (b) the resistivity of the polycrystalline film is inevitably inhomogeneous. When molten zone is moved by the application of a Peltier current, the melting interface does not remain stable since "current channeling" will tend to break up the interface. This current channeling effect on the other hand has a stabilizing effect on the freezing interface as illustrated on an expanded scale in Figure 12. In this figure, for simplicity, the melting interface is assumed to be flat, and the solidifying interface is assumed to be irregular, causing low and high resistivity paths within the interface region. Thus the rate of freezing at the interface varies across it, always tending to make the interface planar. This

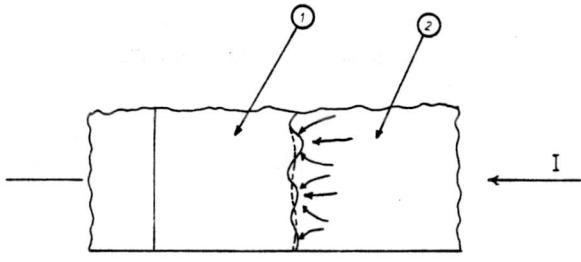

Figure 12. Schematic illustrating the current channelling effect tending to stabilize the interface: (1) liquid, (2) solid

effect is presently being investigated. To conclude, it can be suggested that for the purpose of zoning a thin semiconducting film, the combination of an external source to create a molten zone and Peltier current to stabilize the freezing interface should result in more control over the solidification process, provided that the velocity of zoning is not high enough to allow renucleation before the freezing interface.

Nomenclature

Representative values of the various parameters or Ge are given in parentheses.

a	Length of sample (cm)
A	Amps
A	Cross sectional area of sample (cm^2)
b	Sample thickness (cm)
c	Sample width (cm)
$C_{ps}(C_{pl})$	Specific heat of solid (liquid) [(2.12),(2.3)]/deg/cm^3]
$d_s(d_l)$	Density of solid (liquid) [(5.3) g/cm^3]
$I_s(I_l)$	Current through solid (liquid) (A)
$J_s(J_l)$	Current density through solid (liquid) (A)
J	Peltier current (density) (A)
$K_s, (K_l)$	Thermal conductivity of solid (liquid) [(0.24),(0.24)W/cm]
l (subscript)	Liquid
L	Latent heat of solidification [(2.16 × 10^3)(J/cm^3)]
P	Cross sectional perimeter ($2c + 2b$) (cm)
s (subscript)	Solid
t	Time (sec)
T	Absolute temperature (°K)
T_A	Ambient temperature (°K)
T_m	Melting point (1210°K)

T_s, (T_l)	Absolute temperature of solid (liquid) (°K)
W	Length of the molten zone (cm)
x	Direction along the sample (*see* coordinates in Figure 1) (cm)
α	Absolute difference in seebeck coefficient [7×10^{-5}] (V/deg)
γ	Fraction of the area melted
δ_r	Ratio of the solid-to-liquid resistivity ($\delta = \delta_r - 1$)
ρ_s (ρ_l)	Resistivity of solid (liquid) [(8×10^{-4}), (10^{-4}) ($\Omega - $cm)]
ϵ_s (ϵ_p)	Emissivity of solid (liquid) [(0.2),(0.2)]
σ	Stefan–Boltzmann constant [5.68×10^{-12}]

Acknowledgment

We would like to acknowledge many helpful discussions with E. A. D. White during the preparation of this paper. Thanks are also due to E. Afshari for computer programming, and to S. Alsaee for carrying out the experiments.

Literature Cited

1. Cizek, T. F., Schwuttke, G. H., "Proceedings of the Photovoltaic Power Generation Conference," Hamburg, Deutsche Gesellschaft für Luft-und Raumfahrt e.V., Köhn, Germany, 1974, p. 159.
2. Fang, P. H., "International Congress le Soleil au Service de l'Homme," p. 111, UNESCO House, Paris, 1973.
3. Brissot, J. J., Belouet, C., "Comples International Meeting," Dhahran, Saudi Arabia, 1975. *See also* Brissot, J. J., French Patent PV. 912.050 nr. 1.343.740 (1972).
4. Hammond, A. L., *Science* (1974) **184**, 1359.
5. Wolf, M., "Proceedings of the Photovoltaic Power Generation Conference," Hamburg, 1974, p. 699.
6. Fan, John C. C., Zeiger, H. J., *Appl. Phys. Lett.* (1975) **27** (4), 224.
7. Ocwens, C. Daey, Heigligers, H., *Appl. Phys. Lett.* (1957) **26**, 269.
8. Pfann, W. G., Benon, K. E., Wermich, J. H., *J. Electron.* (1957) **2**, 597.
9. Vojdani, S., Dabiri, A. E., Tavakoli, M., *J. Electrochem. Soc.* (1975) **122**, 1400.

RECEIVED July 27, 1976.

8

Wavelength-Selective Surfaces

JOHN C. C. FAN

Lincoln Laboratory, Massachusetts Institute of Technology,
Lexington, Mass. 02173

Spectrally selective coatings have potentially important applications in solar/thermal/electric conversion, solar heating, and window insulation. These coatings can be divided into two classes: transparent heat mirrors and selective-black absorbers. Transparent heat mirrors transmit solar radiation and reflect thermal radiation; selective absorbers absorb solar radiation and have low infrared emissivity. We have prepared both transparent heat-mirror films ($TiO_2/Ag/TiO_2$, Sn-doped In_2O_3, and Sn-doped In_2O_3 microgrids), and cermet absorbers (MgO/Au) that have excellent wavelength-selective properties. In addition, the cermets promise to be stable at the elevated temperatures required of the absorbers.

Terrestrial solar radiation is a low-intensity, variable-energy source arriving at about 800 W/m². Its economic feasibility depends on efficient collection, conversion, and storage. This paper concentrates on the use of wavelength-selective surfaces to improve the efficiency of solar-energy collectors and to provide insulation to domestic windows.

Figure 1 illustrates the basic principle of a flat-plate solar collector. It consists of a transparent coverplate and an absorber separated by a vacuum or a gas. Solar radiation is transmitted through the coverplate and converted by the absorber into thermal energy, part of which is transferred to a working fluid such as water or air, and part of which is lost. For efficient operation thermal losses from the heated absorber must be reduced. These losses can occur by conduction, convection, and infrared radiation. Wavelength-selective surfaces reduce the radiation losses by taking advantage of the spectral separation of solar radiation and the thermal radiation emitted by objects at terrestrial temperatures.

Figure 2 shows schematically the absorber and coverplate separated by air. Radiation losses, when no wavelength-selective surfaces are used,

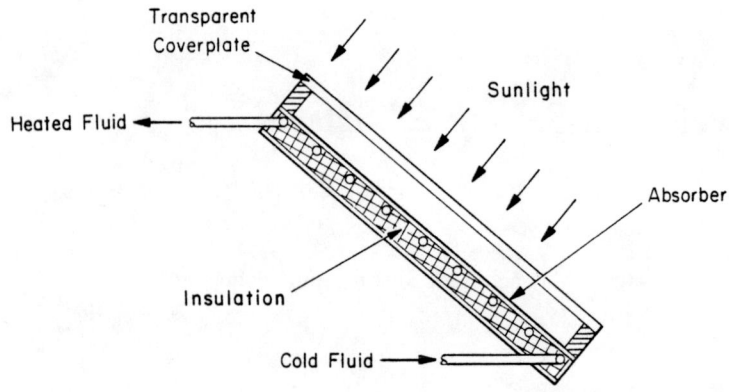

Figure 1. Schematic of a flat-plate solar collector

are comparable with those by convection and conduction. Traditional strategy reduces the radiation losses with a wavelength-selective absorber that has high solar absorptivity but low infrared emissivity. An alternate strategy uses a transparent heat-mirror film coated on the inside of the coverplate. These wavelength-selective films transmit solar radiation to the absorber but reflect the infrared radiation emitted by the absorber back to the absorber. Thermal stability of the heat mirror is easier to achieve since it remains at a much lower temperature than the absorber.

The optical Kirchoff's law states that the sum of transmission, reflectivity, and absorptivity must equal one. Therefore, for a heat mirror a high solar transmission requires a low reflectivity and absorptivity in the solar spectrum. Heat mirrors must also have high infrared reflectivity and hence a low transmission and absorptivity in the infrared. For selective-black absorbers a high solar absorptivity requires a low reflectivity and transmission in the solar spectrum. The absorbers must also have low infrared emissivity to emit minimal heat. A perfect emitter is a perfect absorber; a low infrared emissivity means a low infrared absorptivity.

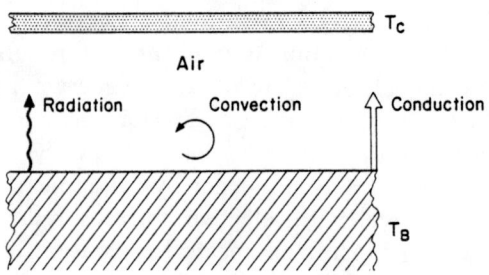

Figure 2. Mechanisms of heat loss in a flat-plate collector

Since absorbers are normally opaque (and hence have no transmission), by Kirchoff's law a low infrared absorptivity is obtained by a high infrared reflectivity. Therefore both the absorber and the transparent heat mirror should have high infrared reflectivity.

It is difficult to tailor simultaneously both the solar and the infrared properties of a film. Nevertheless the infrared losses from a typical solar collector can be reduced well below conduction and convection losses by either transparent heat mirrors or selective-black absorbers. Because the conduction and convection losses are then dominant, it is advantageous to have heat mirrors with a high solar transmission or absorbers with a high solar absorptivity, even at the expense of some reduction in the infrared reflectivity. However, if the collectors are evacuated, the radiation losses are dominant, and a high infrared reflectivity becomes important. In this case, the solar transmission or absorptivity can be somewhat lower.

Figure 3 shows the normalized solar radiation of air mass 2 and the blackbody radiation from a heated body at a temperature of 600°K. The two spectra hardly overlap, and a selective surface having an idealized reflectivity spectrum (dotted line in the figure) would produce almost total wavelength separation. Ideal transparent heat mirrors and selective-black surfaces would both have this reflectivity spectrum. However, the heat mirrors would transmit solar radiation, and the selective-black absorbers would absorb solar radiation.

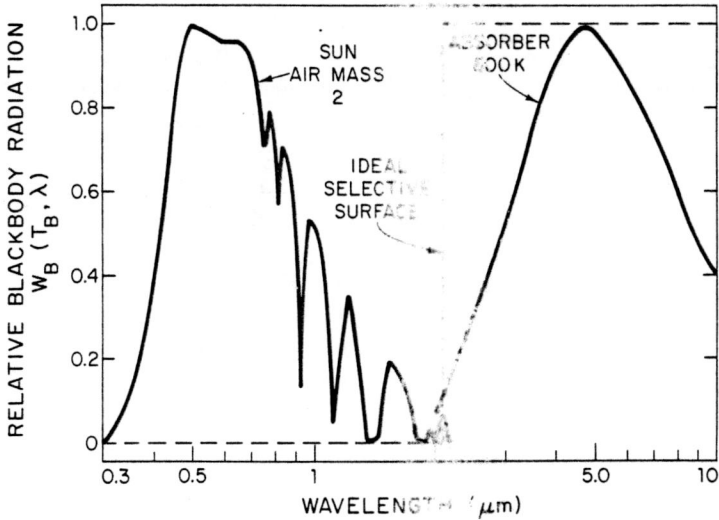

Figure 3. Normalized spectra for solar radiation of air mass 2 and for the radiation from a blackbody at 600°K. Dashed line represents the reflectivity spectrum for an idealized selective surface.

Transparent Heat Mirrors

Transparent heat mirrors can be used in flat-plate collectors for solar heating and cooling of buildings, with receivers of concentrated sunlight for solar/thermal/electric conversion, and on windows as transparent thermal insulation for energy conservation and for heat shields.

Table I shows an estimate of energy savings if heat-mirror films are coated on window panes. The calculations are based on a 90-day heating period per year, an average temperature difference of 20°C between inside and outside, and a 10-mph wind. A single glass pane would lose annually about $2.75 worth of heat per square meter. If a heat mirror reflecting 90% of the thermal radiation were coated on this single glass pane, the heat losses would be cut in half. The heat losses would be equivalent to those of uncoated double-glass panes separated by an air gap of about two inches, as in the storm-window configuration. If double panes are used, a heat-mirror film on the inside surface of the inside pane would reduce the heat loss to only about 90¢/m^2/year. Optimal performance requires, in all cases, placing the heat mirror on the surface facing the inside of the building. Placement of the film on the inside surface of the outside pane is a little less effective, but heat-mirror films on the inside surfaces of both inside and outside panes can reduce the heat loss to about 60¢/m^2/year. According to a study published by the American Institute of Physics (1), ca. 4–5% of the national energy consumption is lost through the windows in residential and commercial buildings. Transparent heat mirrors could save 1–2% of the total national energy consumption.

Transparent heat mirrors can be divided into three types: a multilayer film consisting of a metal film sandwiched between transparent dielectric layers, a single-layer transparent conductor having a controlled concentration of charge carriers, and a conducting microgrid.

The basic principle of the multilayer films follows from Kirchoff's law. It is well known that metals have high infrared reflectivity. Unfor-

Table I. Estimate of Savings with Heat Mirrors on Domestic Windows[a]

Window Heat Losses	Heat Value $/m^2/year	Savings with Heat Mirror
Single glass pane	2.75	1.49
Heat mirror on single pane	1.26	
Double panes	1.45	0.58
Heat mirror on inside pane	0.87	
Heat mirror on outside pane	0.94	
Heat mirrors on both panes	0.62	

[a] Based on: $\Delta T = 20°C$, 10-mph wind, $R_{ir} = 0.90$, 90 days heating per year, and 40¢/gal heating oil.

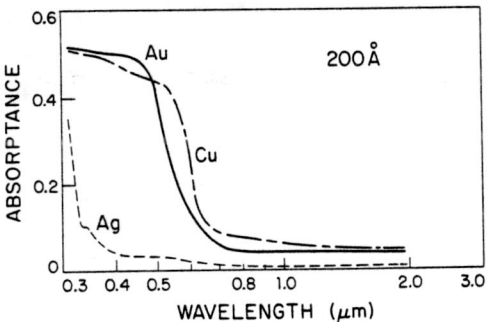

Figure 4. Absorptivity of Ag, Au, and Cu films (each 200 Å thick) vs. wavelength, calculated from bulk optical constants

tunately they also have high solar reflectivity. If their solar reflectivity can be suppressed and if they have low intrinsic absorptivity, then by Kirchoff's law they would have high solar transmission. Figure 4 shows the intrinsic absorptivity of 200-Å-thick layers of three metals: Au, Ag, and Cu. Both Au and Cu have too high absorptance at 0.5 μm, where the solar intensity is a maximum. However, Ag has only a few percent absorptivity in the whole solar spectrum; clearly it is the most promising metal. If the Ag reflectivity in the solar spectrum can be suppressed by antireflection coatings, such as TiO_2, then Ag should indeed be a good transparent heat mirror.

We have prepared such multilayer films. Figure 5 shows the measured optical reflectivity and transmission of a 180-Å TiO_2/180-Å Ag/180-Å TiO_2 film (2, 3) deposited on glass by rf sputtering. The transmis-

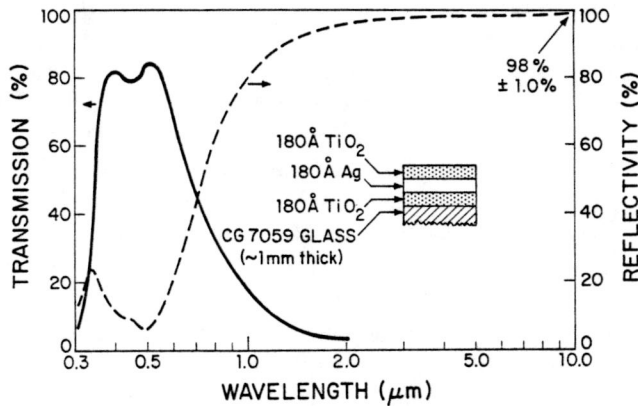

Figure 5. Measured optical transmission and reflectivity of a 180-Å TiO_2/180-Å Ag/180-Å TiO_2 film on Corning 7059 glass

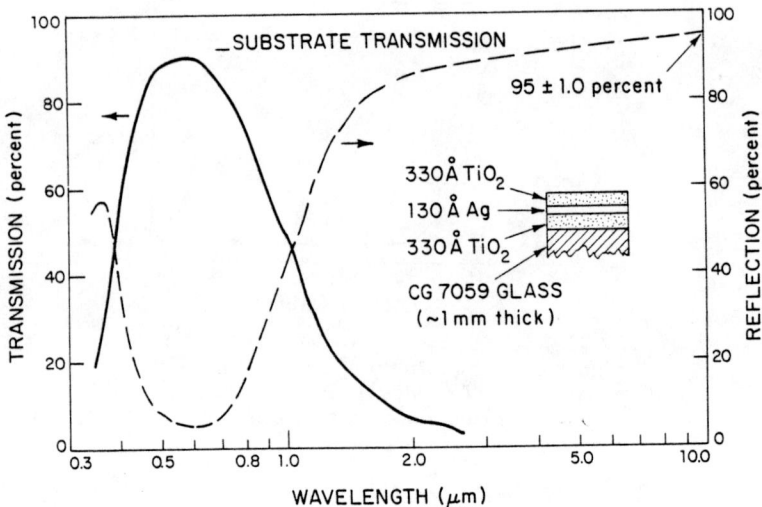

Figure 6. Measured optical transmission and reflectivity of a 330-Å TiO₂/130-Å Ag/330-Å TiO₂ film on Corning 7059 glass

sion in the visible increases above 80% which is excellent since the glass substrate itself transmits only 91%. This multilayer film has a very high infrared reflectivity. At 10 μm, where the room-temperature blackbody radiation peaks, the infrared reflectivity is 98%. Such a multilayer is also inert to water attack and is stable up to at least 200°C. This multilayer film was designed for transparent thermal insulation of a low-temperature furnace; it is indeed an excellent thermal insulator for this purpose. However, it is not optimal for a solar collector. Its visible transmission is high, but its solar transmission, which has a broader wavelength interval, is only about 50%.

The solar transmission of such a multilayer film can be increased easily by depositing different thicknesses of TiO_2 and Ag. Figure 6 shows the measured optical reflectivity and transmission of a 330-Å TiO_2/130-Å Ag/330-Å TiO_2 film. The solar transmission is much larger (over 70%), and the infrared reflectivity at 10 μm decreases to only 95%.

Significantly, this heat mirror scheme can be deposited at low substrate temperatures (lower than 100°C), which makes it possible to coat them on plastics and polyester films. Whereas it would be too expensive to replace standing window panes with coated panes for thermal insulation, transparent heat mirrors deposited on plastics with adhesive backing could be used to retrofit existing windows. We have succeeded in depositing TiO_2/Ag/TiO_2 on Mylar, Lexan, Kapton, and FEP Teflon. Figure 7 shows the visible transparency of bare Mylar and of Mylar coated with a TiO_2/Ag/TiO_2 composite. The black letters are on the back side of the mylar. We do not understand the yellow tint of the coated film; the same

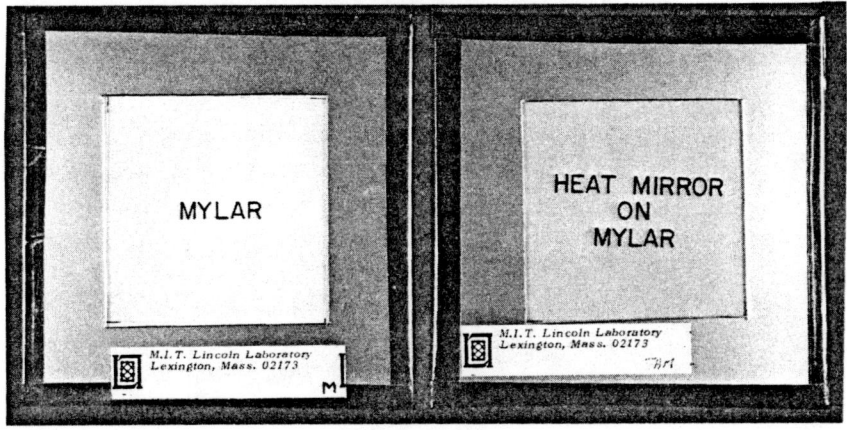

Figure 7. Photographs of bare Mylar and of Mylar coated with a 330-Å TiO_2/ 130-Å Ag/330-Å TiO_2 film

coating on glass is clear (*see* Figure 8). However, for window coatings the slight tint is not detrimental.

Whereas the above films require multilayers of controlled thickness, transparent-conductor films may be made of a single layer whose thickness is not critical. Of several material candidates, Sn-doped In_2O_3 films have the best wavelength-selective properties (*4*). Figure 9 shows the optical transmission and reflectivity of a Sn-doped In_2O_3 film. The solar spectrum and the blackbody radiation of a heated absorber at 600°K are also shown. The Sn-doped film transmits about 85% of the solar energy, and it reflects about 90% of the thermal radiation emitted by the heated

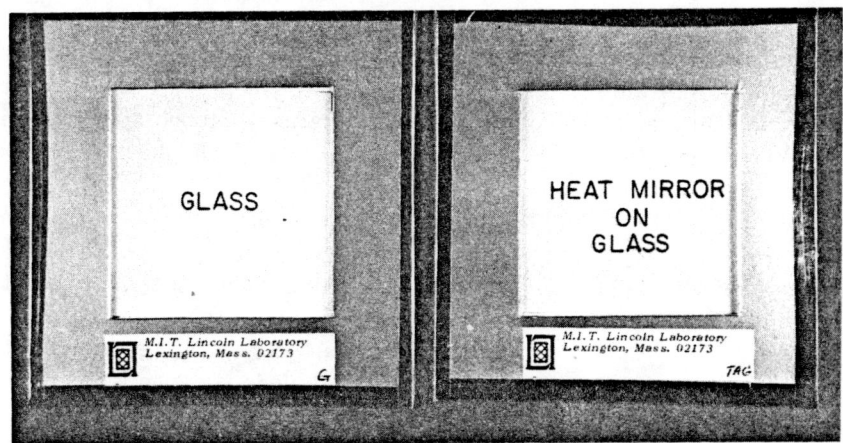

Figure 8. Photographs of a bare glass substrate (Corning 7059 glass) and of a similar glass substrate coated with a 330-Å TiO_2/130-Å Ag/330-Å TiO_2 film

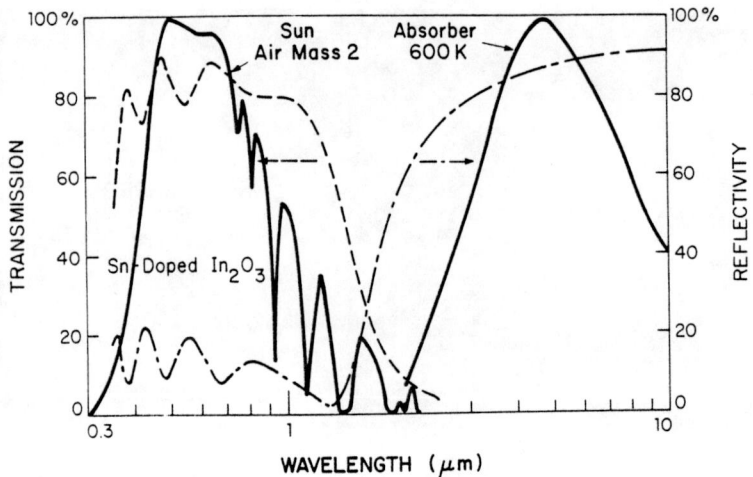

Figure 9. Measured optical transmission and reflectivity of a 0.35-μm-thick Sn-doped In_2O_3 film coated on a Corning 7059 glass substrate. The spectra for solar radiation of air mass 2 and for the radiation from a blackbody at 600°K are also shown.

body. In addition, the film itself has very low intrinsic absorption in the solar spectrum. The transmission losses are caused mostly by reflection losses, which can be suppressed by antireflection coatings.

Figure 10 shows the optical transmission of an Sn-doped In_2O_3 film before and after it was coated with a 1000-Å-thick MgF_2 antireflection coating. The solar transmission, as expected, increased from 85% to 90% without any change in the infrared reflectivity. Sn-doped In_2O_3 films coated with MgF_2 are also stable in air up to at least 300°C for 50 hours. These coatings could be used as transparent thermal insulators for ovens or furnaces. There is, however, a drawback in this material system. Sn-doped In_2O_3 films of good wavelength-selective properties have been deposited only at elevated substrate temperatures (above 400°C); it is therefore not yet possible to deposit on plastics such films with good selective properties.

Figure 11 illustrates the principle of the third type of transparent heat mirror. It was suggested a few years ago by Horwitz (5) that selective surfaces could be fabricated by etching a metal sheet to create openings of the proper size (ca. 2.5 μm) to allow solar radiation with wavelengths < 2.5 μm to pass through but not infrared radiation of wavelengths much larger than 2.5 μm. The whole idea depends, of course, on the width of the metallic lines. If the lines are too narrow, the infrared radiation is not reflected, and the wavelength-selective property disappears. The linewidth effect can be estimated from the reflection coefficient for a wire mesh (6):

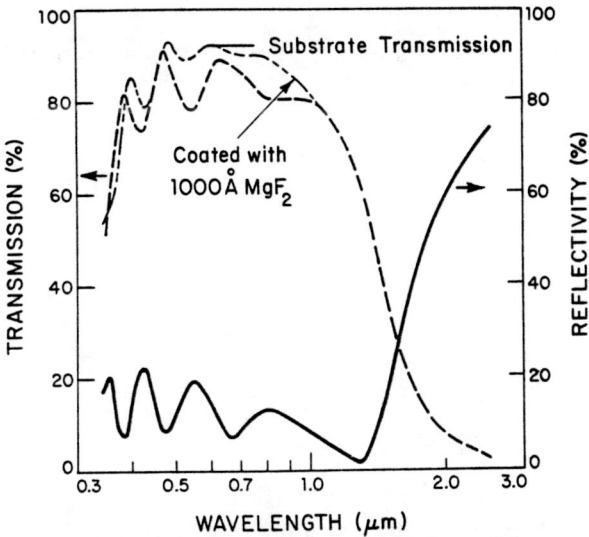

Figure 10. Measured optical transmission of a 0.35-μm-thick Sn-doped In_2O_3 film on Corning 7059 glass before and after it was coated with a 1000-Å-thick MgF_2 antireflection coating

$$\Omega = |R| \exp(i\phi) = \{1 + (\eta k/\cos\theta)(1 - \tfrac{1}{2}\sin^2\theta)\}^{-2}$$

where R is the reflectivity, $\eta = \ln(a/2\pi r)$, $k = 2\pi/\lambda$, and θ is the angle of incidence. This equation is based on the assumption that the metal is a perfect conductor at all frequencies. With an opaque metal, solar transmission of 85% requires a metal linewidth of 0.2 μm. Such a narrow linewidth would drop the infrared reflectivity from 100% to 70%, which is unsatisfactory. However, with a transparent metal that already has a solar transmission of 85% the transmission can be increased to 95% by having a linewidth of 0.6 μm. Such a linewidth would decrease the infrared reflectivity of a perfect conductor by only about 1.5%. This calculation shows that the microgrid technique would be practical only for transparent conductors. We have fabricated such microgrids from Sn-doped In_2O_3 films (7), which are transparent and quite conducting [dc conductivity is about 10^4 $(\Omega\text{-cm})^{-1}$].

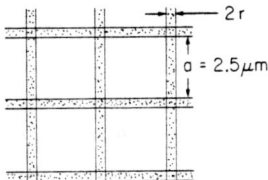

Figure 11. Schematic of a wire-mesh structure

Figure 12. Scanning electron micrograph of microgrid fabricated from a Sn-doped In_2O_3 film

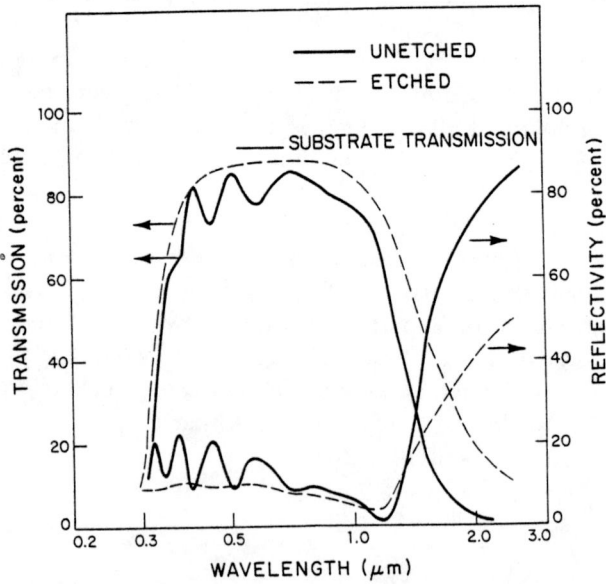

Figure 13. Measured optical transmission and reflectivity of a 0.35-μm-thick Sn-doped In_2O_3 film on Corning 7059 glass before etching (solid lines) and after etching (dashed lines) to form a conducting microgrid

Figure 12 shows a scanning electron micrograph of such a conducting microgrid fabricated by photolithography and chemical etching. The openings are ca. 2.5 μm across; the lines are ca. 0.6 μm wide.

Figure 13 shows the optical transmission and reflectivity of an Sn-doped In_2O_3 film before and after a microgrid has been etched in the film. Before etching the solar transmission is ca. 80%; it increases to 90% after etching. The infrared reflectivity at 10 μm, however, decreases from 91% to 83%. Although the decrease is larger than the 1.5% calculated for perfect conductors, it is still much smaller than the percentage area of Sn-doped In_2O_3 removed, which is about 65%. Therefore, the basic principle is confirmed, and the decrease in infrared reflectivity could be reduced by using a film with higher conductivity. The microgrid technique is just developing; it promises to give transparent heat mirrors with excellent selective properties. However, the cost of fabrication is troublesome.

Selective-Black Absorbers

Selective-black absorbers can also be obtained in three ways. (a) A thin film that absorbs solar radiation but is transparent to infrared radiation may be coated on a metal base. The high infrared reflectivity of the base provides the low infrared emissivity. (b) An opaque metal or compound having a high infrared reflectivity but a low solar reflectivity may have its solar absorption enhanced by antireflection coatings that further suppress the solar reflectivity. (c) Wire mesh may be fabricated to have a special morphology that traps the solar radiation but not the infrared. I am concentrating on the first approach, partly because our own efforts have been on selectively absorbing films on metal and partly because the best known selective-black surfaces are of this type.

The most popular selective-black absorber for low-temperature operation is "Cr-black," sometimes referred to as "black chrome." Cr-black is presently deposited on metal sheets by electroplating. Figure 14 shows the optical reflectivities of two Cr-black coatings. One we electroplated on Cu, the other Mar et al. (8) electroplated on Ni. The coating on Cu has excellent selective properties: a solar absorptivity of 94% and an infrared emissivity of 4%. The Cr-black on Ni also has high solar absorptivity, ca. 94%, but its infrared emissivity is 12%.

Unfortunately the optical properties of Cr-black on Cu degrade at operating temperatures above 200°C in air. However, Cr-black on Ni is stable up to 300°C.

We are now investigating the basic properties of Cr-black absorbers, which are not yet well understood. Our initial x-ray diffraction results suggest that the material is possibly a cermet consisting of polycrystalline Cr_2O_3 and amorphous Cr. If this is the case, the optical properties may

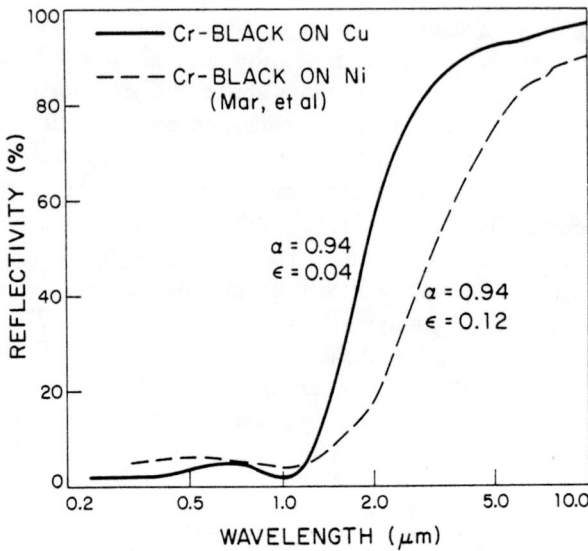

Figure 14. Measured optical reflectivity of Cr-black coated on Cu and Ni

be affected by crystallization of the amorphous chromium at elevated temperatures. In addition, at 350°C Cr oxidizes in air. To circumvent these problems, we are studying a cermet system MgO/Au (9), in which the metal, Au, is not amorphous and does not oxidize.

Figure 15 shows the transmission-electron-microscope results for a MgO/Au film. The electron-diffraction ring patterns belong to MgO and Au, indicating that both are crystalline. The Au and MgO crystallites

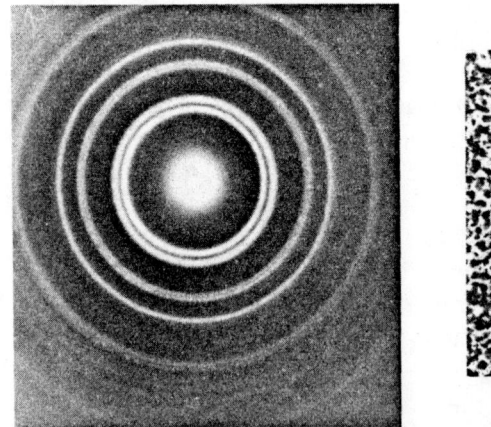
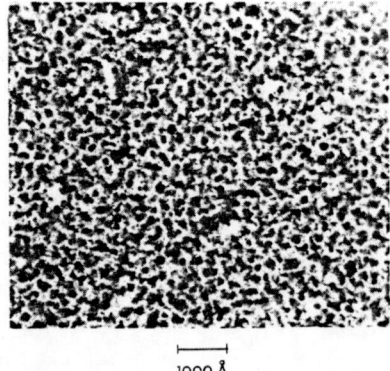

Figure 15. Transmission electron reflectivity of a 1500-Å-thick MgO/Au cermet film

are both smaller than 200 Å, as is shown by the micrograph on the right side of the figure.

We have found that MgO/Au cermet films deposited by rf supttering on metallic substrates are excellent selective-black absorbers. Figure 16 shows the optical reflectivity of a 1500-Å-thick MgO/Au film coated on Mo. (The material cost of films of this thickness is very small). The as-grown surface has a solar absorptivity of 93% and an infrared emissivity of 9%. After this absorber was heated in air at 400°C for 64 hours, the emissivity and solar absorptivity were essentially the same, indicating that such an absorber is quite stable at 400°C. At 500°C, the MgO/Au absorber begins to degrade. In particular, the solar absorptivity decreases while the infrared reflectivity does not change. Examination of degraded films with a scanning electron microscope indicates that degradation takes place by crack formation and subsequent surface aggregation. A better match between thermal expansion coefficients of different layers may stop the degradation at 500°C.

Unlike Cr-black, which is electroplated onto a metal, MgO/Au can be deposited on transparent substrates for measurement of its optical constants. Figure 17 shows the absorption coefficient vs. wavelength for an MgO/Au film (*10*). This curve was obtained from measured reflectivity and transmission spectra. On passing from visible to infrared wavelengths the absorption coefficient decreases. With the film thickness used the cermet is essentially transparent in the infrared, and the

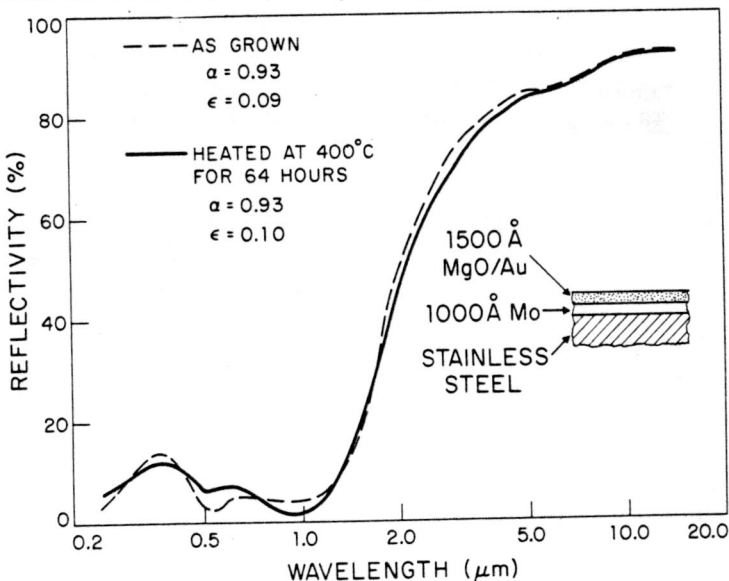

Figure 16. *Measured optical micrograph of a 500-Å-thick MgO/ Au film coated on Mo*

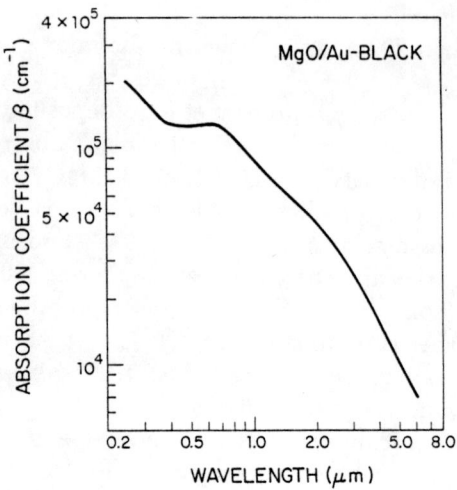

Figure 17. Absorption coefficient vs. wavelength for an MgO/Au film

low infrared emissivity of the MgO/Au absorber is provided by the reflectivity of its metal base. Moreover the solar absorption coefficient is high, and the film absorbs essentially all the solar radiation. These experiments suggest that Cr-blacks most probably have similar optical properties.

Table II summarizes the selective properties of some of our coatings. The α and ϵ for selective-black absorbers as well as α_{eff} and ϵ_{eff} for transparent heat mirrors are defined in the Appendix to this chapter.

In conclusion, wavelength-selective surfaces have important applications in solar-energy collection and energy conservation. Both transparent heat mirrors and selective-black absorbers present challenging material requirements. Although several wevelength-selective films have been developed, materials research in this area continues to be of interest.

Table II. Solar Selective Coatings[a]

Coating	α or α_{eff}	ϵ or ϵ_{eff}	α/ϵ or $\alpha_{eff}/\epsilon_{eff}$
Black chrome on Ni	0.94	0.12	7.8
Black chrome on Cu	0.94	0.04	23.5
Black MgO/Au on Mo	0.93	0.09	10.3
Sn-doped In_2O_3	0.85	0.08	10.6
Sn-doped In_2O_3/1000 Å MgF_2	0.90	0.08	11.3
330 Å/130 Å/330 Å TiO_2 Ag TiO_2	0.72	0.05	14.4

[a] Based on spectrum weightings for solar air mass of 2 and 121°C (250°F) blackbody.

Appendix

For selective black absorbers the solar absorptivity α and the infrared emissivity ϵ are defined as

$$\alpha = \frac{\int_{0.25\,\mu m}^{2.5\,\mu m} [1-R(\lambda)]A_m(\lambda)\,d\lambda}{\int_{0.25\,\mu m}^{2.5\,\mu m} A_m(\lambda)\,d\lambda}$$

$$\epsilon = \frac{\int_{1\,\mu m}^{100\,\mu m} [1-R(\lambda)]W_B(T_B,\lambda)\,d\lambda}{\int_{1\,\mu m}^{100\,\mu m} W_B(T_B,\lambda)\,d\lambda}$$

where $A_m(\lambda)$ is the solar radiation spectrum of air mass 2, $W_B(T_B,\lambda)$ is the blackbody radiation spectrum for a temperature T_B, and $R(\lambda)$ is the reflectivity of the selective absorber at wavelength λ. For transparent heat mirrors we define two parameters α_{eff} and ϵ_{eff} that are analogous to α and ϵ for selective absorbers:

$$\alpha_{eff} = \frac{\int_{0.25\,\mu m}^{2.5\,\mu m} Tr(\lambda)A_m(\lambda)\,d\lambda}{\int_{0.25\,\mu m}^{2.5\,\mu m} A_m(\lambda)\,d\lambda}$$

$$\epsilon_{eff} = \frac{\int_{1\,\mu m}^{100\,\mu m} [1-R(\lambda)]W_B(T_B,\lambda)\,d\lambda}{\int_{1\,\mu m}^{100\,\mu m} W_B(T_B,\lambda)\,d\lambda}$$

where $Tr(\lambda)$ and $R(\lambda)$ are the transmission and reflectivity of the heat mirror at wavelength λ. Where $Tr(\lambda)$ is the transmission of a film-substrate composite, the expression for α_{eff} is divided by the integrated solar transmission of the substrate.

These definitions of α_{eff} and ϵ_{eff} assume that the heat mirror is used in conjunction with a perfect (i.e., black) absorber which absorbs all the solar radiation transmitted by the heat mirror and also the infrared radiation reflected by the heat mirror.

Literature Cited

1. "Efficient Use of Energy," No. 25 in *Proc. Am. Inst. Phys.* (1975) 249.
2. Fan, J. C. C., Bachner, F. J., Foley, G. H., Zavracky, P. M., *Appl. Phys. Lett.* (1974) 24, 693.

3. Fan, J. C. C., Bachner, F. J., *Appl. Opt.* (1976) **15**, 1012.
4. Fan, J. C. C., Bachner, F. J., *J. Electrochem. Soc.* (1975) **122**, 1719.
5. Horwitz, C. M., *Opt. Commun.* (1974) **11**, 210.
6. Kontorovich, M. I., Petrunkin, V. Yu., Yesephkina, A. N., Astrakhan, M. I., *Radio Eng. Electron. Phys.* (1962) **7**, 223.
7. Fan, J. C. C., Bachner, F. J., Murphy, R. A., *Appl. Phys. Lett.* (1976) **28**, 440.
8. Mar, H. Y. B., Lin, J. H., Zimmer, P. B., Peterson, R. E., Gross, J. S., "Optical Coatings for Flat Plate Solar Collectors (Final Report), "Contract No. NSF-C-957 (**AER-74-09104**), Sept. 1975.
9. Fan, J. C. C., Henrich, V. E., *J. Appl. Phys.* (1974) **45**, 3742.
10. Fan, J. C. C., Zavracky, P. M., *Appl. Phys. Lett.* (1976) **29**, 478.

RECEIVED July 27, 1976. Work sponsored by the Department of the Air Force.

9

Thermodynamic Studies of Some Electrode Materials

P. G. DICKENS

Inorganic Chemistry Laboratory, Oxford University, Oxford QX1 3QR, England

The thermochemistry of a class of ternary oxide phases, the "oxide bronzes," A_xMO_n, is investigated. In this formulation MO_n is the highest oxide of a transition metal $M = W$, Mo, or V; A is some other electropositive inclusion element such as Na, K, or H; and x is a variable $0 < x < 1$. The materials chosen for study are chemically inert towards nonoxidizing acid media and are good electronic conductors; some have previously been examined as potential fuel cell electrodes. Enthalpies of formation of a range of tungsten, molybdenum, and vanadium oxide bronzes are determined. The thermodynamic stability of these materials towards oxidation and disproportionation is examined, and their electrochemical characteristics are discussed.

The materials referred to in the title are restricted to the "oxide bronzes" (1, 2) A_xMO_n, a class of ternary oxide phases derived formally by the insertion of an electropositive element A into the oxide matrix MO_n of a transition metal M. Common examples of parent oxides MO_n which form oxide bronzes, and on which special attention is focused here, are WO_3, MoO_3, and V_2O_5. The insertion element A is typically an alkali metal, but analogous compounds H_xMO_n are formed by hydrogen, and these too are of current interest (3, 4). The observed electronic properties of the alkali metal oxide bronzes suggest (1) that the insertion element A is present as a cation, and the average oxidation state of M is lowered accordingly to $< 2n$. Consequently A_xMO_n behaves as an electronic conductor, $[A_x^+MO_n(e_x)]$, either metallic as in the case of Na_xWO_3 ($x > 0.3$) or semiconducting, as for example with β-$Na_xV_2O_5$. Chemically the alkali metal oxide bronzes tend to be rather inert, particularly in the case of A_xWO_3, and resistant to attack by non-oxidizing acids. The combination of metallic conductivity and chemical inertness has stimulated interest in these materials as potential fuel cell electrodes (5).

In A_xMO_n the proportion x of the inserted element may vary continuously over a wide range $0 < x < 1$, and large deviations from classical stoichiometric ratios occur. For example, in the well known cubic sodium tungsten bronzes Na_xWO_3 a single phase exists over the range $0.37 \leqslant x \leqslant 0.95$, which may be regarded as a solid solution of Na in the WO_3 matrix. Provided that the process of dissolution of the insertion element is sufficiently rapid and the range of homogeneity is sufficiently large, such non-stoichiometric solids could function as solid-solution electrodes in appropriate electrochemical cells acting, for example, as battery cathodes. In view of this it is clearly desirable that quantitative thermodynamic data for such materials be available. Prior to 1970 no published thermodynamic data for any oxide bronze system existed. The object of this paper is to present and analyze the thermodynamic data for some oxide bronze phases of vanadium, molybdenum, and tungsten obtained recently by the author and co-workers in Oxford (2). Background for this study is provided by a brief account of the main structural features of the phases examined.

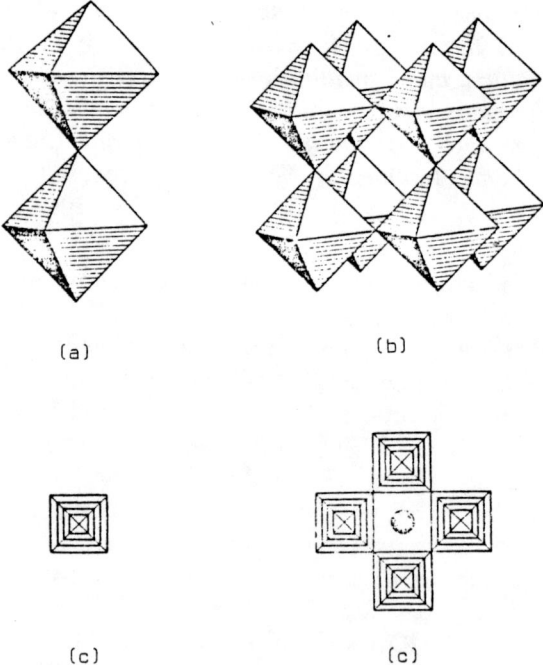

Figure 1. (a) The single ReO_3 chain, (b) perovskite structure, and (c) projection along chain axis

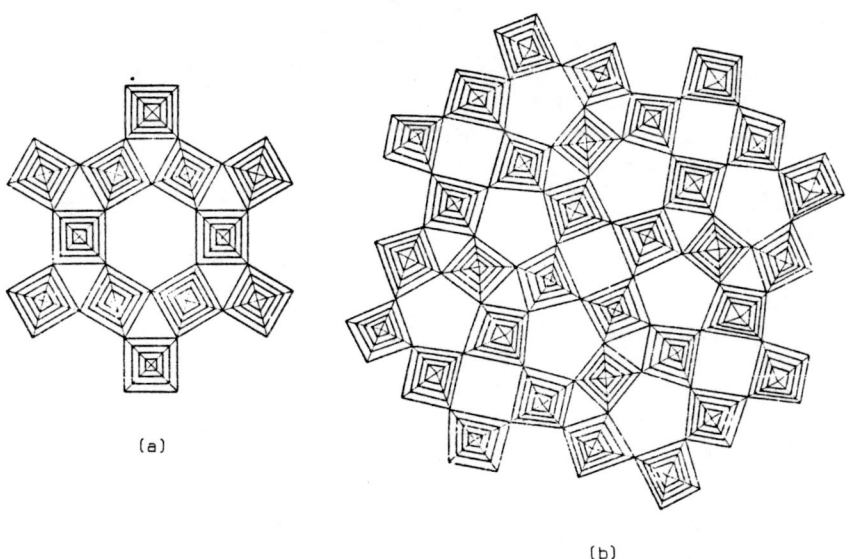

Figure 2. (a) Hexagonal A_xWO_3 structure, and (b) Tetragonal (I) A_xWO_3 structure

Structures of Oxide Bronzes (2)

In A_xMO_n the transition metal M is usually in a high oxidation state and has a small crystal radius. It exerts strong directional bonding effects on its nearest neighbors in the crystal. Accordingly tunnel and layer structures occur, consisting of linked polyhedra. Such structural types can accommodate a large variation of A content either in interstices or between layers. The common structural unit of the bronze-forming parent oxides is the MO_6 octahedron which is essentially regular in WO_3 and MoO_3 but severely deformed in V_2O_5. The bronzes often possess structures closely related to the parent oxide, and these can be described in terms of the arrangement of a few simple groups of octahedra as illustrated in Figures 1 and 2.

Tungsten Bronzes. All the known tungsten bronze structures contain a three-dimensional tunnel framework of the host lattice, of stoichiometry WO_3, consisting of single ReO_3 chains (Figure 1) which share all equatorial vertices with four other chains. Idealized projections of the tungsten-bronze structures are shown in Figures 1 and 2. A atoms occupy tunnel sites level with the apical oxygen atoms. Complete filling of the tunnels in all cases corresponds to the composition AMO_3. The hexagonal tungsten bronze structure (Figure 2) is found in tungsten bronzes with the largest insertion ions K, Rb, Cs, Tl, In, and NH_4. In this structure only the six-sided tunnels are occupied, and the maximum composition corresponds to $A_{0.33}WO_3$. The tetragonal tungsten bronze structure (T_I) shown in Figure 2 contains three types of tunnel: triangular (T), square (S), and pentagonal (P). Complete filling of all the tunnels corresponds to $(T)_{0.4}(S)_{0.2}(P)_{0.4}WO_3$. In the tungsten bronzes only S and P tunnels are occupied, and the structure is found for A ions of medium size. Examples are Na_xWO_3 ($0.21 \leqslant x \leqslant 0.35$), K_xWO_3 ($0.40 \leqslant x \leqslant 0.59$). Distorted perovskite structures (Figure 1) are found in many tungsten bronzes at low x values. As the x content increases, a progressive transition occurs from a lattice of low symmetry to one of higher symmetry, and for certain systems a cubic phase is reached (e.g., Na_xWO_3 is cubic for $0.37 \leqslant x \leqslant 0.95$). Distortions of the WO_3 framework occur in the actual structures, and in the case of the cubic hydrogen tungsten bronze H_xWO_3 the hydrogen atom is displaced from the ideal perovskite site and is randomly attached to oxygen atoms as hydroxyl groups; i.e., this compound is more correctly formulated as $WO_{3-x}(OH)_x$ (4).

Molybdenum Bronzes. MoO_3 itself has a layer structure. The layers consist of double ReO_3 chains sharing edges in the staggered manner shown in Figure 3. The basic framework probably survives intact in the hydrogen molybdenum bronze H_xMoO_3 in which it is suggested (6) that H atoms are inserted as hydroxyl groups between layers. The cathodic battery material Li_xMoO_3 may also be closely related structurally to the parent oxide. In contrast, the well characterized molybdenum bronzes, red $K_{0.33}MoO_3$ and blue K_xMoO_3 ($0.28 \leqslant x \leqslant 0.30$) have structures derived from octahedral clusters, similar to the finite

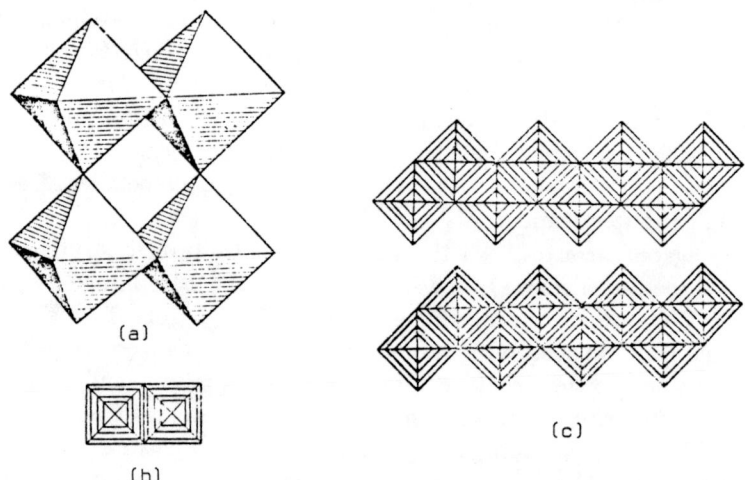

Figure 3. (a) The double ReO_3 chain, (b) projection along chain axis, and (c) MoO_3 layer structure

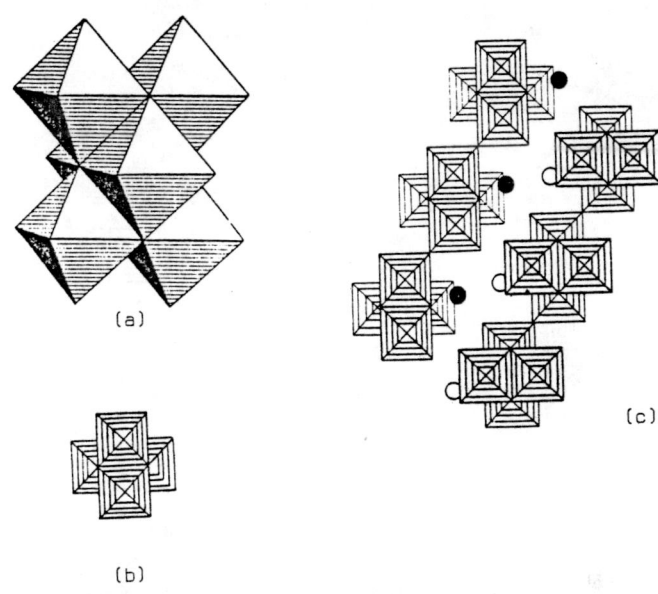

Figure 4. (a) Six-unit cluster in red $K_{0.33}MoO_3$, (b) projection, and (c) red $K_{0.33}MoO_3$ structure

polynuclear groups of the isopolymolybdate anions. The red $K_{0.33}MoO_3$ phase contains a cluster of six octahedra shown in Figure 4. This cluster of composition Mo_6O_{22} forms infinite ribbons in the vertical direction by vertex sharing with similar units. The ribbons then share further vertices horizontally to form infinite two-dimensional sheets (Figure 4). The potassium ions completely occupy positions of irregular eightfold coordination and bond the layers together. This compound has a theoretical composition limit of $K_{0.33}MoO_3$. Blue K_xMoO_3 ($0.28 \leqslant x \leqslant 0.30$) contains a different cluster consisting of 10 edge-shared octahedra ($Mo_{10}O_{30}$) which connect with similar groups to form the layer structure shown in Figure 5. The potassium ions again occupy sites between the layers which are fully occupied at $x = 0.30$. The structure of a lithium molybdenum bronze Li_xMoO_3 ($0.31 \leqslant x \leqslant 0.39$), prepared at high temperatures (8), has not been determined, but the lattice parameters suggest that it too consists of octahedral clusters. The oxygen-deficient molybdenum bronze $NaMo_6O_{17}$ has a distorted perovskite structure.

Vanadium Bronzes. The idealized V_2O_5 structure (Figure 6) can be generated by linking together single octahedral ribbons formed by shearing two ReO_3 chains together along common octahedral edges. Each ribbon connects via its free vertices to four similar ribbons. In the actual structure considerable distortion from this simple representation occurs; the coordination is closer to fivefold (trigonal pyramidal), and the

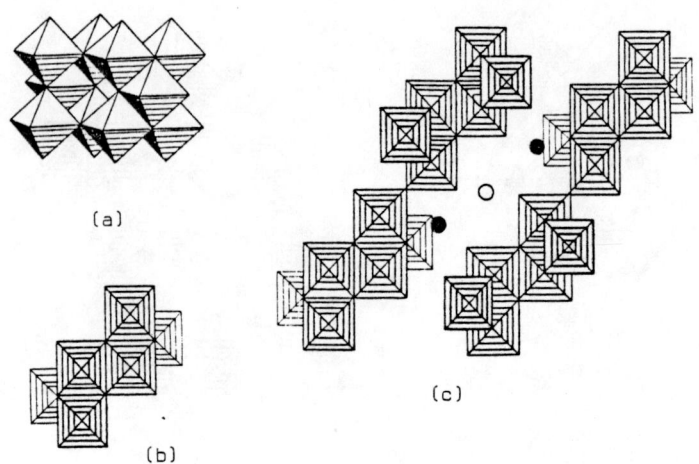

Figure 5. (a) Ten unit cluster in blue $K_{0.30}MoO_3$, (b) projection, and (c) blue $K_{0.30}MoO_3$ structure

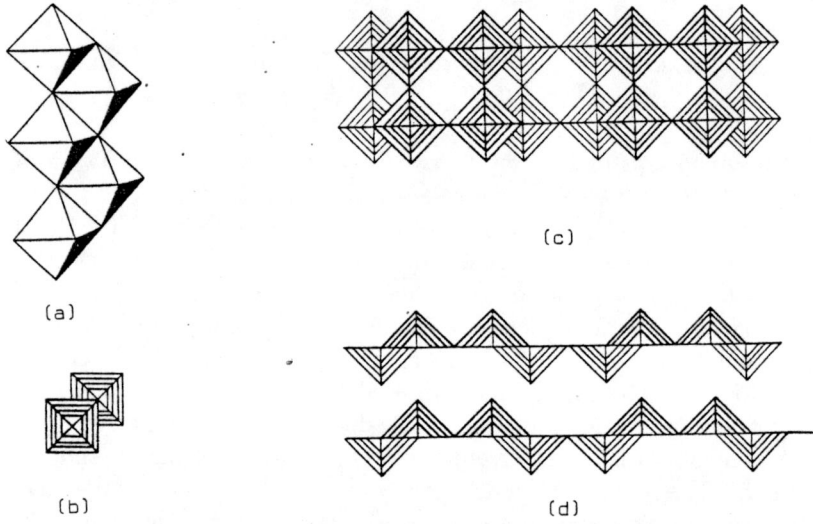

Figure 6. (a) The single octahedral ribbon, (b) projection, (c) V_2O_5 structure (idealized) represented as octahedra, and (d) represented as bipyramids

compound has a layer structure rather than a tunnel structure (Figure 6). One vanadium bronze structure, the α-phase, is based on this structure. It is found for small concentrations of A atoms, e.g. $Na_xV_2O_5$ ($0 < x \leqslant 0.02$) where the inter-layer sites of a more symmetrical V_2O_5 lattice are occupied. The α'-$Na_xV_2O_5$ phase ($0.70 \leqslant x \leqslant 1$) has a similar structure. Low temperature cathodic materials based on V_2O_5 and alkali metal insertion (7) are probably of this type. The β-$Na_xV_2O_5$ structure (0.22

$\leqslant x \leqslant 0.40$) contains irregularly shaped tunnels enclosed by both double (essentially octahedral) and single (trigonal bipyramidal) ribbons (Figure 7). The sodium atoms in the β-phase occupy the tunnel sites M_1, but the proximity of neighboring sites prohibits their simultaneous occupation. The upper composition level is $x = 0.33$ corresponding to a zig-zag arrangement of the A ions down the tunnel. The extra position M_2 and M_3 may be occupied by smaller insertion elements, e.g. in β'-$Li_xV_2O_5$ ($0.44 \leqslant x \leqslant 0.49$) and β-$Cu_xV_2O_5$ ($0.26 \leqslant x \leqslant 0.64$).

Experimental Section

Materials. The oxide bronze phases used in the calorimetric studies were made by standard methods (9). Sodium tungsten and vanadium bronzes were prepared by solid state reactions and the sodium and potassium molybdenum bronzes by electrolysis of the appropriate molybdate and oxide melts. The hydrogen tungsten bronze phases were made (3) by chemical reduction of WO_3 by zinc and hydrochloric acid. A hydrogen molybdenum bronze $H_{0.30}MoO_3$ was prepared by a sealed tube reaction between Mo, MoO_3, and water (10). All products were characterized as single pure phases by x-ray diffraction, and their compositions were confirmed by chemical analysis.

Calorimetry. Enthalpies of solution were measured with an LKB 8700 constant environment solution calorimeter operated at 25°C. Experimental details have been given elsewhere (11). The reaction medium principally used was alkaline hexacyanoferrate(III) in which all the oxide bronze phases dissolved fairly rapidly (< 20 min). Other oxidizing reac-

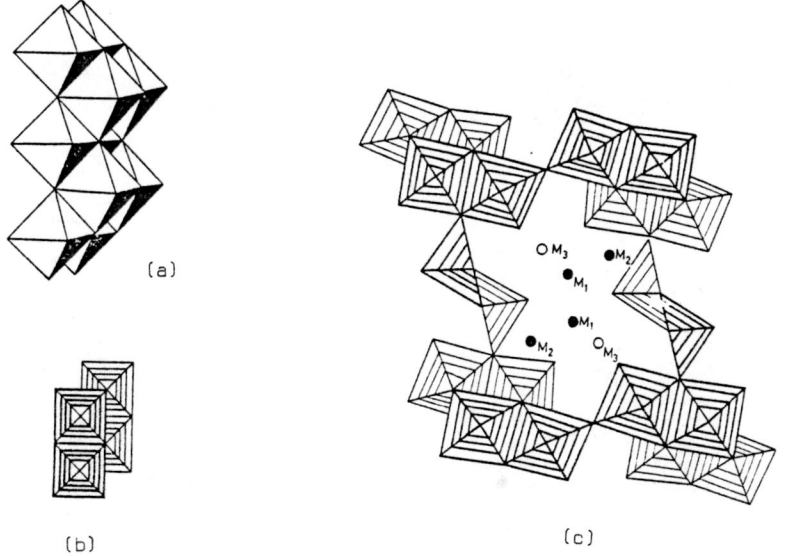

Figure 7. (a) The double octahedral ribbon, (b) projection along chain length, and (c) β-$Na_xV_2O_5$ structure

Table I. Calorimetric Reaction[a]

1. $H_{0.30}MoO_3(s) + 2.30H^-(sol) + 0.3Fe(CN)_6^{3-}(sol) =$
2. $0.3K_3Fe(CN)_6(s) =$
3. $54.841H_2O(l) =$
4. $0.3KCl(s) =$
5. $MoO_3(s) + 20H^-(sol) =$
6. $0.3K_4Fe(CN)_6 \cdot 3H_2O(s) =$
7. $0.3[HCl, 179.80H_2O](l) + 0.3OH^-(sol) =$
8. $H_{0.3}MoO_3(s) + 0.3K_3Fe(CN)_6(s) + 54.841H_2O(l) + 0.3KCl(s) =$

$\Delta H_8 =$
$0.15H_2(g) + MoO_3(s) =$

[a] (sol) means dissolved in solvent of composition [15.73 $K_3Fe(CN)_6$, 923.0 KOH, 14370 H_2O].

tion media were used occasionally for confirmatory purposes. The enthalpies of formation for oxide bronzes of V (11, 12), Mo (14), and W (3, 14) were determined by using appropriate thermochemical cycles; a typical cycle used by J. J. Birtill in current work for the determination of the enthalpy of formation of $H_{0.3}MoO_3$ is shown in Table I.

Results

Measured enthalpy changes (298°K) are summarized in Table II for the following reactions:

Formation, ΔH_1

(1)
$x Na(s) + WO_3(s) = Na_xWO_3(s)$
$x/2 H_2(g) + MO_3(s) = H_xMO_3(s) \quad (M = Mo, W)$
$x A(s) + V_2O_5(s) = \beta\text{-}A_xV_2O_5(s) \quad (A = Na, Ag)$
$x K(s) + MoO_3(s) = K_xMoO_3(s)$
$0.17 Na(s) + MoO_{2.83}(s) = Na_{0.17}MoO_{2.83}(s)$

Oxidation, ΔH_2

(2)
$2/x\, Na_xWO_3(s) + \tfrac{1}{2}O_2(g) = Na_2WO_4(s) + (2-x)/x\, WO_3(s)$
$2/x\, H_xMO_3(s) + \tfrac{1}{2}O_2(g) = H_2O(l) + 2/x\, MO_3(s)$
$\quad (M = Mo, W)$
$2/x\, A_xV_2O_5(s) + \tfrac{1}{2}O_2(g) = 2AVO_3(s) + (2-x)/x\, V_2O_5(s)$
$2/x\, K_xMoO_3(s) + \tfrac{1}{2}O_2(g) = K_2MoO_4(s) + (2-x)/x\, MoO_3(s)$
$2/3\, NaMo_6O_{17}(s) + \tfrac{1}{2}O_2(g) = 1/3\, Na_2MoO_4(s) + 11/3\, MoO_3(s)$

Disproportionation, ΔH_3

(3)
$1/x\, Na_xWO_3(s) = \tfrac{1}{2}Na_2WO_4(s) + (1-x)/x\, WO_3(s) + \tfrac{1}{2}WO_2(s)$
$1/x\, H_xMO_3(s) = \tfrac{1}{2}H_2O(l) + (2-x)/2x\, MO_3(s) + \tfrac{1}{2}MO_2(s)$
$\quad (M = Mo, W)$
$1/x\, A_xV_2O_5(s) = AVO_3(s) + (1-x)/x\, V_2O_5(s) + VO_2(s)$
$1/x\, K_xMoO_3(s) = \tfrac{1}{2}K_2MoO_4(s) + (1-x)/x\, MoO_3(s) + \tfrac{1}{2}MoO_2(s)$
$NaMo_6O_{17}(s) = \tfrac{1}{2}Na_2MoO_4(s) + 4\,MoO_3(s) + 3/2\,MoO_2(s)$

Reaction Scheme for $H_{0.30}MoO_3$

	$\Delta H_{298°K}/kJ\ mol^{-1}$
$1.3H_2O(sol) + MoO_4^{2-}(sol) + 0.3Fe(CN)_6^{4-}$	$-121.0\ \pm 1.0$
$0.9K^+(sol) + 0.3Fe(CN)_6^{3-}(sol)$	13.11 ± 0.20
$54.841H_2O(sol)$	-4.09 ± 0.05
$0.3K^+(sol) + 0.3Cl^-(sol)$	4.47 ± 0.01
$MoO_4^{2-}(sol) + H_2O(sol)$	-83.50 ± 0.30
$1.2K^+(sol) + 0.3Fe(CN)_6^{4-}(sol) + 0.9H_2O(sol)$	13.86 ± 0.10
$0.3Cl^-(sol) + 54.241H_2O(sol)$	-23.54 ± 0.07
$MoO_3(s) + 0.3K_4Fe(CN)_6 \cdot 3H_2O(s)$	
$\quad\quad\quad\quad + 0.3[HCl, 179.805H_2O]$ (1)	
$\Delta H_1 + \Delta H_2 + \Delta H_3 + \Delta H_4 - \Delta H_5 - \Delta H_6 - \Delta H_7$	$-14.3\ \ \pm 1.1$
$H_{0.3}MoO_3(s)$	$-12.1\ \ \pm 2.0^b$

[b] Calculated from ΔH_8 and standard enthalpy data given in Ref. 3.

Values of $1/x\Delta H_1$, the molar enthalpy of solution of A(s) into $MO_n(s)$, are given in Table II (column 2). [For β-$Na_xV_2O_5$ and $Ag_xV_2O_5$, $\overline{\Delta H}_{Na}$, the partial molar enthalpy of solution of A(s) into $MO_n(s)$ was determined and is tabulated here.]

Discussion

Formation of Oxide Bronzes. In all the examples given in Table II the formation reaction xA (s or g) $+ MO_n(s) = A_xMO_n(s)$ is exothermic. For formation reactions in which only solid phases are involved, entropy changes are small, and free energies of reaction ΔG are approximated by the measured enthalpy changes ΔH. Consequently, the corresponding oxide bronze phases A_xMO_n are all thermodynamically stable with respect to the components A(s) and $MO_n(s)$. This assertion is well illustrated in the case of cubic Na_xWO_3 where both calorimetric enthalpy (14) and entropy (15) data are available (298°K).

$$0.68\ Na(s) + WO_3(s) = Na_{0.68}WO_3(s)$$
$\Delta H° = -165\ kJ\ mol^{-1}, \Delta S° = -15.6\ kJ\ mol^{-1}, \Delta G° = -160\ kJ\ mol^{-1}$

ΔG and ΔH, in this typical example, differ by only 5 kJmol^{-1}. For the formation reactions of H_xWO_3 and H_xMoO_3, which involve gaseous H_2, fairly accurate estimates of the free energies of formation can be made by ignoring the contributions made to the total entropy change by the solid phases. Thus we find (at 298°K)

$$1/2\ H_2(g) + 1/xWO_3(s) = 1/xH_xWO_3(s)$$
$$(x = 0.35, \Delta S°\ measured)$$

$\Delta H° = -27.4\ kJ\ mol^{-1}, \Delta S° = -61.1\ kJ\ mol^{-1}, \Delta G° = -9.1\ kJ\ mol^{-1}$
and,

Table II. Enthalpies[a] of Formation and Reaction of Some Oxide Bronzes (298°K)

Oxide Bronze	Formation[b] ΔH_1	$1/x \Delta H_1$	Oxidation ΔH_2 (± 10)	Disproportionation ΔH_3 (± 10)
$Na_{0.53}WO_3$	-129.9 ± 0.9	-243 ± 3	-220	$+20$
$Na_{0.77}WO_3$	-185.8 ± 1.8	-243 ± 3	-220	$+20$
$H_{0.35}WO_3$	-9.6 ± 0.8	-27 ± 4	-230	$+10$
$H_{0.30}MoO_3$	-12.1 ± 2.0	-40 ± 7	-205	-20
$Na_{0.20}V_2O_5$	-72.7 ± 1.0	-360 ± 7	~ 0	$+50$
$Na_{0.22}V_2O_5$	-79.1 ± 1.8	-360 ± 7	~ 0	$+50$
$Na_{0.25}V_2O_5$	-90.0 ± 1.0	-360 ± 7	~ 0	$+50$
$Na_{0.29}V_2O_5$	-104.7 ± 1.3	-360 ± 7	~ 0	$+50$
$Na_{0.33}V_2O_5$	-119.3 ± 1.3	-360 ± 7	~ 0	$+50$
$Ag_{0.30}V_2O_5$	-19.0 ± 2.2	-90 ± 10	~ 0	$+60$
$Ag_{0.33}V_2O_5$	-22.4 ± 2.1	-90 ± 10	~ 0	$+60$
$Ag_{0.40}V_2O_5$	-27.9 ± 2.2	-90 ± 10	~ 0	$+60$
$K_{0.30}MoO_3$	-101.0 ± 0.7	-337 ± 5	-80	$+40$
$K_{0.33}MoO_3$	-115.5 ± 1.0	-346 ± 5	-60	$+50$
$Na_{0.17}MoO_{2.83}$	-52.0 ± 2.0	-310 ± 5	-140	$+30$

[a] kJ mol^{-1}.
[b] Experimental uncertainties expressed as twice the standard error of the mean.

$$1/2\, H_2(g) + 1/x MoO_3(s) = 1/x H_xMoO_3(s)$$
$$(x = 0.30, \Delta S° \text{ estimated})$$
$\Delta H° = -40.3$ kJ mol^{-1}, $\Delta S° = -65.3$ kJ mol^{-1}, $\Delta G° = -20.8$ kJ mol^{-1}

Both phases H_xWO_3 and H_xMoO_3 are thermodynamically stable to dissociation, but H_xWO_3 is only marginally so.

Quantitative confirmation of the accuracy of the calorimetric values obtained for the reaction $xNa(s) + WO_3(s) = Na_xWO_3(s)$ is provided by recent electrochemical measurements made by Whittingham (7) with the cell

$$Na_xWO_3 \,\Big|\, \begin{array}{c} NaI \\ \text{(propylene carbonate)} \end{array} \,\Big|\, Na(s), \Delta G° \text{ (298°K)}$$

values were derived from measurements of $\overline{\Delta G}_{Na}$ for a range of x values ($0.27 \leqslant x \leqslant 0.79$). Figure 8 illustrates the comparison of $\Delta G°$ (cell) and $\Delta H°$ (calorimetric) values. The agreement is excellent.

Partial molar enthalpy changes $\overline{\Delta H}_A$ can be calculated from the data given in Table II for β-$Na_xV_2O_5$ and β-$Ag_xV_2O_5$. In both cases $\overline{\Delta H}_{Na}$ is independent of x as is also implicitly the case for cubic Na_xWO_3 (Figure 8). Presumably in these cases the interactions between the ions of the insertion elements are too small to be detected calorimetrically. The variation of $\overline{\Delta H}_A$ (or $1/x\Delta H_1$) with the various types of host lattices and insertion elements recorded in Table II can be understood qualitatively in chemical terms. Thus in the seqeunce $\beta\text{-Na}_x\text{V}_2\text{O}_5$, $\text{NaMo}_6\text{O}_{17}$, Na_xWO_3, $-\overline{\Delta H}_{Na}$, which measures the binding energy of Na in the oxide matrix, increases in the order $\text{WO}_3 < \text{MoO}_3 < \text{V}_2\text{O}_5$. This is the order of increasing stability of the lower $(2n-1)$ oxidation state of the transition metal concerned. It is also, significantly, the order of increasing voltage generated by solid solution cells of the type $A_x\text{MO}_n(s)|A^+|A(s)$ (7). Hydrogen is more tightly bound in the host lattice MoO_3 than in WO_3. Thus $1/x\Delta H_1$ for H_xMoO_3 is -40 kJ mol^{-1} and -27 kJ mol^{-1} for H_xWO_3.

The variation of $\overline{\Delta H}_A$ for different insertion elements in the same oxide matrix is illustrated by $\beta\text{-Na}_x\text{V}_2\text{O}_5$ and $\beta\text{-Ag}_x\text{V}_2\text{O}_5$ (Table II). Here $-\overline{\Delta H}_{Na} > -\overline{\Delta H}_{Ag}$, an order which is predictable since the lower ionization potential of Na favors the electron transfer process involved in $xA(s) + \text{V}_2\text{O}_5(s) = [A_x^+\text{V}_2\text{O}_5(e_x)](s)$.

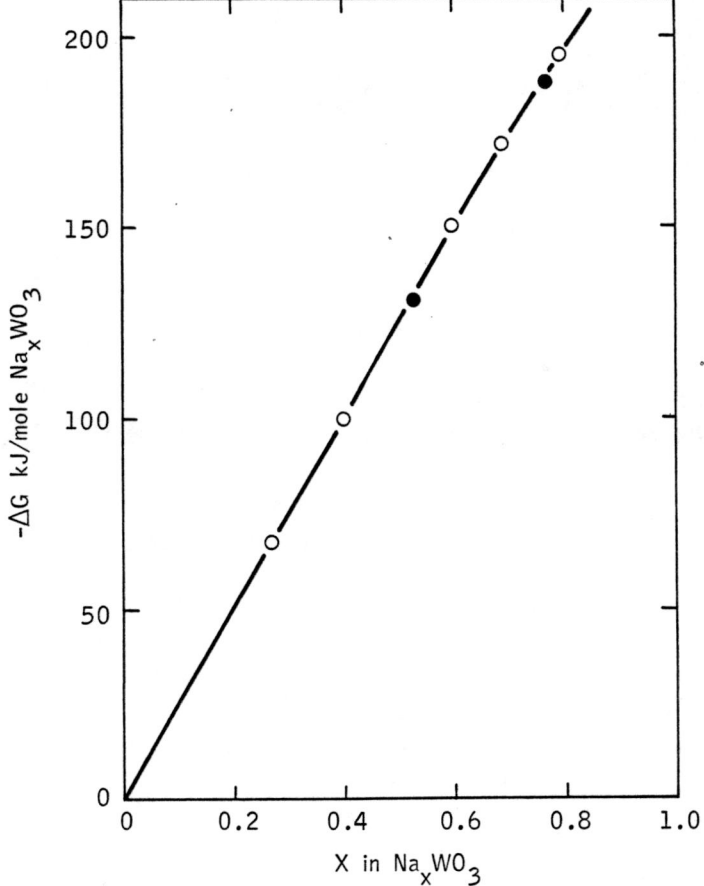

Figure 8. Free energy of formation of $Na_xWO_3(s)$ from $xNa(s)$ and $WO_3(s)$; ○, Ref. 7, ●, calorimetric enthalpy, Ref. 14

Oxidation and Disproportionation. With the exceptions of β-$Na_xV_2O_5$ and β-$Ag_xV_2O_5$, the oxide bronzes listed in Table II should all undergo exothermic reactions with gaseous oxygen. Reasonable estimates of the entropy changes involved establish that, except for β-$Na_xV_2O_5$ and β-$Ag_xV_2O_5$, all the oxide bronzes are metastable with respect to oxidized products under standard conditions at room temperature. In fact, only H_xWO_3 and H_xMoO_3 undergo discernible bulk oxidation under these conditions; for the remainder, low mobility of the inserted cation (16) prevents significant reaction at ambient temperatures.

Equilibrium electrochemical characteristics of oxide bronzes in aqueous media can readily be estimated from the data given in Table II (if $\overline{\Delta G_A}$ is identified with $\overline{\Delta H_A}$).

Thus for Na_xWO_3, a material whose electrochemical behavior as a fuel cell electrode has been closely examined (5, 17), the following relevant standard potential data (298°K) in aqueous acid media can be deduced:

$$Na^+(aq) + 1/xWO_3(s) + e = 1/xNa_xWO_3(s), (x \sim 0.5),$$
$$E° = -0.26 \text{ V}$$

$$1/xWO_3(s) + H^+(aq) + e = 1/xH_xWO_3(s), (x \sim 0.3),$$
$$E° = 0.1 \text{ V}$$

These data may be of assistance in clarifying the complex electrochemical behavior observed for Na_xWO_3 electrodes in acidic media. The depletion of Na from such electrodes after anodic treatment and the formation of a non-metallic surface layer is well attested (5, 17). The process is thermodynamically feasible even at cathodic potentials. The well defined peak found in linear voltammetry studies (17) of Na_xWO_3 electrodes at ~ 0.2 V is reasonably assigned to the oxidation-reduction process,

$$1/xWO_3(s) + H^+(aq) + e = 1/xH_xWO_3(s)$$

The rest potential (18) for the hydrogen evolution reaction on Na_xWO_3 electrodes also lies in this region (anodic, 0.1–0.5 V).

Most of the disproportion reactions included in Table II are endothermic, within the limits of error involved, but the hydrogen tungsten and molybdenum bronzes H_xMO_n appear either metastable or only marginally stable to disproportionation.

The Oxide Bronzes as Battery Cathodes. The usefulness of an oxide bronze A_xMO_n as a battery cathode in a cell

$$A_xMO_n(s) \; \bigg| \; \genfrac{}{}{0pt}{}{A^+}{\text{nonaqueous or solid electrolyte}} \; \bigg| \; A(s \text{ or } l)$$

is limited by the requirements that $\overline{\Delta G_A}$ should be large and negative, that the solubility of A in MO_n should be extensive, and that the mobility of A in the bronze phase should be high. Cubic Na_xWO_3 and β-$Na_xV_2O_5$

satisfy the first two criteria, but the observed low mobilities of the alkali metal cations preclude their use as practical cathodic materials (5, 16).

Viewed in retrospect this result is not surprising since a high temperature is frequently required in the preparation of a stable oxide bronze phase to ensure the dispersal of an insertion element throughout a rearranged oxide matrix. Much more promising practical cathodic materials are substances such as Li_xMoO_3 and $Li_xV_2O_5$ which can be prepared electrochemically at ambient temperatures using nonaqueous solvents as electrolytes (19). Their structures are probably closely related to the layer structures of the parent oxides and distinct from the familiar high temperature phases of the same general formulas. H_xMoO_3 is structurally of this same type and can also be prepared at ambient temperatures. Its application in an electrochemical device is feasible.

Acknowledgment

The author would like to thank the Petroleum Research Fund of the American Chemical Society for financial assistance in attending the New York Meeting of the Society.

Literature Cited

1. Hagenmuller, P., "Progress in Solid State Chemistry," Vol. 5, H. Reiss, Ed., p. 71, Pergamon, New York, 1971.
2. Dickens, P. G., Wiseman, P. J., "MTP International Review of Science," Series Two, Inorganic Chemistry, Vol. 10, L. E. J. Roberts, Ed., p. 211, Butterworths, London, 1975.
3. Dickens, P. G., Moore, J. H., Neild, D. J., J. Solid State Chem. (1973) 7, 241.
4. Wiseman, P. J., Dickens, P. G., J. Solid State Chem. (1973) 6, 374.
5. McHardy, J., Bockris, J. O'M., "From Electrocatalysis to Fuel Cells," G. Sandstede, Ed., p. 109, University of Washington Press, Seattle and London, 1972.
6. Wilhelmi, K. A., Acta Chem. Scand. (1969) 23, 419.
7. Whittingham, M. S., J. Electrochem. Soc. (1975) 122, 713.
8. Reau, J. M., Fouassier, C., Hagenmuller, P., J. Solid State Chem. (1970) 1, 326.
9. Banks, E., Wold, A., "Preparative Inorganic Reactions," Vol. 4, W. L. Jolly, Ed., p. 237, Interscience, New York, 1968.
10. Glemser, O., Lutz, G., Meyer, G., Z. Anorg. Allg. Chem. (1956) 285, 173.
11. Dickens, P. G., Jewess, M., Neild, D. J., Rose, J. C. W., J. Chem. Soc. Dalton Trans. (1973) 30.
12. Randall, P. A., Thesis, Part II, Oxford University, 1971.
13. Neild, D. J., Ph.D. Thesis, Oxford University, 1970.
14. Dickens, P. G., Neild, D. J., J. Chem. Soc. Dalton Trans. (1973) 1074.
15. Gerstein, B. C., Klein, A. H., Shanks, H. R., J. Phys. Chem. Solids (1964) 25, 177.
16. Steele, B. C. H., "Mass Transport Phenomena in Ceramics," A. R. Cooper and A. H. Heuer, Eds., p. 269, Plenum, New York (1975).
17. Randin, J. P., Vijh, A. K., Chughtai, A. B., J. Electrochem. Soc. (1973) 120, 117.

18. Sepa, D. B., Ovcin, D. S., Vojnović, M. V., *J. Electrochem. Soc.* (1972) **119**, 1235.
19. Campanella, L., Pistoia, G., *J. Electrochem. Soc.* (1971) **118**, 1905.

RECEIVED July 27, 1976.

10

New Solid Electrolytes

H. Y-P. HONG

Lincoln Laboratory, Massachusetts Institute of Technology,
Lexington, Mass. 02173

This paper discusses several new solid electrolytes for fast alakli-ion transport in terms of the applicable crystallographic principles. These materials are oxides in which the mobile alkali ions partially occupy a two- or three-dimensionally linked interstice within a rigid three-dimensional network. The networks, which are stabilized by electrons from the mobile ions, are formed from cation-oxygen polyhedra. They are classified into three groups depending on whether the polyhedra are tetrahedra, octahedra, or a combination of the two. Each group is further divided into subgroups depending on the number of network cations to which the oxygen anions are bonded. The first group is represented by the system $K_{2-2x}Mg_{1-x}Si_{1+x}O_4$, where MgO_4 and SiO_4 tetrahedra share corners in such a way that each oxygen is bonded to two cations to form the three-dimensional network. The second group is represented by the systems $NaSbO_3 \cdot 1/6NaF$ and $Na_{1+2x}Ta_2O_5F \cdot O_x$, in which the networks are formed by sharing corners and/or edges of SbO_6 and TaO_5F octahedra, respectively. The third group is represented by the system $Na_{1+x}Zr_2Si_xP_{3-x}O_{12}$, in which the network is formed by corner-sharing of SiO_4 and PO_4 tetrahedra with ZrO_6 octahedra. For one composition in this system, $Na_3Zr_2Si_2PO_{12}$, the Na^+-ion conductivity is 0.3 Ω^{-1} cm^{-1} at $300°C$, which is comparable with that of the best Na β''-alumina.

Solid electrolytes for fast alkali-ion transport are under investigation at many laboratories for possible use in high-specific-energy secondary batteries, electrolytic cells for extracting metals from molten salts, and thermoelectric generation. This chapter summarizes the results of studies at Lincoln Laboratory that have led to the synthesis of a number of new alkali-ion solid electrolytes (*1, 2, 3*). One of these materials is comparable in Na⁺-ion conductivity at 300°C to β''-alumina, which is presently the leading candidate for use in Na–S high-specific-energy batteries, and another has the highest K⁺-ion conductivity so far reported.

The chapter begins with a general discussion of the crystallographic principles relating to alkali-ion transport that have evolved during these studies and have been employed in the search for new solid electrolytes. In the remainder of the chapter the new materials developed are discussed individually in terms of these principles. No attempt has been made to present details concerning the procedures used for synthesis, ceramic fabrication, x-ray diffraction analysis, or electrical characterization. Such details have been given in a previous publication (*3*).

Crystallographic Principles

The essential structural feature of the solid electrolytes considered in this paper is a rigid, three-dimensional network having an interstice connected in at least one dimension and partially occupied by alkali ions (A⁺). The mobility of the A⁺ ions is governed by the transition probability for ion transfer from an occupied interstitial position to a neighboring empty one. In all these materials the rigid networks consist of metal atoms $M = \alpha M_1 + \beta M_2 + \ldots$ and oxygen atoms O. They may be formed from MO_4 tetrahedra, MO_6 octahedra, or from both types of polyhedra. If only alkali ions occupy the interstice, the general chemical formula is $A_x^+(M_yO_z)^{x-}$, where the rigid network $(M_yO_z)^{x-}$ is stabilized by accepting x electrons from the mobile A⁺ ions. It is not uncommon for molecules such as H_2O, A_2O, or AF also to be present in the interstice. Because these molecules tend to act as contaminants that interfere with alkali-ion transport, they will be ignored in this general discussion.

For fast A⁺-ion transport, the A–O polyhedra around adjacent A⁺ positions must share a common face. The smallest diameter of such faces, which act as bottlenecks to ion motion, should be greater than twice the sum of the alkali-ion and oxygen-ion radii. Thus for fast Na⁺-ion transport the smallest diameter should exceed 4.8 Å since the radii of the Na⁺ and O^{2-} ions are respectively 1.0 and 1.4 Å (*4*). For fast transport of K⁺ ions, with a radius of 1.33 Å (*4*), the smallest diameter should exceed 5.46 Å.

In addition to these geometrical constraints, chemical bonding also plays a role in determining ion mobility. In order to increase the alkali-ion mobility, the covalent contribution to the A–O bonds should be minmized and the ionic contribution maximized by polarizing the electron cloud away from the A⁺ ion and toward the rigid network. This polarization can occur in two ways: (a) the anions can bond with more than two M cations of the network and (b) the anions can form strongly covalent M–O bonds within the network. However, bonding to additional cations results in an undesirable limitation on the dimensionality of the A^+-ion transport.

In the extreme case where the anions bond to four or more network cations, the anion array is close-packed, and the volume of the interstice for A^+-ion transport is greatly reduced. If each anion is bonded to three network actions, transport is more likely to be either one- or two-dimensional rather than three-dimensional. Thus, H^+- or Li^+-ion transport occurs in one dimension in the tetragonal rutile structure, while two-dimensional transport occurs in layered compounds such as β-alumina. One-dimensional ion transport is disadvantageous because it is readily blocked by stacking faults and impurities, while in polycrystalline ceramics two-dimensional transport causes a reduction in mobile-ion density and grain-boundary conductance if the transporting layers are widely separated from each other, as in β-alumina. Moreover, anisotropic thermal expansion reduces the mechanical strength and operating life of thermally cycled membranes. Three-dimensional transport is therefore preferable. This is most probable if the number of M–O bonds per anion is limited to two. In this case, polarization of the O^{2-} charge density away from the A^+ ion requires strongly covalent σ and π M–O bonds within the rigid network. To obtain strong σ bonds it is advisable to select such strongly covalent complexes as BO_3^{3-}, CO_3^{2-}, NO_3^-, SiO_4^{4-}, PO_4^{3-}, or SO_4^{2-}, and transition-metal cations having an empty d shell should be used for strong π bonding.

New Solid Electrolytes

Table I summarizes the properties of the new solid electrolytes developed during this study by employing the crystallographic principles discussed above. These materials are divided into three groups depending on whether their rigid networks are formed from MO_4 tetrahedra, MO_6 octahedra, or both tetrahedra and octahedra. These groups are further divided into subgroups depending on the number of M–O bonds per O^{2-} ion. The same classification scheme is used in the following discussion, where the crystallographic principles are applied to these new electrolytes (and related materials, as appropriate).

Network Formed by Linked Tetrahedra. A rigid network M_yO_z formed entirely by linked tetrahedra must have $z/y = 2$, corresponding to two M–O bonds per network anion, since no other ratio can allow a three-dimensional network. All zeolites, for example, have this ratio. However, most of these aluminosilicates have too large an interstice to be effective solid electrolytes; if their cell edge exceeds 10 Å, they are generally stabilized by water—or some other molecule—that fills this space and blocks A^+-ion transport.

Table I. Properties of

3D Network System	A^+-ion Transport	A^+ Ions per cc
Linked Tetrahedra		
O^{2-} bonded to 2 cations		
$K_{2-2x}Mg_{1-x}Si_{1+x}O_4$, $x = 0.05$	3D	16.4×10^{21}
Linked Octahedra		
O^{2-} bonded to 2 cations		
$NaSbO_3 \cdot 1/6NaF$	3D	17.4×10^{21}
$Na_{1+2x}Ta_2O_5F \cdot O_x$, $x = 0.28$	3D	10.9×10^{21}
O^{2-} bonded to 3 cations		
$K_xMg_{x/2}Ti_{8-x/2}O_{16}$, $x = 0.8$	1D	5.4×10^{21}
O^{2-} bonded to 2,3 cations		
$K_2Sb_4O_{11}$	3D	7.6×10^{21}
Linked Tetrahedra and Octahedra		
O^{2-} bonded to 2 cations		
$Na_{1+x}Zr_2Si_xP_{3-x}O_{12}$, $x = 2$	3D	11.1×10^{21}
O^{2-} bonded to 2, 3, and 4 cations		
β''-alumina	2D	5.6×10^{21}

Carnegieite, the high-temperature form of $NaAlSiO_4$, has a cubic $(AlSiO_4)^-$ network with cell edge $a = 7.38$ Å (5), too small for water to be a contaminant. Unfortunately this structure is not stable at room temperature. At this temperature, only Na_2CaSiO_4 and Na_2MgSiO_4 are reported to form stable structures with networks related to that of carnegieite, and these $(M^{2+}SiO_4)^{2-}$ networks are distorted by bonding with the Na^+ ions (6). These networks have distorted-hexagonal bottlenecks with a smallest diameter less than 4.31 Å, significantly less than the value required for fast Na^+-ion transport. Presumably this explains our observation that materials in the system $Na_{2-x}CaSi_{1-x}P_xO_4$ have poor ionic conductivity. The carnegieite $(AlSiO_4)^-$ network has been stabilized to room temperature by the incorporation of Na_2O molecules in the interstice to give: cubic $Na_2Al_2Si_2O_8 \cdot (Na_2O)_x$, where $x = 0.5$ or 1. Apparently these molecules greatly reduce ion mobility since according to our measurements (3) conductivities of these compounds are three orders of magnitude lower at 300°C than that reported for β''-alumina.

Since the Na^+ ion is apparently too small to support the $(AlSiO_4)^-$ network of carnegieite, we prepared analogs with the larger K^+ ion. It

Solid Electrolytes, $A_x^+M_yO_z$

Shortest A^+–A^+ (Å)	Bottleneck A^+–O (Å)	ϵ_a (eV)	$\rho_{300°C}$ (Ω cm)
3.35	hexagon 2.74	0.35	28
2.87	triangle 2.38	0.35	13
3.70	hexagon 2.67	0.40	150
2.94	square 2.34	—	—
3.50	rectangle 2.52	0.40	4×10^5
3.52	hexagon 2.49	0.24	3
3.23	rectangle 2.71	0.16	4

was found that at room temperature the system $K_{2-2x}Mg_{1-x}Si_{1+x}O_4$ has the cubic network with no distortion. A single-crystal structure determination gave space group Fd3m. A projection of the M_yO_z network on the a-b plane is shown in Figure 1, and the properties of the material with $x = 0.05$ are listed in Table I. Each K^+ ion is surrounded by twelve O^{2-} ions at a distance of 3.2 Å. As shown in Figure 1 the bottleneck is a regular hexagon formed by MO_4-tetrahedra edges; its diameter of 5.48 Å is just large enough for fast K^+-ion transport. Because no molecules are introduced into the interstice to stabilize the structure, good K^+-ion transport was anticipated and found. The resistivity measured at 300°C for $x = 0.05$, 28 Ω cm, is the lowest reported for K^+-ion solid electrolytes.

Network Formed by Linked Octahedra. If the M cations in $A_xM_yO_z$ are all coordinated octahedrally by the oxygen anions, formation of a three-dimensional network requires $z/y \leq 3$. Such networks may be divided into three subgroups depending on the value of z/y: (a) $z/y = 3$, corresponding to two M–O bonds per network anion, (b) $z/y = 2$, corresponding to three M–O bonds per network anion, and (c) $2 < z/y < 3$, corresponding to some network anions with two, others with three, M–O

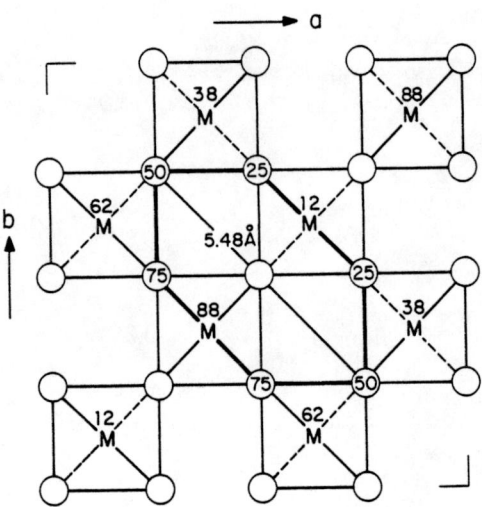

Figure 1. Projection on the a-b plane of the M_yO_z network of cubic $K_{2-2x}Mg_{1-x}Si_{1+x}O_4$. The O^{2-} ions are represented by open circles, the Mg^{2+} and Si^{4+} ions by the letter M. The ions that lie above the plane are identified by numbers giving their height as a percentage of the lattice constant. A similar representation is used in several of the following figures. The bottleneck to K^+-ion transport, which has a diameter of 5.48Å, is the regular hexagon indicated by the heavy lines.

bonds. Examples of each subgroup are listed in Table I and discussed below.

Two M–O Bonds per Network Anion. $NaSbO_3 \cdot 1/6NaF$. This compound has a cubic structure (3) that is stabilized by the NaF molecules. The $(SbO_3)^-$ network of this structure contains pairs of edge-shared octahedra that are corner-shared to form the three-dimensional array illustrated in Figure 2. Each O^{2-} ion is bonded to two Sb^{5+} ions. The $(SbO_3)^-$ network contains tunnels along the $\langle 111 \rangle$ axes that intersect at the origin and body-center positions. The tunnel intersections in the interstice are occupied by F^- ions (3). The tunnel segments between these intersections consist of three face-shared octahedra, squashed along the tunnel axis, which are partially occupied by Na^+ ions. The central octahedron, site 8c of the structure, is bounded by two triangular bottlenecks of O_2 atoms that separate it from the two outer octahedra, site 16f, each of which has a triangular face of O_1 atoms in common with one of the F-polyhedra. The Na^+ ions are distributed between the 8c and 16f sites, with electrostatic forces resulting in vacancies at the 16f sites that are adjacent to an occupied 8c site. The Debye–Waller factor determined by x-ray diffraction analysis shows that the Na⁺ ions in 16f sites have anomalously large thermal motions along ⟨100⟩ directions, indicating excellent ion transport between 16f sites of two neighboring tunnel segments around the F⁻ ion occupying the intersection; in this case the added NaF molecules apparently do not have a strong effect on alkali-ion mobility. The x-ray data also reveal the presence of large number of Na⁺ ions at 8c positions, indicating that these ions do not differ too much in potential energy from those at 16f sites. We conclude that the principal bottleneck to Na⁺-ion motion is the O_2 triangle separating the 8c and 16f positions. The distance from the center of this triangle to an O_2 position is 2.38 Å, just large enough for fast Na⁺-ion transport. This relatively tight bottleneck and the absence of π-bonding orbitals at the Sb^{5+} ions may account for the relatively large activation energy (0.35 eV) observed for this compound. Nevertheless, a relatively large pre-exponential factor in the conductivity expression—caused by a high Na⁺-ion concentration of 17.4×10^{21} cm^{-3}—lowers the resistivity measured (3) at 300°C to 13 Ω cm (the actual value may be even lower since no correction was made for contact resistance).

Defect Pyrochlore $NaTa_2O_5 \cdot xNa_2O$. The cubic pyrochlore structure (type formula $A_2B_2X_6X'$, with A the larger of the two cations) has a

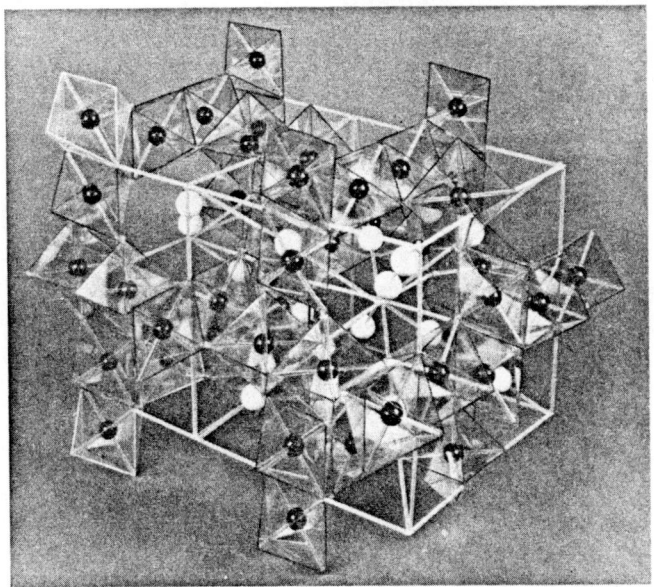

Figure 2. Model of cubic structure of $NaSbO_3 \cdot 1/6\ NaF$. Dark spheres represent Sb^{5+} ions, light spheres Na⁺ ions. The O^{2-} ions are at the corners of the octahedra around the Sb^{5+} ions.

rigid B_2X_6 network of corner-shared octahedra (7). A projection of the structure on the (110) plane is shown in Figure 3. With space group Fd3m, assignment of the B cations to the 16c positions places the A cations at 16d and the X' anions at 8b. The three-dimensional interstitial A_2X' array consists of corner-shared $X'A_4$ tetrahedra.

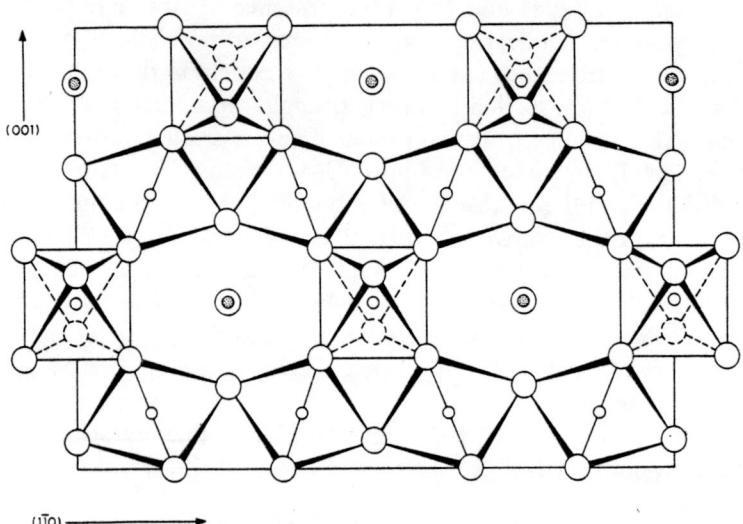

Figure 3. *A (110) projection of cubic pyrochlore $A_2B_2X_6X'$. The linked octahedra represent the B_2X_6 network, the shaded small circles represent the interstitial A^+ cations in positions 16d, and the large open circles projected onto the shaded circles represent the two near-neighbor X' ions in 8b sites on either side of an A^+ cation.*

Two types of defect pyrochlores have been identified previously. In one the 8b positions normally occupied by X' anions are vacant; in the other the 16d positions are vacant; and a large A^+ cation replaces the X' anion at 8b. The first type is illustrated by $Ag_2Sb_2O_6$ (8), the second by $RbTa_2O_5F$ (3). [In the latter compound the F^- ions are randomly substituted for O^{2-} ions of the $(B_2X_6)^-$ network.] In both cases one set of interstitial positions is occupied by cations, and the other is vacant; however, since the 16d and 8b positions are not crystallographically equivalent, fast-ion transport requires either the presence of both cations and vacancies on the same set of positions or an adjustment of the cell size that makes the potential energies nearly equal for A^+ ions in 16d and 8b positions. The Madelung energy tends to stabilize A^+ ions on 16d sites, while elastic forces tend to stabilize large A^+ ions on 8b sites, where the A–O distances exceed 3.2 Å.

Stable $A^+(B_2X_6)^-$ defect pyrochlores can be prepared with the large Rb^+ and Cs^+ ions, and also with K^+ ions if the size of the B_2X_6 array is small enough, as in $KMgAlF_6$. In the case of $A^+(Ta_2O_5F)^-$, $RbTa_2O_5F$ can be synthesized directly, but the K^+ analog can be made only by ion exchange with the Rb^+ compound. Furthermore, the ion-exchanged product is hygroscopic, with water occupying 8b sites and displacing the K^+ ions to 16d sites in $KTa_2O_5F \cdot xH_2O$, $x \leq 1$. When the K^+ compound is ion exchanged with Na, Na_2O' molecules are introduced to form $NaTa_2O_5F \cdot xNa_2O'$, $x = 0.28$, with Na ions partially occupying the 16d positions and the O' ions occupying 8b positions. This compound is thus a third type of defect pyrochlore, which has the same basic structure as an ideal pyrochlore but has both the 16d and 8b positions only partially occupied. Its properties are listed in Table I. Transport of Na^+ ions from an occupied 16d position to a vacant one is partially blocked by the O' ions. In addition, the short 16d–8b distance of 2.27 Å indicates strong $Na-O'$ bonding, which is probably responsible for the high activation energy, 0.40 eV, observed for ion transport in this material. Nevertheless, at 300°C the resistivity is only 150 Ω cm.

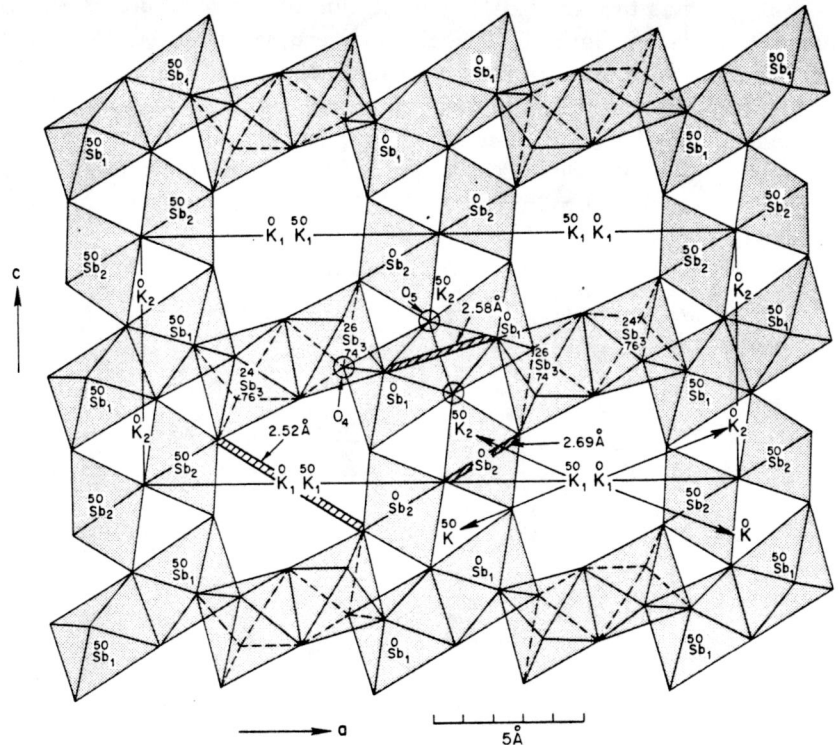

Figure 4. Projection on the a-b plane of the M_8O_{16} network of priderite, $K_xMg_{x/2}Ti_{8-x/2}O_{16}$. (Ions lying in the plane are identified by zeroes). The bottleneck to K^+-ion transport in the one-dimensional tunnels of the structure is a square indicated by the dashed lines.

THREE M–O BONDS PER NETWORK ANION. Priderite, $K_xMg_{x/2}Ti_{8-x/2}O_{16}$, $x = 0.8$, has the hollandite structure. In this structure an M_8O_{16} network is formed by MO_6 octahedra sharing corners and edges, with each O^{2-} ion bonded to three M ions. A projection of this network on the a-b plane is shown in Figure 4. The interstice partially occupied by K^+ ions consists of one-dimensional tunnels that are isolated from one another. In their equilibrium positions, the K^+ ions occupy large sites surrounded by twelve O^{2-} ions at a distance of 3.3 Å. As shown in Figure 4, the bottleneck between adjacent positions in a tunnel is a square with a diameter of only 3.6 Å, much too small for fast K^+-ion transport. Accordingly, no dc conductivity is detected (9). High frequency ac conductivity with an activation energy of 0.2 eV has been observed (9), apparently for K^+-ion displacements within the large tunnel sites.

TWO AND THREE M–O BONDS PER NETWORK ANION. The compound $K_2Sb_4O_{11}$ has a rigid network composed of SbO_6 octahedra that share corners and edges in such a way that some O^{2-} ions are bonded to two Sb^{5+} ions, others to three (10). A projection of the structure on the a-c plane is shown in Figure 5. The K^+ ions occupy interstitial positions in one-dimensional tunnels running parallel to the b and c axes. As shown in Figure 5, the shortest radii of the b-tunnel, c-tunnel, and cross-tunnel bottlenecks are 2.58, 2.52, and 2.69 Å, respectively, all of which are less than the minimum value of 2.73 Å required for fast K^+-ion transport. This explains the high resistivity value, $4 \times 10^5 \Omega$ cm, that we have measured at 300°C.

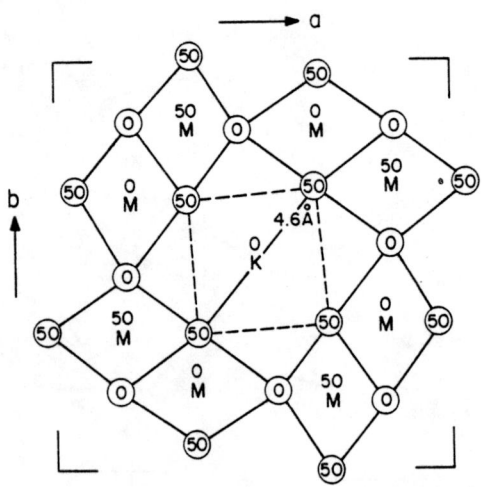

Figure 5. Projection on the a-c plane of the structure of $K_2Sb_4O_{11}$. Diameters of the three types of bottlenecks to K^+-ion transport are indicated by the hatched bars.

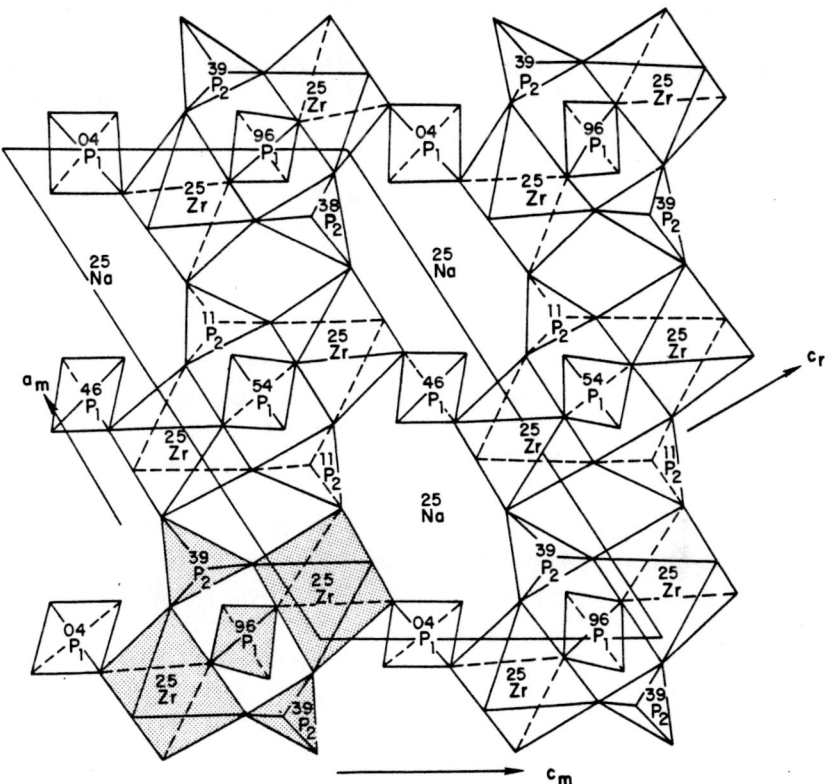

Figure 6. Projection of the hexagonal structure of $NaZr_2P_3O_{12}$. A $Zr_2(PO_4)_3$ unit of the three-dimensional network is indicated by shading in the lower left corner.

Network Formed by Linked Octahedra and Tetrahedra. Two M–O Bonds per Network Anion. The system $Na_{1+x}Zr_2Si_xP_{3-x}O_{12}$ ($0 \leq x \leq 3$) has a three-dimensional network of PO_4 and SiO_4 tetrahedra corner-shared with ZrO_6 octahedra. Every O^{2-} ion is bonded to one Zr^{4+} ion and to one Si^{4+} or P^{5+} ion. As shown in Figure 6, the network is built up of $Zr_2Si_xP_{3-x}O_{12}$ units, and the $(1+x)$ Na^+ ions occupy a three-dimensionally linked interstice. The two end members of the system have hexagonal symmetry, while in a composition range about $Na_3Zr_2Si_2PO_{12}$ the structure is distorted to monoclinic symmetry. In both symmetries the interstice contains four possible Na positions per formula unit. In the hexagonal phases these consist of one Na(1) position and three Na(2) positions. The Na(2) positions form close-packed layers, to which the Na(1) positions are octahedrally coordinated; each Na(2) position has two Na(1) neighbors. In the monoclinic symmetry, the Na(2) layers become Na(2) + 2Na(3) layers, and each Na(1) position is octahedrally coordinated by two Na(2) and four Na(3) positions, as shown in Figure 7. The bottlencks between Na(1) and Na(2) or Na(3) sites are puckered hexagons formed by three ZrO_6 octahedral edges alternating with three SiO_4 and PO_4 tetrahedral edges, as illustrated in Figure 8. The shortest diameter across these hexagons is 4.95 Å, which is significantly greater than the minimum required for fast Na^+-ion transport. Moreover, each O^{2-} ion forms a strong, covalent σ bond with a Si^{4+} or P^{5+} ion and also forms a strong π bond with the Zr 3d orbitals. Therefore, the anion charge should be well polarized into the network, away from the Na^+ positions.

Although the $Na_{1+x}Zr_2Si_xP_{3-x}O_{12}$ structures are seen to be favorable for fast Na^+-ion transport, both end members of the system have poor ionic conductivity. For $NaZr_2(PO_4)_3$, with $x = 0$, the Na(1) positions are all filled and the Na(2) positions all empty, showing that the potential energy is significantly lower for occupation of the former. Since the positions are not energetically equivalent, the absence of vacancies on the Na(1) positions and the absence of cations on the Na(2) positions strongly inhibit fast-ion transport. As x is increased, the additional Na^+ ions begin to occupy Na(2) positions—in the monoclinic modification, both Na(2) and Na(3) positions. This causes the ionic conductivity to increase rapidly, presumably caused by transport by a double-jump process in which a vacant Na(2) or Na(3) position is filled by transfer of a Na^+ ion from a neighboring Na(1) position, which is filled in turn by an ion from a previously occupied Na(2) or Na(3) position. The conductivity reaches a maximum for $Na_3Zr_2Si_2PO_{12}$ ($x = 2$), in which about two-thirds of the Na(2) and Na(3) positions are occupied. For this composition, the resistivity measured at 300°C is only 3 Ω cm (3), comparable with that of the best Na β''-alumina (11), and the activation energy is about 0.24 eV (3). As x increases still further, the conductivity decreases because of the decrease in vacancies on the Na(2) and Na(3) sites. For $Na_4Zr_2(SiO_4)_3$, with $x = 3$, all the interstitial positions are filled with Na^+ ions, and the conductivity again becomes very low.

Two, Three, and Four M–O Bonds per Network Anion. This subgroup is represented by β-alumina, $(Na_2O)_{1+x} \cdot 11\,Al_2O_3$, which is well known because of its potential application as a solid electrolyte for Na–S batteries. Although this is not one of the new solid electrolytes that we have investigated, we discuss it here because it provides an additional illustration of the principles of fast-ion transport.

Figure 9 is a schematic of the structure of β-alumina (12). Its principal elements are Al_2O_3 layers with spinel structure, which are stacked along the z direction. Adjacent layers are held together by O^{2-} ions (represented by the shaded circles in Figure 9) that are bonded to two Al^{3+} ions, one in each layer. The O^{2-} ions at the boundaries of the

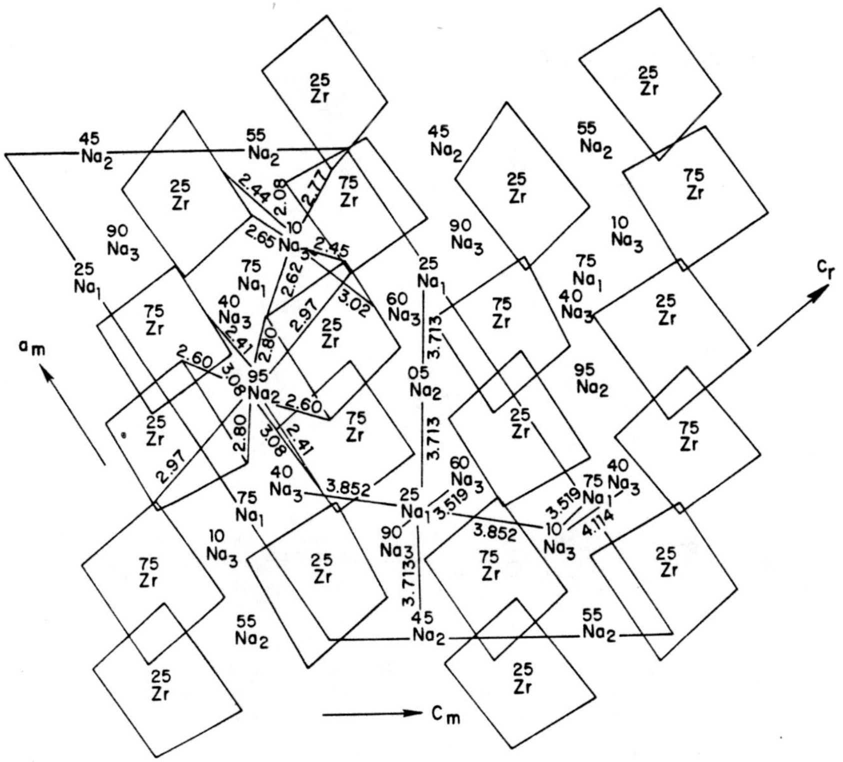

Figure 7. Projection of monoclinic structure of $Na_3Zr_2Si_2PO_{12}$

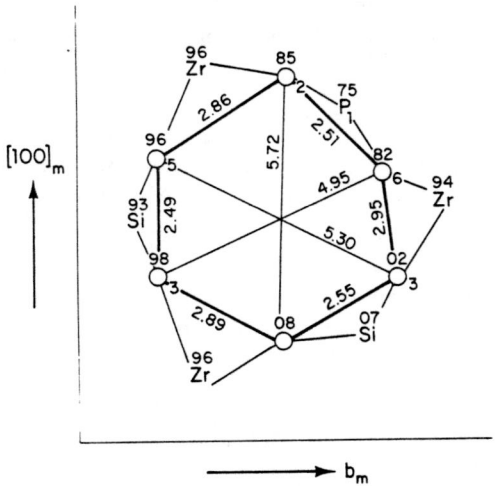

Figure 8. Bottleneck to Na^+-ion transport in $Na_3Zr_2Si_2PO_{12}$

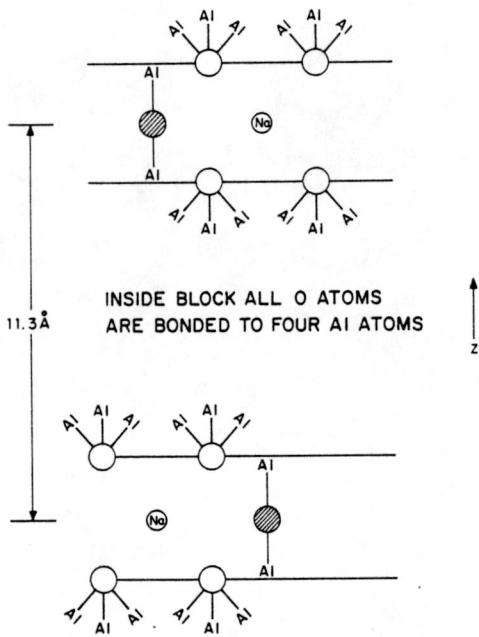

Figure 9. Schematic of β-alumina layer structure. Open and shaded circles represent O^{2-} ions in Na_2O layers between spinel blocks.

layers (represented by the open circles) are bonded to three Al^{+3} ions, while those within the layers (not shown in Figure 9) are bonded to four Al^{3+} ions. No Na^+ ions are present within the Al_2O_3 layers, but $2(1 + x)$ Na^+ ions per formula unit are present in the interstice between adjacent layers.

Transport of Na^+ ions in β-alumina is limited to two dimensions. It takes place within the interstices, in the plane normal to the z direction, but it cannot occur in the z direction. Within the spaces there are two types of Na^+ position, known as Beevers–Ross and anti-Beevers–Ross sites. Most of the Na^+ ions occupy Beevers–Ross sites, but the presence of x excess Na_2O molecules requires that the anti-Beevers–Ross sites be partially occupied. Transport of Na^+ ions may take place by a double-jump process, analogous to the one in $Na_{1+x}Zr_2Si_xP_{3-x}O_{12}$, that involves a filled Beevers–Ross site and two anti-Beevers–Ross sites, one intially empty and the other initially filled.

The bottlenecks to transport between the Beevers–Ross and anti-Beevers–Ross sites are rectangles formed by the O^{2-} ions bounding the Al_2O_3 layers. The shortest diameter of these rectangles is 5.42 Å, much larger than twice the sum of the Na^+ and O^{2-} ionic radii. This large distance, together with the polarization of the charge of the boundary O^{2-} ions toward the Al_2O_3 layers because they are bonded to three cations, explains the remarkably low value of 0.16 eV observed (13) for the activation energy of Na^+-ion transport.

The ionic conductivity of β-alumina can be increased significantly by replacing some of the Al^{3+} ions with cations of lower valence, e.g., Li^+ or Mg^{2+}, to form β''-alumina. The resulting change in symmetry makes all the Na^+ positions in the interstice equivalent, eliminating the necessity for a double-jump transport process. The substitution also causes a decrease in the number of O^{2-} ions present in the interstice that interfere with Na^+-ion transport.

Conclusion

In this chapter we have discussed a number of new solid electrolytes for alkali-ion transport. These include $Na_3Zr_2Si_2PO_{12}$, with Na^+-ion conductivity at 300°C comparable with that of the best Na β''-alumina, and $K_{1.90}Mg_{0.95}Si_{1.05}O_4$, with the highest K^+-ion conductivity at 300°C so far reported. In this discussion we have presented crystallographic principles that can be used to account qualitatively for the differences in ionic conductivity exhibited by these and other solid electrolytes. In all cases transport takes place by the two- or three-dimensional motion of alkali ions between crystallographic positions within the interstice inside a three-dimensional $(M_yO_z)^{x-}$ network. High conductivity therefore requires that these positions be partially but not completely occupied, and also that the minimum dimension of the bottlenecks between them exceed twice the sum of the alkali ion and anion radii. Ion mobility is increased if the charge on the anions is polarized away from the alkali ions by the formation of strongly covalent bonds between the anions and the network cations and also if the anions are bonded to more than two cations, although the latter tends to inhibit three-dimensional transport. Application of these principles should assist in the search for new structures permitting fast-ion transport and also in achieving higher conductivities by optimizing the chemical compositon for these and other structures already identified.

Acknowledgments

This paper reports the results of a collaborative effort by the author, John B. Goodenough, and James A. Kafalas. The expert technical assistance of Carl H. Anderson, Jr. and David M. Tracy is gratefully acknowledged.

Literature Cited

1. Hong, H. Y.-P., Kafalas, J. A., Goodenough, J. B., *J. Solid State Chem.* (1974) **9**, 345.
2. Hong, H. Y-P., *Mater. Res. Bull.* (1976) **11**, 173.
3. Goodenough, J. B., Hong, H. Y-P., Kafalas, J. A., *Mater. Res. Bull.* (1976) **11**, 203.
4. Shannon, R. D., Prewitt, C. T., *Acta Crstallogr.* (1969) **B25**, 925.
5. Barth, T. F. W., Posnjak, Z., *Z. Kristallogr.* (1932) **81**, 135.
6. Barth, T. F. W., Posnjak, Z., *Z. Kristallogr.* (1932) **81**, 370.
7. Gaertner, H. R. V., *Neues Jahrb. Mineral. Geol. Beilageband* (1930) **61A**, 1.
8. Schrewelius, N., *Z. Anorg. Allg. Chem.* (1938) **238**, 241.
9. Singer, J., private communication.
10. Hong, H. Y-P., *Acta Crystallogr.* (1974) **B30**, 945.
11. Ford Motor Company Semi-Annual Report, Jan. 1974, RANN, Division of Advanced Energy Research and Technology, Wash., D.C. (Contract No. NSF-C805).
12. Beevers, C. A., Ross, M. A. S., *Z. Kristallogr.* (1937) **97**, 59.
13. Whittingham, M. S., Huggins, R. A., NBS Spec. Pub. No. 364, R. S. Roth, S. J. Schneider, Eds. (1972) 139.

RECEIVED July 27, 1976. This work was sponsored by the Defense Advanced Research Projects Agency and NSF/RANN. The views and conclusions contained in this document are those of the contractor and should not be interpreted as necessarily representing the official policies, either expressed or implied, of the United States Government.

11

Polarizability Enhancement of Ionic Conductivity for A^{1+} in $A^{1+}M_2X_6$ Series

A. W. SLEIGHT, J. E. GULLEY, and T. BERZINS

Central Research and Development, E. I. du Pont de Nemours & Co. Experimental Station, Wilmington, Del. 19898

The ionic conductivity of A^{1+} is studied in the isostructural series of the general formula $A^{1+}M_2X_6$ where A^{1+} is Rb, Cs, or Tl; M is Ta, Nb, or W; and X is O or F. High polarizability of either A^{1+} or the $(M_2X_6)^{1-}$ framework enhances the ionic mobility of A^{1+}. Thus although Tl^{1+} is essentially the same size as Rb^{1+}, the mobility of Tl^{1+} is much higher in a given $(M_2X_6)^{1-}$ framework. This is attributed to the higher polarizability of Tl^{1+}. The polarizability of the $(M_2X_6)^{1-}$ framework is related to the polarizability of M–X bonds. The influence of polarizability in β-alumina and anion conductors is also discussed.

In a search for new materials exhibiting fast ion transport, various thoughts have been expressed concerning the structural and chemical requirements for such materials. The structural features have generally been stressed; however, there are other important considerations.

The difficulty in assessing factors other than structural is that one must subdue the geometric effects which generally appear to dominate. This is done by comparing the ionic conductivity of ions with nearly identical sizes (e.g., Li^+ vs. Cu^+ or Rb^+ vs. Tl^+) in the same structure. Although some data exist on the relative mobilities of Tl^+ and Rb^+ in β-alumina, a comparison in this structure is hindered by the complex behavior of Rb-β-alumina (1). Another approach is to change the chemistry of a host structure without significantly altering its geometry. The effect of this nongeometric change could then be directly established.

A family of compounds with the general formula AM_2X_6 having a pyrochlore-related structure was discovered by Babel et al. (2) in 1967. This is an $(M_2X_6)^-$ framework structure formed by corner-sharing MX_6

octahedra. There are infinite tunnels in this framework which frequently intersect. In this AM_2X_6 family, X can be O and/or F. The M cation can be practically any cation which will readily accept octahedral coordination to O or F. The A^+ cation can be nearly any univalent cation, but if this cation is too small, water also enters the framework as in zeolites. Table I presents some examples of compounds in this AM_2F_6 family.

Table I. Ionic Conductivity Data for AM_2X_6 Phases

Compound	a (Å)	E (kcal)	σ_0 (ohm^{-1} cm^{-1})	$\sigma_{300°C} \times 10^8$ (ohm^{-1} cm^{-1})
CsTaWO$_6$	10.39	18	18	0.42
CsNbWO$_6$	10.40	13	55	110
RbTaWO$_6$	10.35	21	832	1.4
RbNbWO$_6$	10.37^a	20	1250	5.1
TlTaWO$_6$	10.36	7.1	20	680
TlNbWO$_6$	10.36^a	5.4	13.1	20000
TlTa$_2$O$_5$F	10.48	6.9	2.0	810
TlNb$_2$O$_5$F	10.49	4.4	1.0	3700
RbTa$_2$O$_5$F	10.49	16	213	29
RbNb$_2$O$_5$F	10.49	15	140	46
CsTa$_2$O$_5$F	10.51	16	459	63
CsNb$_2$O$_5$F	10.52	11	212	2400

^{a} These compounds are not cubic at 25°C; see text.

The ideal formula for the pyrochlore structure is $A_2M_2X_6X'$, e.g., $Cd_2Nb_2O_7$. Thus, compounds of the AM_2X_6 family may be formulated as $A\square M_2X_6\square'$ or $\square_2M_2X_6A$ where the empty lattice sites are indicated by $\square$. Therefore this structure would appear to be one in which for AM_2X_6 compounds there are numerous available empty lattice sites for the diffusion of A^+ cations. Although large A^+ cations are found on the "anion site" (i.e., $\square'$), electrostatic energy calculations by Pannetier (3) indicate that this is not a good cation site. Pannetier's calculations suggest that the electrostatic potential along the tunnels could be very smooth when the size of the A cation is taken into account. Thus this $(M_2X_6)^-$ framework may be very favorable for high mobility of A^+ cations in the tunnels when the sizes are properly matched.

Recent papers on compounds of this AM_2X_6 family have expanded the chemistry and demonstrated ion exchange properties for the A cations (4–22). Crystallographic papers have shown that, in at least some cases, the A cations are not well localized on particular lattice sites (23, 24, 25, 26). The search for very high ionic conductivity in this family has been somewhat disappointing (27). However, this class of compounds seems ideal for a study of factors which influence ionic conductivity. In par-

Experimental

Polycrystalline samples were prepared in the same manner as indicated in the literature using temperatures of 800°–900°C. Oxides were generally prepared in air using gold containers; however, all thallium-containing samples were sealed in evacuated gold tubes. Oxyfluorides were prepared in evacuated platinum or gold tubes.

Crystals of pyrochlore-type $KWNbO_6$ were easily grown from a potassium tungstate flux. However, these crystals are difficult to handle since they readily hydrate to $KWNbO_6 \cdot H_2O$. Although this hydrate has the same basic structure, hydration and dehydration destroy the quality of the large crystals. Crystals of $RbWNbO_6$ and $CsWNbO_6$ were grown from fluxes of rubidium and cesium tungstates. However, these phases have small stoichiometry ranges of the type $A_{2x}W_{2-2x}Nb_{2x}O_6$, and x varies somewhat from region to region in a single crystal. This inhomogeneity limits the usefulness of such crystals even though they do not hydrate.

X-ray diffraction patterns at 25°C were obtained on all samples both with an automatic Picker powder diffractometer using $CuK\alpha$ radiation and a graphite monochromator and with a Hägg–Guinier camera using $CuK\alpha_1$ radiation and an internal standard of high purity KCl ($a = 6.2931$ Å at 25°C). Cell dimensions (Table I) are based on a least-squares refinement of the Guinier data.

The centric or acentric character of the samples was determined by measuring the second harmonic radiation generated in an apparatus previously described ([28]). NMR measurements on Tl^{205} were carried out at $-100°-+250°C$ using a broadline cw spectrometer.

The AC conductivity was measured at $10^3–10^6$ Hz and 300°–700°C. The data were used only if there was no frequency dependence to the conductivity. Measurements were made on sintered pellets, and contacts were generally sputtered gold. The electronic component of the conductivity was estimated by a low potential DC method using blocking (Au) electrodes. For the phases in Table I, the electronic conductivity was found to be negligible. In the case of Tl^+ conductors, the DC conductivity was determined with Tl metal electrodes. These measurements were in excellent agreement with the AC conductivities determined with gold electrodes.

Results

All the phases given in Table I are reported to be cubic and centric at room temperature (space group Fd3m). However, we find that in fact $TlNbWO_6$, $TlTaWO_6$, $RbNbWO_6$, $RbTaWO_6$, and $CsNbWO_6$ give significant second harmonic signals at room temperature. Therefore, these phases are not centric at 25°C. Only in the case of $TlNbWO_6$ and $RbNbWO_6$ could we detect departure from cubic symmetry at room temperature. Tetragonal symmetry is observed for RbNbWO$_6$ ($a =$ 10.360 Å, $c =$ 10.379 Å), and the symmetry is even lower for TlNbWO$_6$. Despite certain departures from the ideal structure at room temperature, high temperature studies (*29*) suggest that all phases in Table I are cubic and centric at temperatures greater than 150°C.

The ionic conductivity data are summarized in Table I. The relationship $\sigma T = \sigma_0 e - (E/RT)$ was used to determine σ_0 and E. Because of the large variations in both σ_0 and E, comparisons of conductivities cannot be made unless the temperature is specified. The chosen temperature of 300°C is somewhat arbitrary, but most generalizations concerning the relative merits of these conductors hold over a considerable temperature range. The data of Table I cannot be accepted with complete confidence since these data were obtained on pellets rather than single crystals. However, the activation energies are likely to be meaningful even if the use of pellets has resulted in some error to the values of σ_0 and $\sigma_{300°C}$.

The comparisons of Rb$^+$ with Tl$^+$ are particularly revealing. Here the geometric influence must be very small, as evidenced by the cell dimensions in Table I. The activation energies of the Tl phases are consistently much smaller than the analogous Rb phases regardless of whether the formula is ATaWO$_6$, ANbWO$_6$, ANb$_2$O$_5$F, or ATa$_2$O$_5$F. Although the prefactor (σ_0) is also greater for the Rb phases, this is insufficient in this temperature range to offset their high activation energies. Thus the Tl phases are the better ionic conductors.

The comparison of the Nb vs. Ta phases also reveals a trend. Again there is little chance that geometric factors play a significant role in view of the very similar size of Ta^{5+} and Nb^{5+} evidenced by the cell dimensions in Table I. Although activation energies are consistently lower for the niobates relative to tantalates regardless of whether one considers the TlM^{5+}WO$_6$, RbM^{5+}WO$_6$, CsM^{5+}WO$_6$, CsM^{5+}O$_5$F, or TlM$_2^{5+}$O$_5$F phases, the margin of difference is not always convincing. Again prefactors are generally working against the activation energies, but the niobates are generally better conductors than tantalates at 300°C.

The conductivity of TlWNbO$_6$ is plotted in Figure 1 at 158°–500°C. The AC (Au electrodes) and DC (Tl electrodes) measurements agree well. The activation energy from 158°C to ca. 300°C is 14 kcal. However, near 300°C the slope changes, and the temperature dependence of the conductivity becomes small at higher temperatures. This type of behavior is apparently characteristic of many good ionic conductors (*30*). At 300°C the conductivity of TlNbWO$_6$ is smaller than that of Tl-β-alumina (*31*) by a factor of about four.

The NMR line widths for Tl^{205} in TlNb$_2$O$_5$F and TlTa$_2$O$_5$F are plotted vs. temperature in Figure 2. At low temperatures the line widths

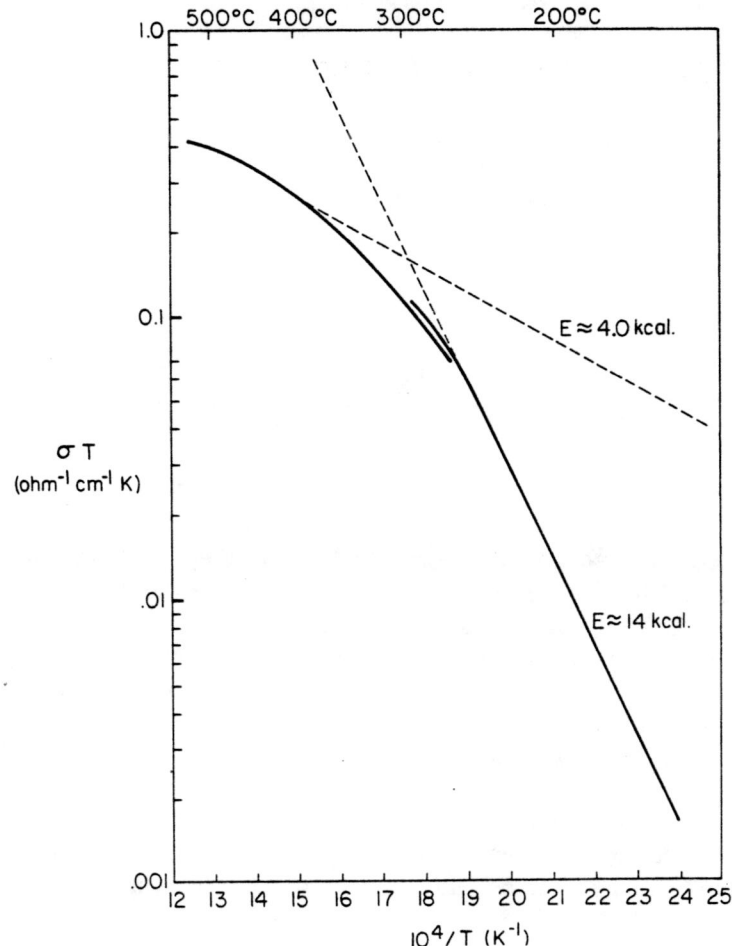

Figure 1. Conductivity data (solid lines) for TlWNbO$_6$. Low temperature data for Tl electrodes (DC), and high temperature data for Au electrodes (AC).

are those expected for a rigid lattice. There is line narrowing near room temperature indicating a significant increase in the motion of the Tl$^+$ cations. Although the line widths have leveled off by 150°C, this residue line width suggests that the motion of the Tl$^+$ cations is still somewhat restricted. Perhaps a further line narrowing event exists at temperatures above 250°C. Line narrowing commences at a lower temperature for TlNb$_2$O$_5$F than for TlTa$_2$O$_5$F. This is consistent with our conclusion that the A cations move more freely in a niobate host than in an analogous tantalate host.

For different ionic conductors, the prefactor (σ_0) decreases as the activation energy decreases (*32, 33*). Good ionic conductors are good

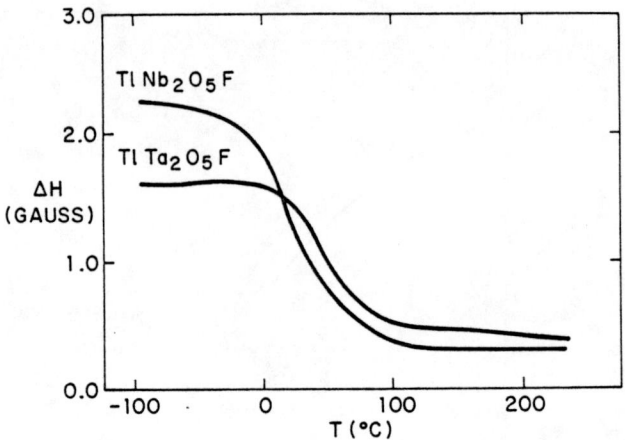

Figure 2. NMR data for Tl^{205} in $TlNb_2O_5F$ and $TlTa_2O_5F$

not because of large prefactors but because they have low activation energies. The same general trend is found within the AM_2X_6 family of the present study. The reason for this trend is not entirely clear because of the number of different factors which contribute to the prefactor. The jump frequency is expected to decrease with the activation energy (33), and this will cause a decrease in the prefactor. However, there is evidence that ion–ion interactions can cause departures from a random-hopping model (34). These interactions can also decrease the prefactor, and they may be attributable to large ordered regions (35). Some of the prefactors in the AM_2X_6 phases are especially small (Table I), and this may then indicate strong A^+–A^+ interactions.

Discussion

The higher mobility of Tl^+ than Rb^+ in analogous AM_2X_6 phases should probably be attributed to polarizability effects. Despite the nearly identical size of Tl^+ and Rb^+, the polarizability of Tl^+ is more than twice that of Rb^+ (36). Although one could argue that the structures of the Tl^+ and Rb^+ phases differ from each other in the manner of distribution of these cations over the various possible sites, this should not affect the activation energies which should be a measure of the "bottleneck barrier." The nature of the barrier would be essentially the same for both Rb and Tl regardless of possible different distributions. However, the more polarizable cation will pass through the barrier more easily because it can readily change its shape. Recent calculations by Wang et al. (37) support this view in that they indicate "that high polarizability of the charge carrier ion is beneficial in lowering the potential energy barrier heights."

Armstrong, Blumber, and Dickinson (32) have attributed the special ability of Cu^+ and Ag^+ as mobile charge carriers to the ability of these cations to adopt both four- and three-coordinated configurations. However, the ability of these cations to accept such low coordination numbers (even two!) could also be attributed to their polarizability which is high relative to their size when compared with the alkali cations. In addition to adopting low coordination numbers, highly polarizable cations can also easily adopt very assymetric coordinations, and this should also enhance their mobility.

High polarizability of anions is frequently supposed to be beneficial for high mobility of cations (32). Certainly the highest mobilities for Ag^+ and Cu^+ are found with the highly polarizable anions, i.e., I^-, Br^-, Cl^-, S^{2-}, Se^{2-}, and Te^{2-}. However, the evidence on this point is presently less convincing when one considers the mobility of the alkali cations. Furthermore, the calculation of Wang et al. (35) suggests that "large polarizability of fixed oxygen ions is detrimental to Na^+ mobility in β-alumina."

The effect of anion polarizability on cation mobility is difficult to establish experimentally because our supply of anions is much more limited than our supply of cations. There are no two different anions with the same size and charge. Thus it becomes difficult to separate polarizability effects from the important geometric factors. However, the polarizability of oxygen is known to vary considerably depending on its environment (36). Thus we might hope to alter the polarizability of oxygen in a given host without significantly altering the geometry of the host. For example, the polarizability of oxygen might differ between ATa_2O_5F and ANb_2O_5F phases, and here the geometric factor would be constant.

One can also consider the polarizability of the host $(M_2X_6)^-$ without resorting directly to consideration of the polarizability of the individual M and X ions. A niobate structure is always more polarizable than its tantalate counterpart (28, 38). Although oxygen is probably more polarizable in a niobate than a tantalate, the movement of the cores relative to each other is also significant. Thus one can distinguish lattice polarizability from the polarizability of individual ions despite some connection between the two. The higher polarizability of the niobate relative to a tantalate can be attributed to the greater covalence of the Nb–O bond relative to the Ta–O bond and to the consequent greater tendency toward double bond formation in a niobate (28). This difference in polarizability between niobates and tantalates is evident in comparing their behavior as ferroelectrics. The higher polarizability for niobates also appears to favor high mobility of the cation charge carrier. In this case, a "softer" host framework presents a lower energy barrier to diffusion of the A^+ cations.

Since the highest reported Na⁺ conductivity in solid electrolytes is for β-alumina phases, the importance of polarizability should be examined for such phases. Structure determinations for Na-β-alumina (*39*) and Ag-β-alumina (*40*) show high thermal vibrational parameters for the mobile cations (i.e., Na and Ag). Such high thermal parameters are characteristic of mobile ions in the better ionic conductors. However, the oxygen atoms in the plane of these mobile cations also have high thermal parameters, and these oxygen atoms are not mobile. The high thermal parameters for these oxygen atoms should be attributed instead to the high polarizability of an Al–O–Al linkage when the bond angle is 180°. (Other oxygen atoms in the β-alumina structure are coordinated to three or four aluminum atoms; the Al–O–Al bond angles are not close to 180°, and the thermal parameters are normal.) A 180° Al–O–Al or Si–O–Si linkage represents a structural instability just as Ti^{4+} or Nb^{5+} at the center of a regular octahedron of O^{2-} represents a structural instability. This structural instability can lead to a highly polarizable Al–O–Al bond which is reflected by high thermal motion perpendicular to this bond. In the β-alumina structure these highly polarizable bonds may well enhance the mobility of the mobile cations.

Polarizability also appears to have important consequences for anion conductivity. The fluoride ion conductivity in PbF_2 is high relative to other fluorides with the fluorite structure. This should probably be attributed to the high polarizability (*36*) of Pb^{2+} relative to Ca^{2+}, Sr^{2+}, Ba^{2+}, and Cd^{2+}. The highest O^{2-} conductivity observed in solid oxides is for phases containing the highly polarizable Bi^{3+} cation (*41*). The lone-pair cations Tl^{1+}, Pb^{2+}, and Bi^{3+} owe their high polarizability to easy hybridization between 6s (formally filled) and 6p (formally empty) orbitals. Although one could attribute high ionic conductivity to the ease of hybridization, high polarizability results from facile hybridization. Thus facile hybridization and high polarizability are essentially inseparable.

Despite high O^{2-} conductivity, bismuth-containing oxides are not useful in fuel cell applications because of the easy reduction of bismuth. However, certain selective oxidation catalysts rely on high mobility of lattice oxygen for their selectivity. Bismuth is a necessary component of many such catalysts, and one of its roles would appear to be the promotion of high O^{2-} mobility (*42, 43*).

Conclusions

Geometric considerations are important for ionic conductors. However, polarizability is also an important consideration. Although one can do little to influence the polarizability of a given mobile ion, one does have considerable flexibility with the polarizability of the host structure. High polarizability can be introduced into a host structure even when high polarizability is not a basic characteristic of any of the individual constituents of the host structure, e.g., β-alumina. Thus, the discovery or design of new host structures with high polarizability is a challenge to crystal chemists.

Literature Cited

1. Kummer, J. T., *Prog. Solid State Chem.* (1972) **7**, 141.
2. Babel, D., Pausewung, G., Viebahn, W., *Z. Naturforsch.* (1967) **22**, 1219.
3. Pannetier, J., *J. Phys. Chem. Solids* (1973) **34**, 583.
4. Raveau, B., Thomazeau, J., *C. R. Acad. Sci. Ser. C* (1968) **266C**, 540.
5. Michel, C., Raveau, B., *C. R. Acad. Sci. Ser. C* (1969) **268C**, 323.
6. Darriet, B., Rat, H., Galy, J., *C. R. Acad. Sci. Ser. C* (1970) **271C**, 1324.
7. Hervieu, M., Raveau, B., *C. R. Acad. Sci. Ser. C* (1970) **271C**, 1568.
8. Fourquet, J. L., Ory, G., Courtier, G., DePape, R., *C. R. Acad. Sci. Ser. C* (1970) **271C**, 773.
9. Jacoboni, C., Courbion, G., Fourquet, J. L., Gauthier, G., DePape, R., *C. R. Acad. Sci. Ser. C* (1970) **270C**, 1455.
10. LeFlem, G., Salmon, R., *C. R. Acad. Sci. Ser. C* (1970) **271C**, 1182.
11. Courbion, G., Jacoboni, C., Pape, R., *C. R. Acad. Sci. Ser. C* (1971) **273C**, 809.
12. Darriet, B., Rat, M., Galy, J., Hagenmuller, P., *Mater. Res. Bull.* (1971) **6**, 1305.
13. Hervieu, M., Michel, C., Raveau, B., *Bull. Soc. Chim. Fr.* (1971) 3939.
14. Groult, D., Michel, C., Raveau, B., *C. R. Acad. Sci. Ser. C* (1972) **274C**, 374.
15. Sabatier, R., Baud, G., *J. Inorg. Nucl. Chem.* (1972) **34**, 872.
16. Groult, D., Michel, C., Raveau, B., *J. Inorg. Nucl. Chem.* (1973) **35**, 3095.
17. Muller, M., Hoppe, R., *Z. Anorg. Allg. Chem.* (1973) **395**, 239.
18. Michel, C., Groult, D., Raveau, B., *Mater. Res. Bull.* (1973) **8**, 201.
19. Michel, C., Groult, D., Raveau, B., *J. Inorg. Nucl. Chem.* (1975) **37**, 247.
20. Michel, C., Groult, D., Deschanvres, A., Raveau, B., *J. Inorg. Nucl. Chem.* (1975) **37**, 251.
21. Groult, D., Michel, C., Raveau, B., *J. Inorg. Nucl. Chem.* (1975) **37**, 1957.
22. Ory, G., Fourquet, F., Jacoboni, C., Miranday, P., DePape, R., *C. R. Acad. Sci. Ser. C* (1971) 273.
23. Allais, G., Michel, C., Raveau, B., *C. R. Acad. Sci. Ser. C* (1972) **274C**, 1625.
24. Babel, D., *Z. Anorg. Allg. Chem.* (1972) **387**, 161.
25. Michel, C., Raveau, B., *Mater. Res. Bull.* (1973) **8**, 451.
26. Fourquet, J. L., Jacoboni, C., DePape, R., *Mater. Res. Bull.* (1973) **8**, 393.
27. Goodenough, J. B., Hong, H. Y-P., Kafalas, J. A., *Mater. Res. Bull.* (1976) **11**, 203.
28. Sleight, A. W., Bierlein, J. D., *Solid State Commun.* (1976) **18**, 163.
29. Sleight, A. W., Zumsteg, F. C., Barkley, J. R., Gulley, J. E., to be published.
30. O'Keefe, M., in "Fast Ion Transport," W. van Gool, Ed., p 233, North Holland, Amsterdam, 1973.
31. Whittingham, M. S., Huggins, R. A., *National Bureau of Standards Special Publication 364* (1972) 139.
32. Armstrong, R. D., Bulmer, R. S., Dickinson, T., *J. Solid State Chem.* (1973) **8**, 219.
33. Huggins, R. A., in "Diffusion in Solids," A. J. Novich, Ed., Academic, New York, 1975.

34. McWham, D. B., Allen, S. J., Remeika, J. P., Dernier, P. D., *Phys. Rev. Lett.* (1975) **35**, 953.
35. O'Keefe, *in* "The Chemistry of Extended Defects in Non-Metallic Solids," L. Eyring and M. O'Keefe, Eds., p 609, North Holland, Amsterdam, 1970.
36. Tessman, J. R., Kahn, A. H., Shockley, W., *Phys. Rev.* (1953) **92**, 890.
37. Wang, J. C., Gaffari, M., Choi, S., *J. Chem. Phys.* (1975) **63**, 772.
38. Bergman, J. G., Crane, G. R., *J. Solid State Chem.* (1975) **12**, 172.
39. Peters, C. R., Bettman, M., Moore, J. W., Glick, M. D., *Acta Crystallogr.* (1971) **B27**, 1826.
40. Roth, W., *J. Solid State Chem.* (1972) **4**, 60.
41. Takahashi, T., Iwahara, H., *J. Appl. Electrochem.* (1973) **3**, 65.
42. Sleight, A. W., *in* "Advanced Materials for Catalysis," J. J. Burton and R. L. Garten, Eds., Academic, New York, 1976.
43. Sleight, A. W., Linn, W. J., *Ann. N. Y. Acad. Sci.* (1976) **272**, 22.

RECEIVED July 27, 1976.

12

The Sodium–Sulfur Battery: Problems and Promises

S. A. WEINER

Research Staff, Ford Motor Co., Dearborn, Mich. 48121

> *The current status of work on the sodium–sulfur battery is reviewed, with emphasis on the ceramic electrolyte and container and electrode materials for the sulfur electrode. The baseline studies for the cell testing program are run on cells constructed of carbon, glass, and β''-alumina and containing no metal other than sodium. In sodium–sodium test cells ceramic life has exceeded 1000 A-h/cm^2 one way. Sodium–sulfur cell life still remains short of sodium–sodium cell life. Separate cells designed to maximize energy and power density, respectively, were studied. The high energy cell #89 delivered 2.3 Wh/cm^2 at 64% efficiency. The high power cell delivered 0.35 W/cm^2 at 62% electrical efficiency and with long life. Cost studies indicate the ceramic to be the high cost item with a materials cost of 11.6 cents/cm^2. While problems still remain, there is no known fundamental obstacle that precludes the commercial development of the sodium–sulfur battery.*

The sodium–sulfur battery consists of two liquid electrodes, sodium and sulfur, and a ceramic electrolyte membrane allowing the transport of sodium ions (1). The sodium electrode is well characterized and does not present material problems. Excess sodium is used to keep the ceramic electrolyte completely covered at all times. The use of excess sodium together with a stainless steel sodium container eliminates the need for an electrical feed-through. The β''-alumina electrolyte consists of Na_2O, Al_2O_3, stabilized by Li_2O. Typically it has a strength on the order of 20 kpsi and a resistivity of 5 ohm-cm at 300°C.

The operation of the sulfur electrode is quite complex. Because elemental sulfur is an electronic insulator, graphite felt is added to provide a large area electrode. On discharge from sodium and sulfur, the sodium polysulfide formed is not soluble in sulfur. Thus the sulfur electrode contains two liquid phases throughout some 60% of the discharge. Beyond this point essentially no elemental sulfur remains, and all of the polysulfides are miscible, forming one phase. To keep this phase liquid throughout its compositional range (Na_2S_5 to Na_2S_3) it is necessary to operate above 270°C with typical operating temperatures falling at 300°–375°C. A schematic of a cell with a cylindrical ceramic electrolyte is shown in Figure 1.

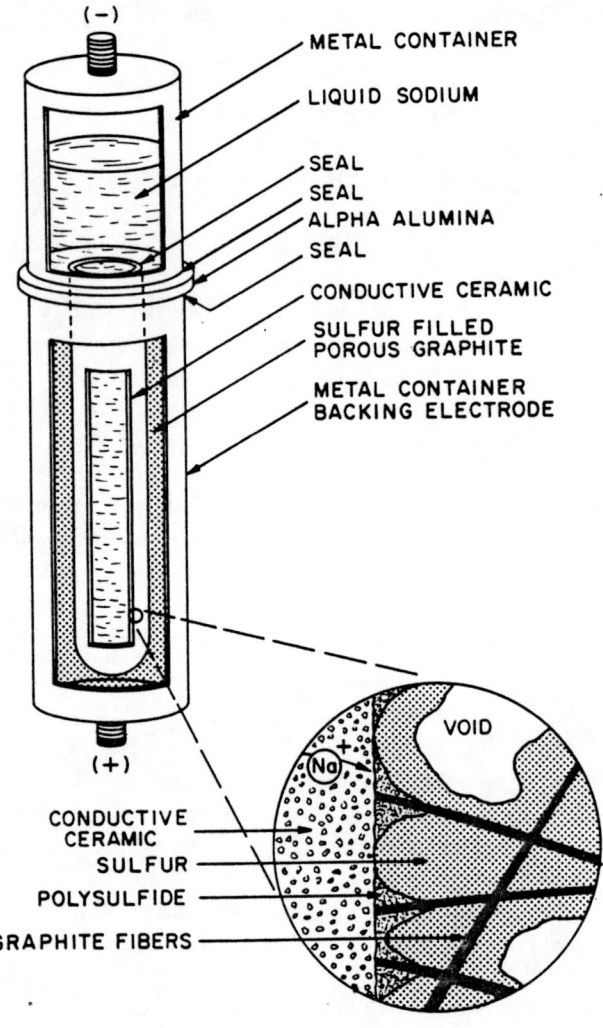

Figure 1. *Schematic of a sodium–sulfur cell*

The two major applications currently envisioned for the sodium–sulfur battery are electric utility load leveling and automotive propulsion. For load leveling the sulfur electrode must meet stringent electrical efficiency requirements with less importance placed on achieving high utilization of reactants since weight and volume are not as critical. In contrast, high reactant utilization is more important than even the operating efficiency of the vehicular battery. Furthermore, the battery for automotive propulsion must have a higher power density than the battery used for load leveling. Our program has two goals: the development of an efficient high energy battery and the development of a low weight, high power battery.

In order to compare current laboratory achievement with overall program goals, the goals of the program have been translated from units of W/kg and Wh/kg to W/cm² and Wh/cm² where the unit of area is the surface area of the β''-alumina ceramic electrolyte. The goals are given in Table I. The translation from units of weight to units of electrolyte area was necessary because the bulk of the laboratory results were obtained using cells constructed mainly from carbon and glass to avoid the effects of corrosion products originating from metallic electrode containers or current collectors in contact with the sulfur/polysulfide melt.

Table I. Cell Goals

Variable	High Energy Cell	High Power Cell
Energy density (Wh/cm²)	2 (265 Wh/kg)	0.15 (60 Wh/kg)
Average power density (W/cm²)	0.2–0.4 (55–110 W/kg)	0.35 (140 W/kg) [a]
Utilization of reagents (%)	50	25
Electrical efficiency (%)	65	70
Capacity (A-h/cm²)	1.0	0.1
Discharge time (h)	5–10	0.4
Durability (A-h/cm²)	2500	100
Cycle life	2500	1000

[a] The goal for peak power density is 0.7 W/cm² or 280 W/kg.

Ultimately the use of the sodium–sulfur battery will depend on its ability to compete economically with the alternate means available for load leveling and automotive propulsion. Presently the limits of both cell durability and cost are set by the β''-alumina electrolyte. Accordingly this paper emphasizes the ceramic electrolyte.

Results of Cell Testing

For ceramic evaluation sodium–sodium test cells (Figure 2) are constructed and run at relatively high current densities of 0.75–1.25 A/cm^2 so that substantial ionic currents can be passed through the electrolyte in a reasonable period of time. During testing cell polarities are reversed periodically. This subjects each surface of the ceramic electrolyte to both a charging operation in which sodium ions are converted to sodium metal and a discharge operation in which sodium metal is converted to sodium ions.

The characterization of individual sodium–sulfur cells involves two distinct testing programs—endurance testing and performance testing (2). The purpose of the endurance test program is to establish the durability of the cell and its components by monitoring the electrical performance at fixed operating conditions as a function of time and conditions of use. In addition to the time to failure, the rates of deterioration of cell performance (e.g., capacity, internal resistance) are obtained. The goal of performance testing is the characterization of the electrical behavior of a cell at various operating conditions (e.g., temperature, charge, and discharge rates) during the early stages of cell life. Specifically these tests involve determining the capacity vs. rate of charge and discharge

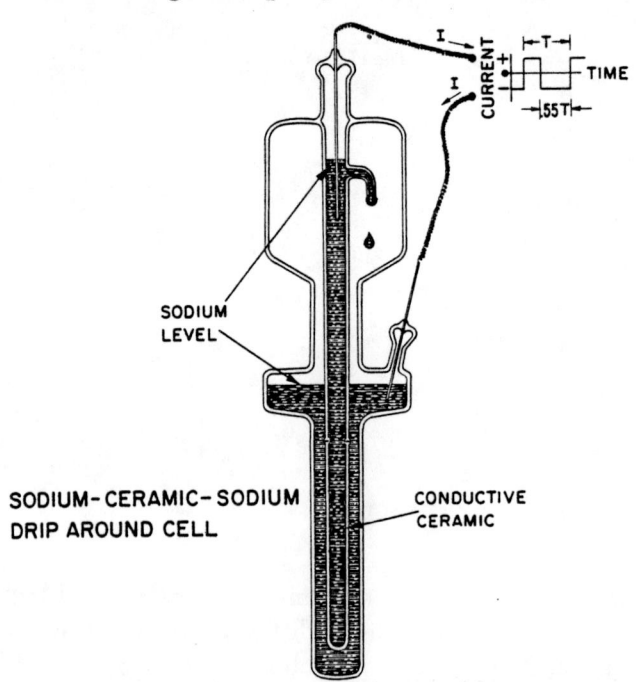

Figure 2. Schematic of a sodium–sodium cell

and the measurement of ohmic and concentration polarizations as a function of temperature, rate, and state of charge of the cell. At the conclusion of the electrical test program each cell is dissected and examined visually. Cell components are prepared for further examination as appropriate.

Sodium–Sodium Cells

One of the major uses of the sodium–sodium cell test program has been to evaluate different ceramic compositions. Several of the compositions tested passed over 1000 A-h/cm^2 in one direction, whereas others showed clear evidence of electrolytic degradation. The major factor in the electrolytic degradation of β''-alumina when subjected to high current densities in a charging mode is the Li_2O concentration. While the Na_2O content may vary within certain limits, β''-alumina compositions containing $Li_2O \leq 0.8\%$ appear to be significantly more resistant to electrolytic degradation at high current densities than β''-alumina compositions containing $Li_2 \geq 0.9\%$.

As a further result of this study the composition 9.0% Na_2O/0.8% Li_2O is being tested in sodium–sulfur cells. Cells 1723-1 and 1723-2 each passed over 1000 A-h/cm^2 undirectionally without deterioration at a current density of 1.25 A/cm^2. The resistivity of the material (5.3 Ω-cm at 300°C) is comparable with that of the 8.7% Na_2O/0.7% Li_2O previously used (5.0 Ω-cm at 300°C), while its strength is greater (19,000 psi vs. 16,000 psi). Furthermore, β''-alumina of the 9.0% Na_2O/0.8% Li_2O composition is easier to process than β''-alumina of the 8.7% Na_2O/0.7% Li_2O composition. Comparison of the performance in Na–Na cells of ceramic of these two compositions is given in Table II.

Testing of cells 1266-1, 1266-2, 1266-3, and 1487-1 was discontinued because of failure of the outer glass envelope. Testing of cells 1269-3,

Table II. Summary of Data from High Current Density Na–Na Test Cells

Cell Number	Current Density (A/cm^2)	Composition		Approx. Time on Test (Mo)	Specific Capacity (A-h/cm^2) One Direction
		% Na_2O	% Li_2O		
1266-1	0.75	8.7	0.7	1.3	377
1266-2	0.75	8.7	0.7	6	1512
1266-3	0.75	8.7	0.7	1.5	378
1578-1	1.00	8.7	0.7	1.9	636
1269-3	1.25	8.7	0.7	1.3	525
1723-1	1.25	9.0	0.8	2.1	1155
1723-2	1.25	9.0	0.8	3.2	1575

1723-1, and 1723-2 was discontinued because of malfunction of the cell test controller. After cell termination most of the ceramic membranes were examined by a variety of methods including light microscopy, scanning electron microscopy (SEM), x-ray diffraction, and x-ray fluorescence. Only the ceramic from cell 1578-1 was found to be cracked. Some anomalies were observed, however, in the examination of other ceramics. The presence of K and Cl was found by SEM in cracked areas of tube 1578-1, but not in undamaged sections. There were several indications of seal damage that could not properly be called seal failures in the four cells built with ceramic of composition 9.0% Na_2O/0.8% Li_2O. In these cases the α-Al_2O_3-glass-β''-alumina seals were badly discolored or pitted but not broken.

Typically ceramic electrolytes that have failed on high current testing exhibit multiple crack patterns and appear weakened even in areas where there are no visible crack patterns. There have also been indications of failure caused by deterioration of the β''-alumina in the vicinity of the seal.

In one case cracks were formed in a β''-alumina tube adjacent to the β''-glass-α seal but not in other portions of the tube. This raised the possibility of stress corrosion caused by the seal. To test this possibility perpendicular pairs of strain gauges were mounted on β''-alumina tubes at distances of 0.5 cm and 2 cm from previously formed β''-glass-α seals. After initial readings were taken, the tubes were cut with a low-speed diamond saw at points between the seals and the strain gauges closest to them. The change in strain was then measured and the stress calculated using the expression:

$$G_x = \frac{E}{1-\nu^2}(\epsilon_x + \nu\epsilon_y)$$

$$G_y = \frac{E}{1-\nu^2}(\epsilon_x + \nu\epsilon_x)$$

where

G_x = circumferential stress in psi
G_y = axial stress in psi
E = Young's modulus $\approx 28.04 \times 10^6$ psi
ϵ_x = strain in ppm, circumferential
ϵ_y = strain in ppm, axial
ν = Poisson's ratio ≈ 0.259

The values of E and ν were determined for the composition 9.0% Na_2O/0.8% Li_2O. The value of E differs from that reported for the composition 8.7% Na_2/0.7% Li_2O, i.e., 18×10^6 psi (3). This degree of variation is not unreasonable, however, as the value of E is influenced by sample porosity and does not affect the conclusion significantly.

All tubes were sealed to α-Al_2O_3 tubes with sealing glass in the customary manner. Tubes of the following compositions were examined:

(a) 9.5% Na_2O/0.9% Li_2O. The tube had been degraded in a Na–Na cell. Cracks were observed near the seal, but no cracks were found 2 cm or farther away from the seal.

(b) 9.5% Na_2O/0.9% Li_2O. The tube was new.

(c) 9.5% Na_2O/0.8% Li_2O. The tube had been subjected to current of 1.25 A/cm² in a Na–Na cell. It was undamaged.

(d) 8.7% Na_2O/0.7% Li_2O. The tube was freshly prepared.

These data are summarized in Table III. There is an apparent correlation between strain and ceramic degradation, and it would be tempting to ascribe the observed ceramic degradation to stress corrosion. However, the stresses calculated from the measured strains normally would not be expected to contribute significantly to stress corrosion. More recent work by A. Virkar, University of Utah, has shown that β''-alumina is subject to stress corrosion in liquid sodium. In these experiments, the K-V diagram was generated. The stress corrosion effects are small and are somewhat a function of composition of ceramic.

More recently the Na–Na test cell program has been used to evaluate the ceramic electrolyte produced at the University of Utah. Of the four

Table III. Stress and Strain on β''-Alumina Tubes Near Seals

	% Composition Na_2O/Li_2O			
	9.5/0.9	9.5/0.9	9.5/0.8	8.7/0.7
History	Na–Na cell degraded near seal	Fresh	Na–Na cell undamaged	Fresh
0.5 cm from seal:				
ϵ_y (ppm)	22	28	4	−13[a]
G_y (psi)	856	1075	159	−157
0.5 cm from seal:				
ϵ_x (ppm)	25	30	5	30
G_x (psi)	922	1119	181	800
2 cm from seal:				
ϵ_y (ppm)	2	13	0	8
G_y (psi)	146	523	31	263
2 cm from seal:				
ϵ_x (ppm)	11	17	4	3
G_x (psi)	346	612	120	152

[a] Negative values indicate compression.

cells containing Utah-produced β''-alumina ceramic of composition 9.0% Na_2O/0.8% Li_2O one has failed after passing 1415 $A\text{-}h/cm^2$ one way, and the others are still in operation. Present plans call for using Na–Na cells to test the effects of process and raw material changes made by the University of Utah. Although Na–Na cell testing remains a useful tool for ceramic evaluation, we have found that the crack patterns exhibited by β''-alumina electrolytes after Na–Na testing are different from those observed on Na–S cell testing. Furthermore, while we have established our preferred composition on the basis of Na–Na cell tests, we have not established why one composition behaves differently from another similar composition nor why and by what mechanism(s) β''-alumina electrolytes degrade.

Sodium–Sulfur Cells

Cell design involves creating an overall cell configuration which is compatible with the β''-alumina electrolyte, seals, container materials, and assembly procedures. The cell components must be sized to be consistent with (a) the electrical requirements of capacity, power, and efficiency, and (b) the mechanical requirements of strength, ruggedness, and simplicity of assembly set by operational and fabrication loads.

The sodium–sulfur cell testing program is directed toward improving the electrical performance of cells, developing an understanding of those factors which control cell performance, establishing cell durability, and identifying factors which limit cell life. While the present limit to cell life for those cells constructed mainly of carbon and glass is set by the durability of β''-alumina electrolytes, the fact that cells built using other

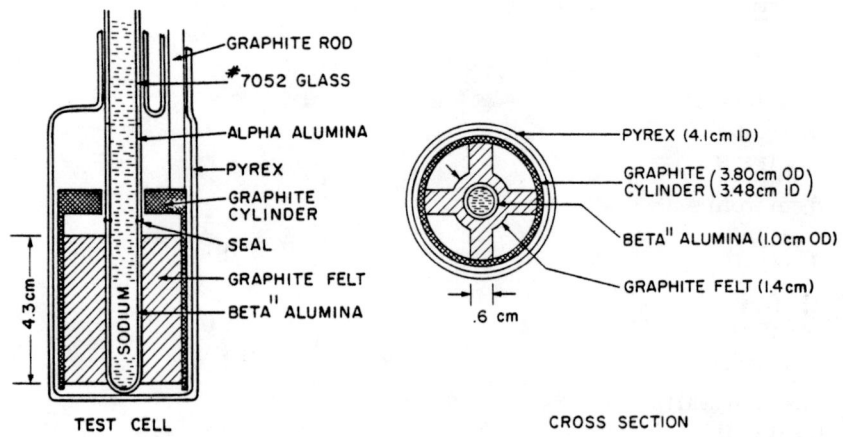

Figure 3. Schematic of Cell 89

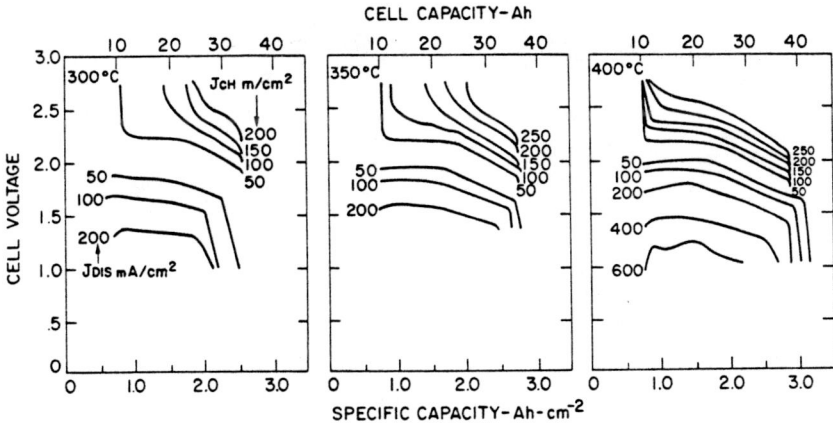

Figure 4. Performance of Cell 89

materials of construction have shorter lives underlines the importance of factors such as the presence of metal ions whose influence on β''-alumina ceramic electrolyte durability is not yet understood.

A cell incorporating a shaped graphite felt electrode designed for very high energy storage (Figure 3) gave low internal losses and high utilization of reactants (Figure 4). In attempts to obtain further improvements in cell operation at temperatures of about 350°C and to aid in developing an understanding of the effects of electrode shape a series of three cells, designated cells 93, 94, and 95 in Figure 5, was constructed to compare the effect of open volumes and the location of the open channels relative to the ceramic surface. The results obtained with cell 94 are shown in Figure 6. This cell shows good discharge performance and fair chargeability—utilizing a sizeable fraction of the reactants and operating moderately deep into the two phase region of the Na–S phase diagram. The results obtained with cell 93 are shown in Figure 7. It is clear that the cell with uncovered ceramic is capable of recharging much more completely, returning to nearly pure sulfur, although with a steadily increasing polarization as charging continues. The losses for this cell are about double those in cell 94.

These results can be interpreted in terms of mass transfer by convection, wicking, and diffusion. Cell 94 was capable of discharging well because the reaction product—polysulfides—formed in the thin ring was able to diffuse to the edge of the felt and react or convect away. Large-scale convection in the open channels brought sulfur up to the felt to supply it to the reacting zone. Because all the β''-alumina ceramic was utilized and the reaction zone was close to the ceramic, the internal losses were low. On charge, however, the sodium pentasulfide was prevented from reaching the ceramic at a sufficient rate because the graphite fibers

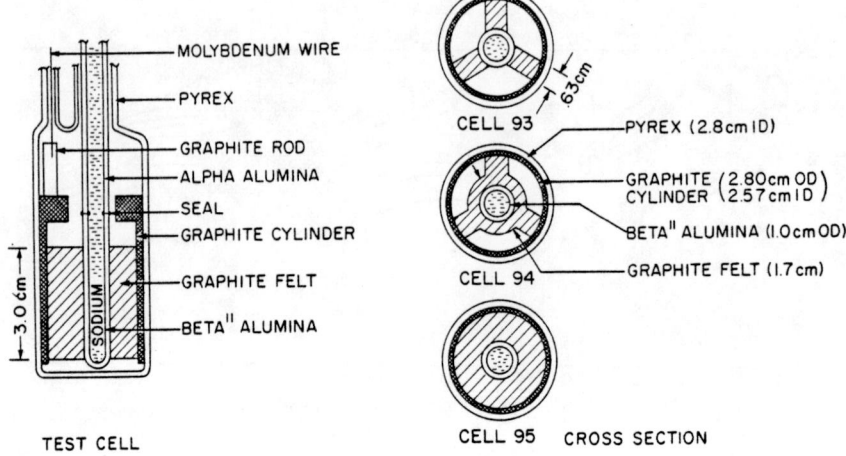

CELL DESIGN IV

Figure 5. Schematic of Cells 93–95

are wetted better by sulfur. The charge is limited thereby to a small portion of the two-phase region.

For cell 93 this interpretation would suggest that on discharge, the losses would be higher because only about half of the ceramic is fully active. The portion of the ceramic area not covered by felt is inactive, since it is covered by sulfur initially. As polysulfide is formed during discharge and fills the open channels, the portion of ceramic covered by

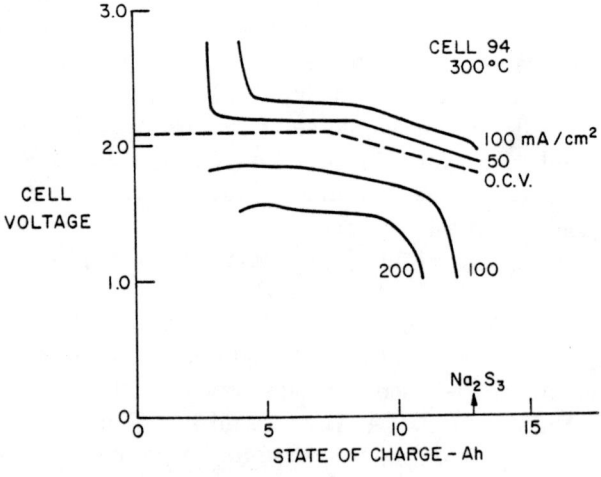

Figure 6. Performance of Cell 94

polysulfide becomes active. In this region the ionic path extends through the ceramic and through the polysulfide melt in the open channel and terminates on the graphite fiber surfaces on the edges of the shaped felt. The contribution of this conduction path is proportional to the height of the polysulfide in the open channels and is probably a small factor throughout the discharge cycle, although it is responsible for the improvement in chargeability of the cell. As with most cells, there are no problems associated with charging through the one-phase region because of diffusion and chemical reactions. Once pentasulfide has formed, however, the portion of ceramic covered by felt is expected to become inactive because of sulfur film formation which blocks the felt surface adjacent to the ceramic. The only remaining ionic path is through the uncovered ceramic surface. We believe that the graphite fiber surfaces at the edge

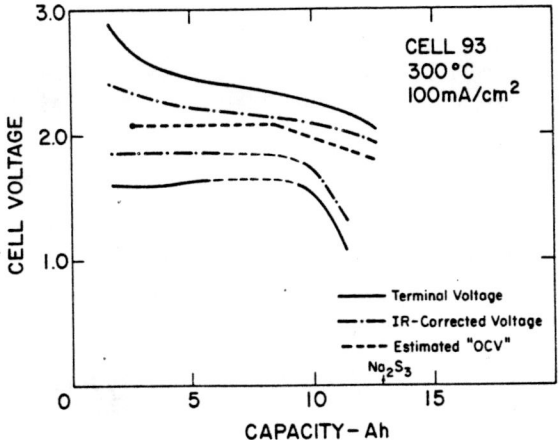

Figure 7. Performance of Cell 93

of the felt remain active because these surfaces are exposed to a freely convecting liquid phase that can remove the sulfur film by convection. According to this model we would expect the cell conductance to decrease in proportion to the remaining height of polysulfide in the open regions, in qualitative agreement with data from cell 93.

A cell designed for a more qauntitative test of this concept is shown in Figure 8. The inner hole in the electrode was enlarged to provide a 1-mm gap between the ceramic surface and the electrode. The charge/discharge characteristics of this cell (cell 102) are given in Figure 9. The results can be interpreted in terms of changes in geometry associated with the varying level of pentasulfide as the state of charge is varied. If phase separation occurs rapidly, the area of ceramic covered by the ionically conducting polysulfide varies in proportion to the amount of polysulfide present in the cell. To a first approximation the cell conductance should vary linearly with the state of charge as shown in Figure 10.

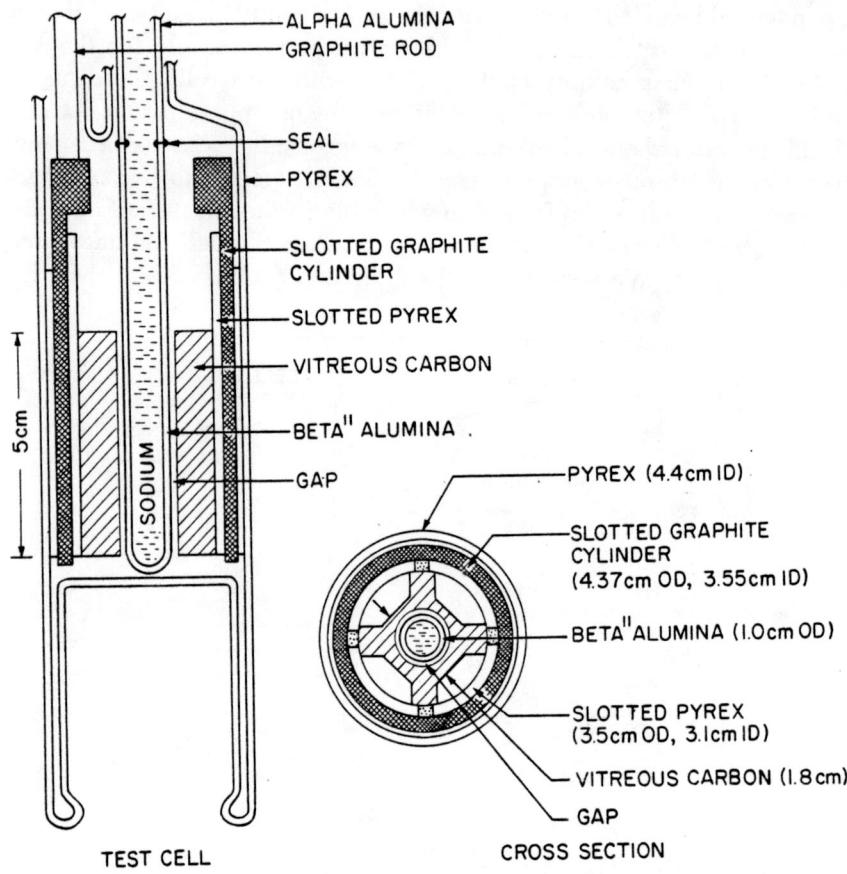

Figure 8. Schematic of Cell 102

Examination of Sodium–Sulfur Cells after Testing

After high-temperature testing was completed, cell 89 was returned to 300°C. Its performance had degraded significantly. The charge cycle appeared to have become limited to the single-phase region even at low rates of charge. The cell was taken out of service after three months of operation and examined. The major finding was that the "cemented" felt arms had become detached from an eroded graphite current collector, thus reducing the electrical contact between the current collector and the electrode to that provided by a few pressure contacts. Under this

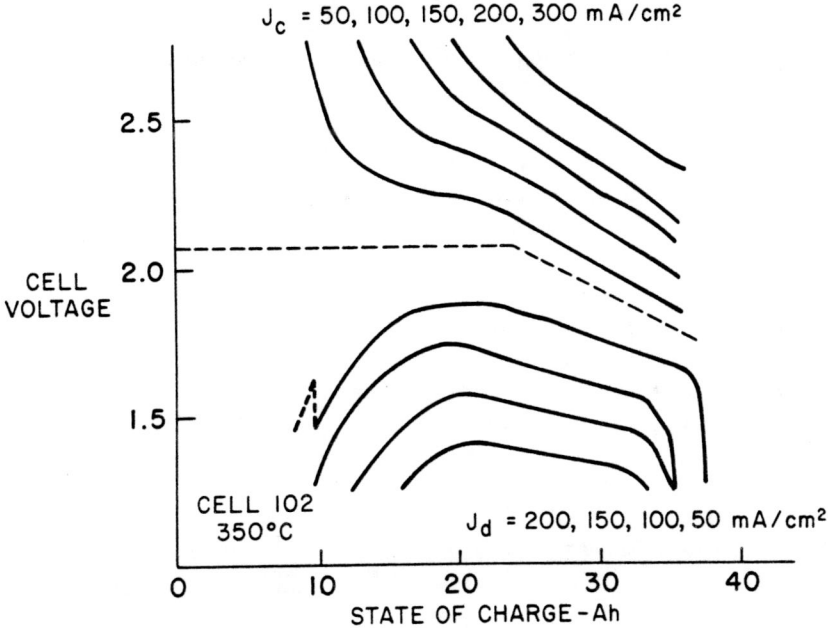

Figure 9. Performance of Cell 102

condition, only the inner wall of the cylindrical current collector remains active as the electrode. The ceramic electrolyte was found to be intact. Examination of fracture surfaces using scanning electron microscopy did not reveal any signs of degradation.

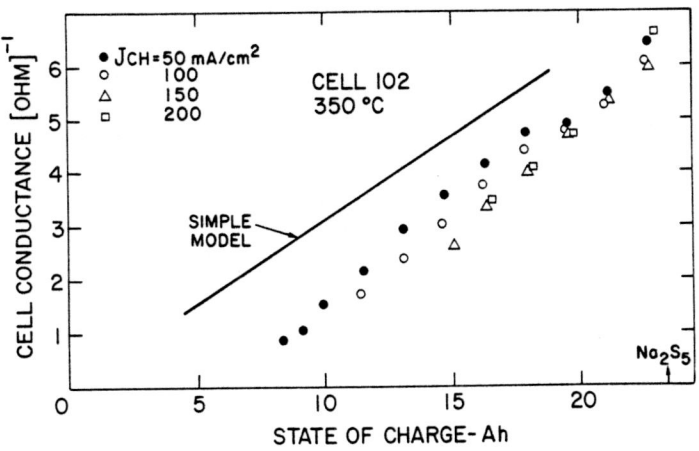

Figure 10. Plot of cell conductance vs. state of charge

The performance of cell 93 deteriorated slowly. Examination of the cell after its failure showed that the "cemented" graphite felt slabs had become detached, as had occurred in cell 89.

Cell 94 failed before extensive electrical testing could be accomplished. Only the initial charge/discharge characteristics were obtained. Unexpectedly, cell 95 also charged far into the two-phase region, contrary to all previous results on such cells having thick full ring electrodes. The cell became non-coulombic, however, before its charge/discharge characteristics could be established fully.

Ceramic tubes were removed from 93 and 94 and, after cleaning, were cut into segments. When feasible, rings were cut from undamaged portions of the tube for diametrical strength tests, microstructures were determined, and surfaces were analyzed by scanning electron microscopy (SEM). In selected cases electron microprobe, Auger, and x-ray fluorescence were also used. In addition, ceramic tubes also were removed from cells E 5, E 16, and E 23. These cells were built with an electrode shape similar to cell 95 (*see* Figures 5 and 11) and used only carbon and glass materials of construction. Cells E 5 and E 16 had been cycled some 6000 and 2200 times respectively in the single-phase region. Cell E 5 had passed 925 A-h/cm^2 of sodium ion one way prior to being terminated while still functioning properly. Cells 89, 93–95, E 16, and E 23 used ceramic of composition 8.7% Na_2O/0.7% Li_2O. Cell E 5 used ceramic of composition 9.25% Na_2O/0.25% Li_2O.

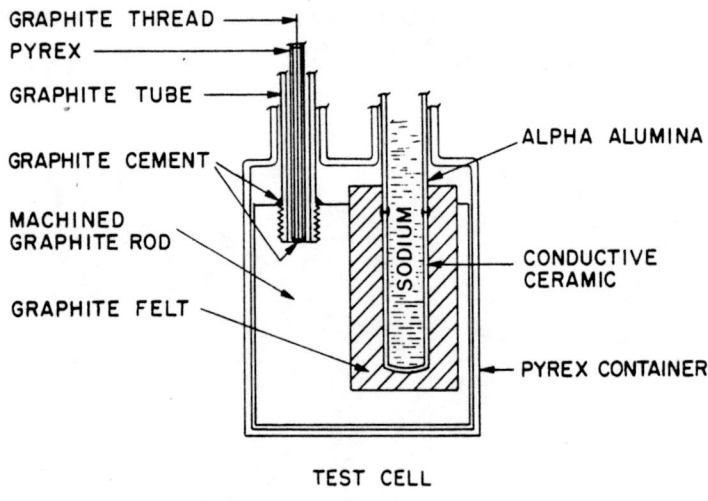

Figure 11. Schematic of "metal-free" cell

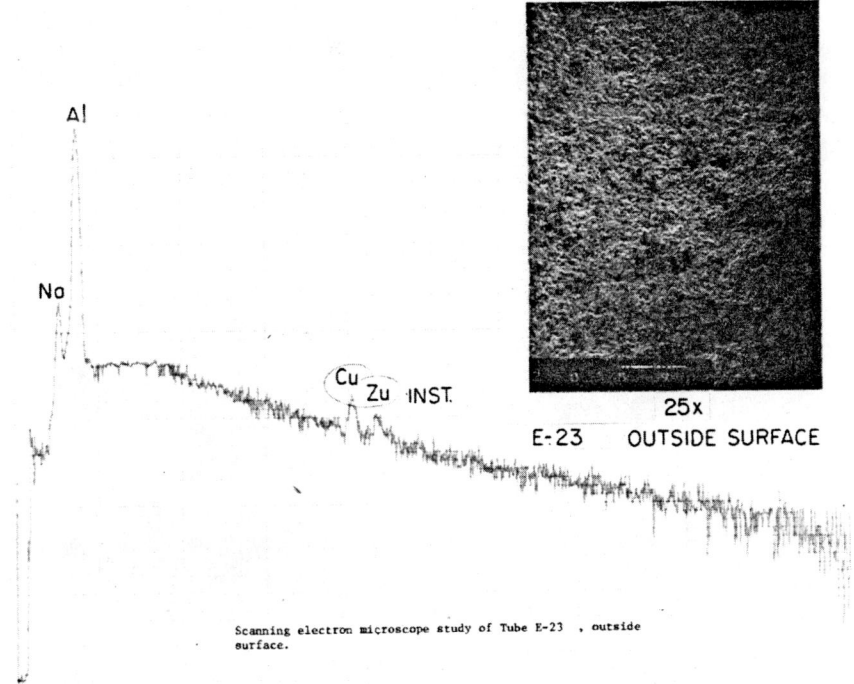

Figure 12. *SEM scan of outside surface of ceramic from Cell E 23*

The β''-alumina electrolyte from cell E 23 was undamaged, and no impurities were found on either the inner surface (which had been in contact with Na) or the outer surface (which had been in contact with sulfur). Diametral tests indicated no deterioration in strength. This sample was used as a standard, and subsequent reference to contamination or impurity levels are made relative to E 23 (Figure 12). This reduces the probability of misinterpreting the presence of very low level impurities resulting from normal cell construction. The appearance of tubes from cells E 5, E 16, 93, and 94 was quite different from that of a typical ceramic degraded in a Na–Na cell. Tubes from cells E 5, 93, and 94 displayed a single long crack, with some branching in the lower portion of the tube. The tube from cell E 16 divided into two parts by a uniform circular crack in the upper portion of the tube. There was some erosion at the edges of the cracks, but this was found only on the outer surfaces. The erosion was probably caused when the crack formed and sodium came into explosive contact with sulfur. The areas away from the crack appeared undamaged, and diametral strength tests indicated no loss of strength for tubes E 16, 93, and 94. Degradation is usually manifested in Na–Na cells as multiple unconnected cracks.

The SEM techniques used for most of the reported surface analysis involves the analysis of points on a surface. This raises the possibility of any one point being atypical because of random contamination. Therefore, many points were analyzed for each sample. Some typical data are shown in Figures 12–15. Cell E 5 had been in service for 15 months prior to failure. Analysis of both inner and outer surfaces using SEM indicated small but real amounts of potassium. Cross-sections 25μ from the outer surface, however, had potassium levels 25% higher than that found at the other points.

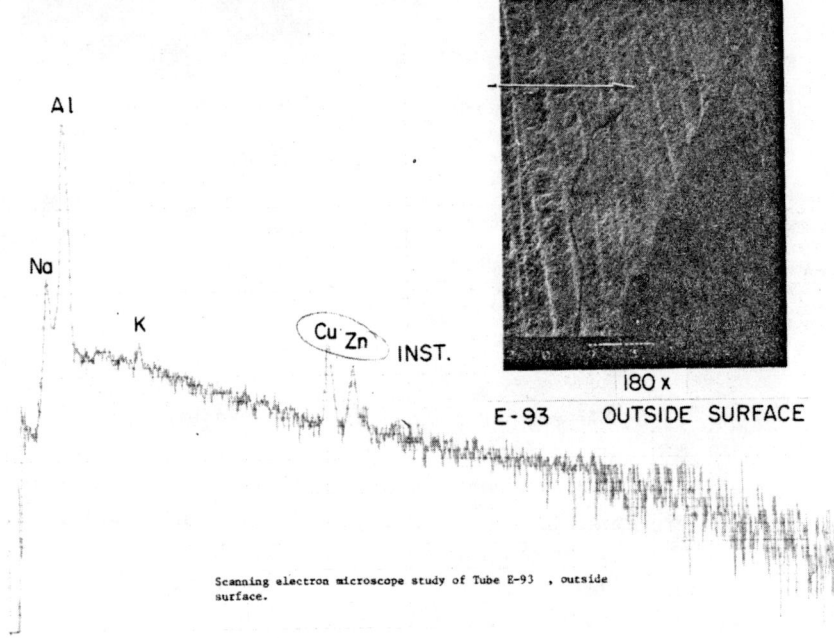

Figure 13. *SEM scan of outside surface of ceramic from Cell 93*

The major impurity in all of the degraded samples was potassium. Calcium was often present with the potassium but always at lower levels. Calcium was never found alone. Traces of iron, silicon, and chlorine were also found in several cases. The potassium impurities were found on the outer surfaces more frequently than on the inner surfaces. The potassium levels found on outer surfaces were often higher, but never lower, than the potassium levels found on inner surfaces. Apparently potassium is not present in the original ceramic and diffuses into the ceramic during cell operation. At this stage it would be improper to conclude that potassium caused the ceramic degradation, but there are enough indications to warrant further study. More recently, some pre-

12. WEINER *The Sodium–Sulfur Battery* 221

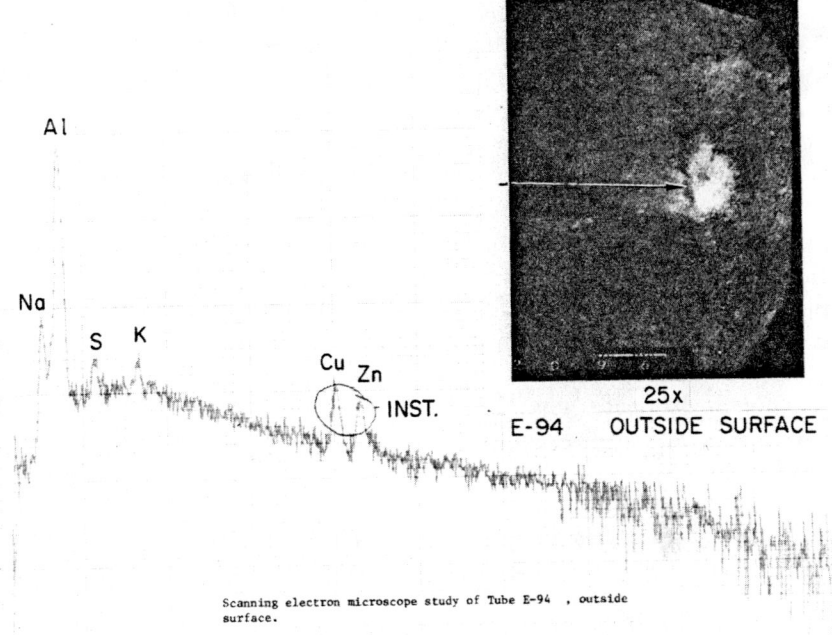

Figure 14. *SEM scan of outside surface of ceramic from Cell 94*

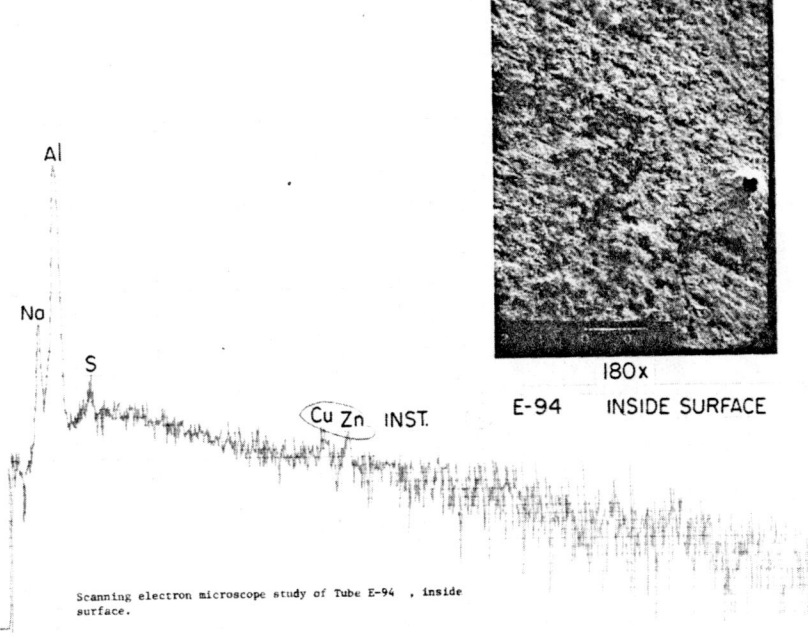

Figure 15. *SEM scan of inside surface of ceramic from Cell 94*

liminary experiments in which samples of β''-alumina of composition 9.0% Na_2O/0.8% Li_2O were immersed for 16 hours at 350°C in a $NaNO_3$–KNO_3 melt containing 0–4.5 mol % potassium suggest that small amounts of potassium reduce the strength of β''-alumina by some few thousand psi.

Recently we have shifted our emphasis from construction of glass and carbon cells to construction of more realistic prototypes of larger size using metal sodium and sulfur containers. We are continuing our efforts, however, to define the effects of such cell variables as temperature, metal ions, current density, electrode shape, and the extent of two-phase (sulfur + sodium polysulfide) operation on β''-alumina durability. It will be necessary for some time to perform experiments in an attempt to delineate how these cell variables interact with β''-alumina electrolytes of varying composition, internal stress levels, and various microstructures.

Materials Costs Estimate for a Sodium–Sulfur Cell

A cost target of $20/kWh has been chosen for the load leveling application based on published data (4). The cost of construction of present laboratory cells will provide an upper limit for cell costs. The direct cost elements can be broken down into four major categories: (a) the costs of raw materials, (b) the costs of component fabrication, (c) the costs of cell filling and assembly, and (d) the costs of external components such as leads, etc. Presently we shall concern ourselves with only the actual costs of raw materials used in construction of laboratory cells.

For purposes of calculation we have assumed a materials usage efficiency of 100% and have based all of our costs in terms of unit area of ceramic electrolyte (cm^2). To convert from area of electrolyte to kWh we have used a factor of 2.3 Wh/cm^2 or 435 cm^2 ceramic electrolyte per kWh delivered. This energy density was achieved in cell 89 which

Table IV. Major Materials Costs for Laboratory Cells

Material	Cost ($/lb)	Amount (g/cm^2 of β''-Alumina)	Cost ($/kWh)
Graphite felt	75	0.091	6.57
Stainless steel	2	2.53	4.86
α-Alumina header	10	0.35	3.18
β''-Alumina	—	0.32	50.46
Sodium	0.4	2.06	0.79
Sulfur	0.1	3.71	0.36
		Total	62.22

was designed specifically for load leveling. Using these assumptions the major raw materials costs were determined (Table IV).

The cost of $3.18/kWh for α-Al_2O_3 headers can be expected to decrease at least one order of magnitude in terms of production costs. A cost reduction of a factor of four can be achieved simply by using 1-in long headers rather than the 4-in long headers currently used in fabrication of laboratory cells.

It is worthwhile to take a somewhat closer look at the costs associated with producing units of β''-alumina electrolyte joined to an α-Al_2O_3 insulator. In the past we have used a cost target of $0.01/cm² of ceramic electrolyte. Using the conversion factor of 2.3 Wh/cm², this translates to a cost of $4.35/kWh of energy delivered.

The raw materials costs associated with β''-alumina fabrication are shown in Table V.

Table V. Ceramic Materials Costs

Material (Unit)	Amount (g/cm²)	Cost ($/unit)	Cost (¢/cm²)
α-Al_2O_3 (lb)	0.07	3.60	0.06
Na_2CO_3 (lb)	0.19 [a]	17.50	2.6
$LiNO_3$ (lb)	0.06 [b]	200.00	8.0 (39) [b]
Pt (troy ounce) [b]	0.02	8.90	0.04
Polyurethane (boots) [a]	0.41	10.00	0.9
		Total	11.6

[a] The conversion factor is 1.5 × 10⁻³ boots/cm².
[b] The cost of the Pt is $0.39/cm², but 80% of the cost is recovered.

The cost of the polyurethane boots (molds) can be decreased by extending boot life which is now on the order of 15 pressings. The cost associated with the platinum used during sintering is shown as 20% of the cost of the formed platinum, the remaining 80% taken as scrap value.

The results given in Tables IV and V indicate that the highest cost item by far is the β''-alumina ceramic electrolyte. To meet target costs the cost of β''-alumina raw materials must be reduced by more than an order of magnitude.

Research is continuing toward our goal of cost reduction. We are exploring the use of raw α-Al_2O_3 powders which cost less than $1.00/lb or 0.09¢/cm², the elimination of platinum encapsulation, and the use of lubricants to extend the life of the polyurethane boots. Although the progress at Utah in forming ceramic electrolyte by isostatic pressing has exceeded that made using extrusion, we may still be able to use an extrusion process which would be one way to eliminate the costs associated with polyurethane boots.

Summary

Results on testing laboratory sodium–sulfur cells continue to demonstrate the potential of this system to meet the goals required for load leveling and automotive propulsion. While much valuable research has been done, the rapid advance of this technology would be aided greatly by answers to some of the fundamental questions that remain. The need to understand the causes of cell and ceramic failure are perhaps demonstrated most clearly by stating that load leveling systems should last at least 10 years.

Acknowledgment

I thank my colleagues at Ford—Tischer, Minck, Gupta, Ludwig, Mikkor, Lingscheit, Tennenhouse, Oei, Winterbottom, and Seaver—for allowing me to cite their work and helping me to prepare this manuscript.

Literature Cited

1. Kummer, J. T., Weber, N., "A Sodium–Sulfur Secondary Battery," *SAE Transactions* (1967) 76, paper 670179.
2. Weiner, S. A., "Research on Electrodes and Electrolyte for the Ford Sodium–Sulfur Battery," Annual Report to the National Science Foundation under Contract No. NSF-C805, July 1975.
3. Virkar, A. V., Tennenhouse, G. J. ,Gordon, R. S., *J. Am. Ceram. Soc.* (1974) 57, 508.
4. Yao, N. P., Birk, J. R., "Battery Energy Storage for Utility Load Leveling and Electric Vehicles: A Review of Advanced Secondary Batteries," 10th Intersociety Energy Conversion Engineering Conference, Newark, Delaware, August 18–22, 1975, paper 759166.

RECEIVED July 27, 1976. Work supported in part by the National Science Foundation—RANN Program under contract NSF-C805.

13

Chemistry of Hot Corrosion

JOHN F. ELLIOTT

Massachusetts Institute of Technology, Cambridge, Mass. 02139

> *The current view of the chemical aspects of the hot corrosion of ceramic and superalloy materials is discussed. The theory based on the acid/base concept is considered by many in the field to account for the corrosion effects that have been observed. It is proposed, however, that there may be other phenomena related to physical imperfections and defects in the protective oxide layer on a ceramic or metal part that may be important. Surface and electrochemical effects must be considered also in the development of a comprehensive treatment of hot corrosion.*

The problem of high-temperature corrosion of highly stressed metallic and ceramic parts of fossil-fuel energy conversion systems can place serious limitations on the life span and operating efficiency of these systems. Examples of parts are turbine blades and vanes of open-cycle gas turbines (1, 2) and boiler tubes (3). The problem of hot corrosion will be important in the design and operation of these systems in the future as a result of higher operating temperatures and the use of more impure fuels. In particular, this problem will require greater attention when coal and residual oil are the fuels.

The metallic and ceramic materials chosen for use in these highly stressed parts are designed to develop a tightly adherent oxide coating when exposed to the hot combustion gases of a system (4, 5). This oxide layer then protects the part from further oxidation and from attack by some of the corrosive agents in the gases. Typically a superalloy that is chosen for use as a turbine blade or combustion pot will contain a significant concentration of chromium so that a thin layer of Cr_2O_3 will form on the exterior surface of the part. For very high-temperature service in aircraft turbines, the alloy may contain aluminum so that a thin layer of Al_2O_3 will form on the surface. More often a metallic coating containing a high concentration of aluminum may be applied to a part, thus assuring the formation of a layer of alumina. Parts made from ceramic materials such as SiC and Si_3N_4 will form a coating of SiO_2 when exposed to hot oxidizing gases. Other ceramic materials considered also form protective oxide coatings, such as Al_2O_3. To provide the protection required, all of these surface layers should form spontaneously, be dense and nonporous, and be tightly adherent to the base material. Also, they should repair themselves if damaged locally and reform if stripped off. Application of a layer of oxide to a part by methods such as flame spraying, etc. has not yet produced satisfactory protective coatings because layers applied this way are easily removed and are not self-healing, nor do they replace themselves.

The principal problem of high-temperature corrosion of metallic and ceramic parts arises when the hot combustion gases contain agents that can deposit as a liquid on a part which in turn causes destruction of the protective qualities of these adherent oxides. These agents may prevent the formation of the oxide layer, actually destroy it, disrupt its continuity and adherence, or possibly prevent the formation of a new, tightly adherent layer to replace the old. Elements that are known to form corrosive agents in the combustion system are principally Na, K, V, Pb, and Si. These agents can react with each other and with oxygen and sulfur to form volatile compounds such as oxides, alkali sulfates, and vanadates when the fuel is burned. In passing through a system the gases cool, the pressure may fall, and the oxygen pressure may rise. As a result, liquid and solid deposits may form on the surfaces of operating parts of the system. Serious problems may also arise if particles of ash and liquid or solid residues from a desulfurization step are carried in the gas stream. The corrosive vapor species may condense on the surfaces of these particles; this will increase the size of the particles and possibly alter their compositions so that they become liquid at the prevailing temperature.

There is good evidence that the problem of serious corrosion of parts at high temperatures is encountered when a liquid phase that includes corrosive agents is present on the surfaces of highly stressed parts (*1, 2, 3*). The destructive action of such agents is often called "hot corrosion," or "sulfation attack" if the destructive agents are sulfates of metals. Such a liquid can also cause corrosion of other metallic and ceramic parts of a system. The principal issue of concern here is the nature of the process of corrosion arising from the presence of the liquid phase on the hot metallic or ceramic part. There is also the broader question of what factors determine the amounts and forms of corrosive species carried by the gas stream and what controls the formation of the liquid corrosive phase. That question is considered only briefly here.

Corrosive Agents

Although a number of agents are known to cause serious hot-corrosion of turbine parts and boiler tubes, the behavior of sodium sulfate has received the greatest attention, and the phenomenology of the nature of the attack has been explored extensively (6–17). Sodium sulfate is considered to be instrumental in the catastrophic failure by corrosion of high-performance gas turbines. Sodium can enter the combustion gases as sodium in the fuel and as sodium chloride from sea water that may be brought into the system with the fuel or combustion air (18). Virtually all fuels contain some sulfur which will be oxidized to SO_2 during combustion. Sodium from sea salt in the fuel and air or from fuel ash, the SO_2, and oxygen in the gas stream can react to form sodium sulfate vapor (18):

$$2Na(v) + SO_2(g) + O_2(g) = Na_2SO_4(v) \tag{1}$$
$$\Delta G_1° = -925{,}300 + 390.7\, T, J$$

Several other reactions can be written to describe the formation of sodium sulfate, but this is the most likely one in a system with high oxygen potential. In combustion gases containing only a few parts per million of sodium and an equivalent level of sulfur, the sodium will be present predominantly as the sulfate as shown in Figure 1. The vapor pressures of both solid and liquid Na_2SO_4 are very low in the operating temperature range of gas turbines as shown in Figure 1, for which the following equations apply (19).

$$Na_2SO_4(v) = Na_2SO_4(l) \tag{2}$$
$$\Delta G_2° = -297{,}500 + 112.17\, T, J$$

$$Na_2SO_4(v) = Na_2SO_4(s) \tag{3}$$
$$\Delta G_3° = -320{,}500 + 132.05\, T, J$$

Thus the concentration of the vapor species in the gas phase at which the liquid or solid phase will precipitate on parts of the energy conversion system is also very low. For example, the gas stream at 900°C containing 0.1 ppm by volume of Na_2SO_4 will tend to deposit the liquid sulfate on a surface at that temperature. Conversely, combustion gases containing 1 ppm Na_2SO_4 will not precipitate the liquid until the temperature of the gas drops to ca. 1050°C. The result is that even relatively low concentrations of sodium and sulfur in the fuel in the range of 0.02% and 0.01%, respectively, can lead to sulfate deposits in locations in a gas turbine where the temperature is below 1050°C. A usual condition for

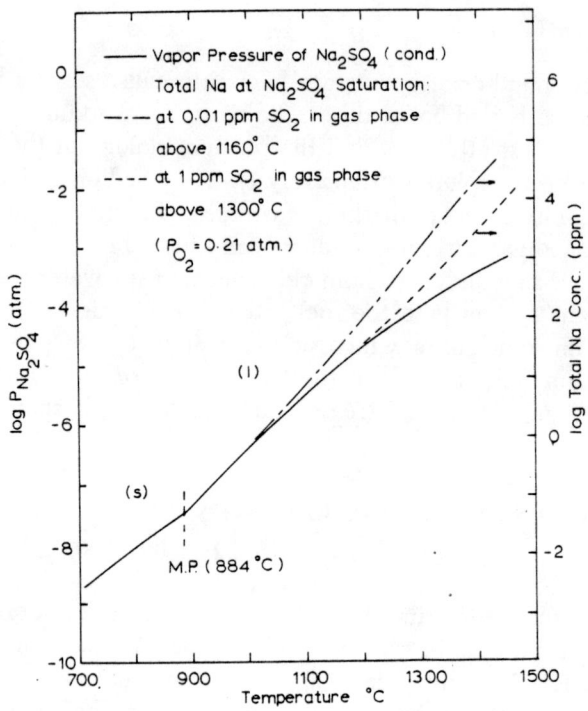

Figure 1. Vapor pressure of sodium sulfate and total sodium content of the equilibrium gas phase

clean fuels is 0.5% sulfur and 1–10 ppm sodium. However, the sodium content of the gases may be raised appreciably by small amounts of sea water in the fuel and sea salt and sea spray in the air.

With the limitations of space here, it is not possible to discuss in detail the many aspects of the hot corrosion process arising from the presence of sodium, potassium, vanadium, and lead in the fuel and air, and from particles of ash carried in the gas stream. There are both kinetic and thermodynamic factors in the corrosion process which are not well understood. Several properties of materials which can lead to corrosion are important. For example, the vapor pressure of potassium sulfate is almost two orders of magnitude higher than that of sodium sulfate (19, 20). Hence proportionately higher concentrations of potassium and sulfur in the fuel are required for the formation of potassium sulfate at a given temperature. One comment that is important to the discussion to follow on the corrosive behavior of sodium sulfate is that the presence of other compounds will most likely lower the melting point of sodium sulfate. Thus if special methods of cooling are used to keep the temperature of the surface of a part below the melting point of the potentially corrosive deposit, greater cooling is needed if several corrosive agents are present rather than only sodium sulfate. However, the addition of magnesium compounds to a gas turbine burning residual fuel oil containing vanadium can substantially reduce the rate of corrosion of turbine parts. The effect is to form magnesium vanadates and magnesium sulfate, all of which apparently form solid deposits on the turbine parts (21, 22).

Phenomenology of Hot Corrosion

The effects of corrosion of highly alloyed steels and superalloys by sodium sulfate have been studied and described extensively (6–17). A wide range of conditions and alloy compositions have been investigated. Some general conclusions are that severe attack of a part may result when a deposit of liquid sodium sulfate forms on that part, as noted earlier. If the temperature is sufficiently low so that the deposit is solid, apparently there is no severe attack (18). In some cases, molybdenum in the metal will enhance the attack (14). Examination of the corroded materials shows that the oxide layer is usually disrupted, dislodged, and possibly destroyed. Elements in the substrate alloy may form sulfides—sulfides of chromium, nickel, and aluminum having been observed. With nickel–chromium alloys that are protected by alumina, laboratory studies at 900°C have also shown the presence of oxides of chromium and nickel in regions of serious attack resulting from a coating of sodium sulfate on the specimens.

Measurement of the change in weight of a specimen of an alloy exposed to a flowing stream of oxygen, air, or synthetic combustion gases in the range of 600°–1200°C is also used in studying the resistance of the alloy to hot corrosion (sulfation). Often the change of weight of a specimen coated with a thin layer of sodium sulfate is compared with that of an uncoated specimen. The uncoated specimen in an oxidizing environment tends to show an increase in weight because of the gradual increase in thickness of the oxide layer, the growth rate often being parabolic. The coated specimen may be expected to show a gain in weight, a gain followed by a loss in weight, or even essentially no change in weight. These seeming inconsistencies arise because the weight gain associated with the formation of an oxide layer can be countered by a loss in weight arising from the volatilization of compounds that may be formed because of reactions among elements in two or more of the phases present: the gas, the sulfate, the metal, and the oxide layer. Bornstein and DeCrescente (10) observed the whole range of behavior of chromium–nickel alloys containing 5–17% chromium which were coated with a thin layer of sodium sulfate and then exposed to flowing oxygen at

1 atm pressure and 1000°C: weight gain, weight gain followed by almost no change in weight, and rapid weight loss following a short period of weight gain. In some cases the total weight lost exceeded the weight of sodium sulfate applied to the specimen. Volatilization of sodium chromate was considered to be responsible for the loss in weight. Bornstein and DeCrescente also observed that the weight-gain/weight-loss behavior of specimens coated with sodium sulfate was similar to that observed for specimens coated with the sulfate (10).

Experiments conducted in the laboratory to study corrosion phenomena should be carried out under carefully controlled conditions to make it possible to interpret the experimental observations. As noted earlier, the classic use of thermogravimetric analysis that is employed to study the oxidation behavior of alloys may not produce easily interpretable results with hot corrosion. Most experimentalists also recognize that subjecting a specimen uncoated or coated with sodium sulfate to a flowing stream of gas uncontaminated with sodium and sulfur does not reproduce conditions in a real combustion system where the gases contain these contaminants. However, useful information can be forthcoming from such experiments with careful interpretation of the results.

Acid–Base Hypothesis

The concept of acidic and basic behavior has long been utilized in interpreting the behavior of fused oxides in the field of extractive metallurgy. In simple terms, "basic oxides," i.e., Na_2O, MgO, and CaO, react with "acidic oxides," i.e., SiO_2, Cr_2O_3, SO_3, etc., and neutralize each other. Applied to hot corrosion, sodium oxide in sodium sulfate can react with chromium oxide to form sodium chromate (23):

$$(Na_2O) + \frac{3}{4} O_2(g) + \frac{1}{2} Cr_2O_3(s) = (Na_2CrO_4) \qquad (4)$$

Species in parenthesis are dissolved in the liquid sulfate. The activity of sodium oxide in the sodium sulfate is controlled by the prevailing pressures of SO_2 and O_2 (or SO_3) in the system in accordance with the equation:

$$Na_2SO_4(l) = (Na_2O) + SO_2(g) + \frac{1}{2} O_2(g) \qquad (1a)$$

The stability diagram for the Na–O–S system at 927°C in Figure 2 shows the isoactivity lines for sodium oxide in the sodium sulfate field. Equation 4 describes the "basic" attack of chromium oxide by sodium sulfate which

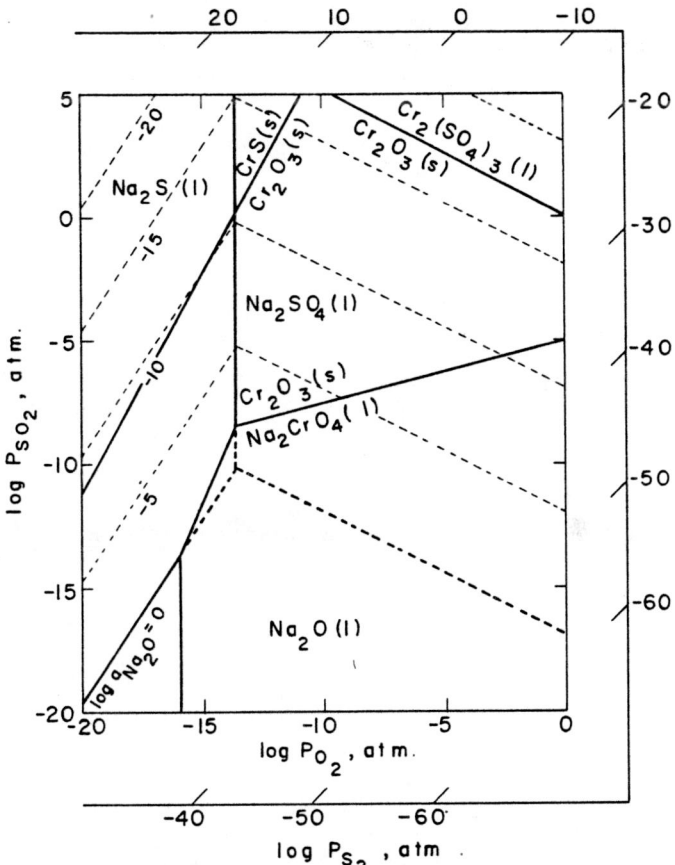

Figure 2. Stability diagram of Na–O–S system at 1200°K (927°C). Also shown are the regions of stability of Cr_2O_3 and CrS.

is favored by low pressures of SO_2. Conversely, "acid" attack is described by the reaction:

$$3Na_2SO_4(l) + Cr_2O_3(s) = [Cr_2(SO_4)_3] + 3(Na_2O) \qquad (5)$$

The conditions for Reactions 4 and 5 assuming unit activity of sodium chromate and chromium sulfate are also shown in Figure 2. Note that $Cr_2O_3(s)$ is stable in the system for conditions represented by the field between the two lines. Equations similar to Equations 4 and 5 can be written to describe the basic and acidic attack of other oxides such as alumina, silica, and oxides of tungsten, vanadium, and molybdenum. Some examples of the basic reactions are:

$$(Na_2O) + Al_2O_3(s) = 2(NaAlO_2) \tag{6}$$

$$(Na_2O) + SiO_2(s) = (Na_2SiO_3) \tag{7}$$

$$(Na_2O) + WO_3(s) = (Na_2WO_4) \tag{8}$$

$$\frac{3}{2}(Na_2O) + \frac{1}{2}V_2O_5(s) = (Na_3VO_4) \tag{9}$$

$$(Na_2O) + MoO_3(s) = (Na_2MoO_4) \tag{10}$$

Several other compounds could be considered to be reaction products for each element. An example of an acidic reaction involving alumina is:

$$3Na_2SO_4(l) + Al_2O_3(s) = [Al_2(SO_4)_3] + 3(Na_2O) \tag{11}$$

Similar reactions may be written for the other metal oxides discussed earlier.

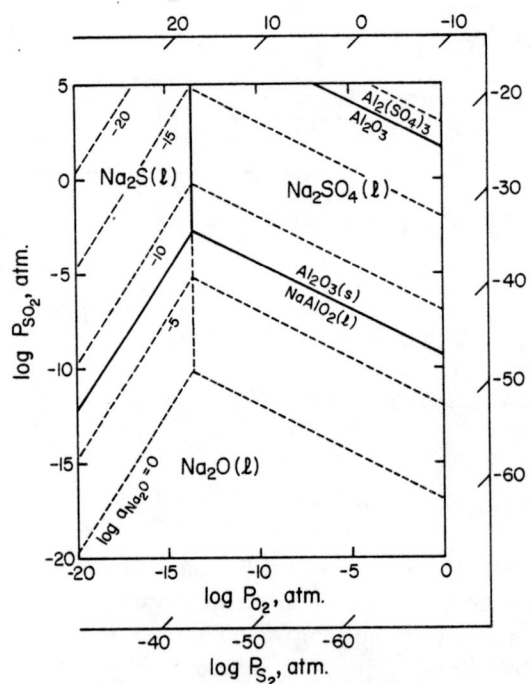

Figure 3. Stability diagram of Na–O–S system at 1200°K (927°C) showing region of stability of Al_2O_3

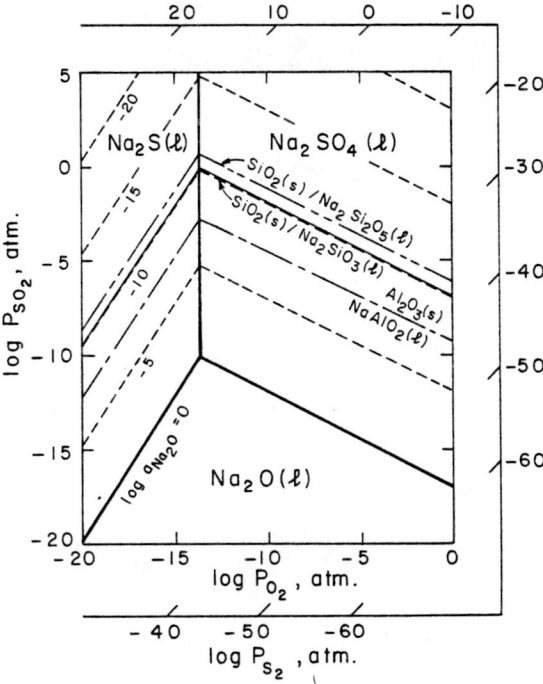

Figure 4. Na–O–S stability diagram at 1200°K (927°C) showing regions of stability of SiO_2. Region of stability is above each designated line.

A somewhat more sophisticated view of the acid–base hypothesis uses the concept that the basicity of a solution is governed by the thermodynamic activity of oxygen ions in a molten oxide or sulfate (*10, 11, 13, 14*). The oxygen ion is assumed to be (O^{2-}), and a basic reaction involves the consumption of oxygen ions. For example:

$$(O^{2-}) + \frac{3}{4} O_2(g) + \frac{1}{2} Cr_2O_3(s) = (CrO_4^{2-}) \tag{4a}$$

An acidic reaction involves the production of oxygen ions:

$$Cr_2O_3(s) = 2(Cr^{3+}) + 3(O^{2-}) \tag{5a}$$

Equations equivalent to these two can be written as alternatives to Equations 6, 7, 8, 9, 10, and 11. The estimated locations of lines representing such equations on the stability diagram for sodium sulfate are shown in Figures 3, 4, and 5. The region of stability for alumina is repre-

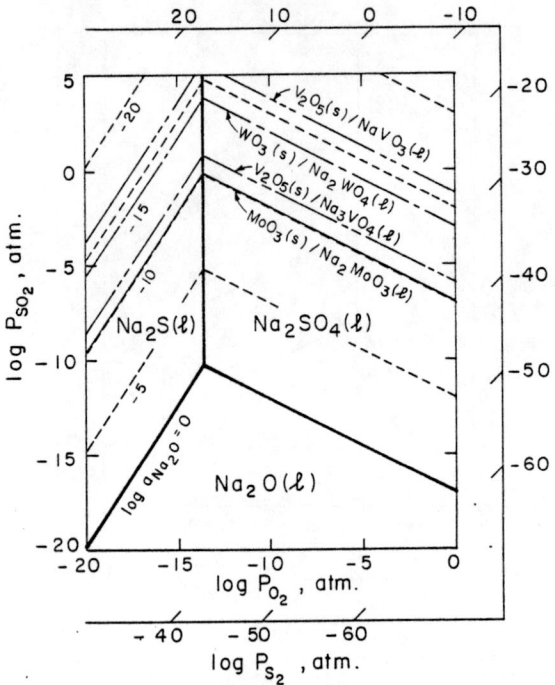

Figure 5. Regions of stability of oxides of tungsten, molybdenum, and vanadium in equilibrium with sodium sulfate at 1200°K (972°C). Region of stability of each oxide lies above the designated line. Unit activity of the compounds is assumed.

sented by the area between the lines designated $Al_2O_3/NaAlO_2$ and $Al_2(SO_4)_3/Al_2O_3$ in Figure 3. The region of stability of silica lies above the lines for silica in Figure 4, and for MoO_3, WO_3, and V_2O_5 it lies above the appropriate lines in Figure 5.

The application of the hypothesis is relatively simple. When a deposit of liquid sodium sulfate forms on a part, the protective oxide on the part will be destroyed if conditions in the system fall outside of the region on the Na–O–S stability diagram for which that oxide is stable.

Because CrS and Cr_2O_3 are observed together as corrosion products in sulfation attack, the line for the CrS/Cr_2O_3 equilibrium is also shown in Figure 2. CrS is stable for conditions represented by the area to the left of the line. In drawing the diagrams it is assumed that the compounds are present at unit activity (pure substance, 1 atm pressure).

Electrochemistry

The concept of corrosion by sulfation being dependent on the oxygen ion content of sodium sulfate (*10, 13*) has prompted electrochemical investigations of the matter. Several studies have shown that it is possible to obtain reversible oxygen and SO_3 electrodes in liquid sulfates (*24–29*). The cell

$$W, WS_2, Na_2S \;\left|\; \begin{array}{c} Na^+ \\ \beta\text{-Alumina} \end{array} \;\right|\; Na_2SO_4(l), SO_2, O_2, Pt \qquad (I)$$

is being used to measure the activity of Na_2O in sodium sulfate (*30*) at 1200°K. A similar cell has been used to measure α_{Na_2O} in sodium chromate (*31*). The activity of sodium oxide at the right electrode in cell(I) is fixed by the prevailing pressures of SO_2 and O_2, and at the electrode at the left the activity of sodium can be determined from the known thermodynamic properties of WS_2 and Na_2S. Accordingly, it is possible to determine the activity of sodium oxide in the sodium sulfate relative to pure liquid sodium oxide for a variety of experimental conditions. An addition of MoO_3 to the liquid sodium sulfate at relatively high values of P_{SO_2} and P_{O_2} caused a small increase in the potential of the cell which indicates that MoO_3 is acidic in behavior, an observation that is in agreement with observations reported by others (*32, 33*).

Several studies (*12, 13, 33*) of the acid–base reactions between sodium sulfate and various metal oxides have utilized the cell

$$Ag, \begin{array}{c} Ag_2SO_4 \\ Na_2SO_4 \end{array} \;\left|\; \begin{array}{c} \text{Electrolyte} \\ \text{(Mullite)} \end{array} \;\right|\; Na_2SO_4, O_2, Pt \qquad (II)$$

or a variation of it, for lower temperatures in which the eutectic mixture of lithium, potassium, and sodium sulfates was substituted for the sodium sulfate. This type of cell can be used to determine the relative change in the activity of the oxygen ion in the liquid at the right eletcrode. However, since no standard electrode has been employed, the measurements of various investigators cannot be compared directly, nor is the result of a measurement readily related to the thermodynamic properties of oxides or of sodium sulfate. Brown, Bornstein, and DeCrescente (*12*) assumed that there was equilibrium between SO_2, SO_4^{2-} ions in the salt, and the established pressure of oxygen. In turn this assumption permitted them to select a reference potential for the cell.

The use of Cell II in studying the effects of additions of oxides to the sodium sulfate of the right electrode is complicated by possible reactions with the mullite. For example, an addition of Li_2O or Na_2O resulted initially in a sharp decrease in the cell potential indicating a rise in the oxygen ion content. However, a gradual increase in potential followed, showing that there may have been a reaction between the Li_2O and the mullite (12, 14). Apparently more work with cells of various kinds is needed to establish clearly the acid/base behavior of sodium sulfate towards the various oxides.

Several potentiokinetic studies have been carried out on superalloys and stainless steels (33, 34). A reasonably good correlation was obtained between the behavior of alloys when anodically polarized and the results of sulfation attack of the alloys that were measured on burner test rigs (33). The measurements were very limited in scope, and much more work is needed if we are to utilize electrokinetic measurements for understanding the physicochemical aspects of hot corrosion.

Commentary

The phenomenological and electrochemical studies described provide us with considerable information on the process of hot corrosion, or sulfidation, of metals and ceramics. However, several additional issues must be considered in the effort of developing a clear understanding of the thermodynamic and kinetic aspects of the sulfation attack by sodium sulfate. The first is that there is an enormous difference in potentials of oxygen and sulfur across a layer of oxide on the surface of a metal or ceramics such as SiC and Si_3N_4. This is illustrated in Figure 6. At the gas–oxide interface at 1000°C the oxygen pressure is ca. 2×10^{-1} atm for

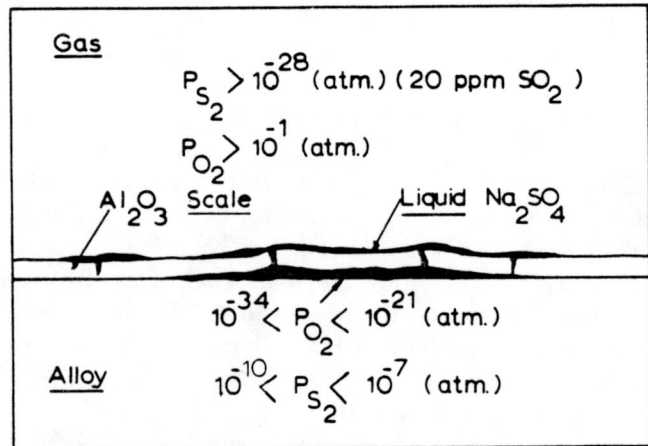

Figure 6. Schematic showing approximate equilibrium pressures of oxygen and sulfur for alloy with oxide coating in contact with $Na_2SO_4(l)$ and combustion gases

well oxidized combustion gases. At the metal–oxide interface it is approximately 10^{-21} atm if the Cr_2O_3/Cr equilibrium is controlling, and possibly as low as 10^{-34} atm if the Al_2O_3/Al equilibrium is controlling. These values generally apply even though chromium and aluminum are not at unit activity in the alloy. Conversely, the sulfur pressure is in the range of 10^{-28} atm in the gas phase for oxidized gases and between 10^{-7} and 10^{-21} atm at the oxide–metal interface. The former figure represents the NiS/Ni equilibrium and the latter the CrS/Cr equilibrium.

Comparison of Figure 6 with Figure 2 shows that sodium sulfate which is brought into contact with the oxide and metal, but out of contact with the highly oxidizing gas phase, will be decomposed with the formation of sulfides and oxides. In addition, the activity of sodium oxide will be very high so that attack of the protective oxide layer by Equations 4, 6, or 7 is highly probable.

The nature of the protective oxide layer must also be considered. Cracks, fissures, and other imperfections are probably present in most protective oxide layers on parts of operating systems. Such imperfections can result from impurities, growth strains, and thermal and mechanical stresses. If sodium sulfate wets the oxide, capillary forces can draw the sulfate to the oxide–metal interface. It may then be possible for the liquid sulfate to penetrate along the interface and destroy the oxide-metal bond. This would lead to loss of the protective oxide and probably would prevent the reformation of an adherent dense oxide layer. Such conditions would cause catastrophic destruction of the part, and the mechanism of oxide solution need not have been the primary cause of the failure.

The effects of localized electrochemical reactions should also be considered. For example, at a location where the liquid sulfate and metal are exposed to the gas phase, the cathodic reaction (Equation 12)

$$SO_3(g) + \frac{1}{2} O_2(g) + (2e^-) = (SO_4^{2-}) \qquad (12)$$

can occur. At a location where the liquid sulfate is in contact with the metal at a low oxygen potential, the anodic reaction (Equation 13)

$$M(s) = (M^{2+}) + 2e^- \qquad (13)$$

may occur with the overall reaction being the formation of the metal sulfate, MSO_4. Also, a wide variety of other reactions is possible. Localized reactions like 12 and 13 would permit very rapid sulfation if there is a path for the ready transfer of electrons from the site of the anode to that of the cathode. The metal phase would provide that path.

Recent work has developed a great deal of information about the phenomenology of hot corrosion of alloy steels and superalloys. The acid–base hypothesis, alternatively the oxygen–ion hypothesis, has been used by a number of investigators to describe the physical chemistry of the phenomena. Further phenomenological studies and work on materials testing are needed, especially as applied to the corrosive action of other liquids which include the alkali vanadates, plumbates, and silicates, as well as mixed potassium and sodium sulfates and chlorides. However, such work alone will not readily advance our understanding of the fundamental physicochemical aspects of hot corrosion by agents that are present in fuels such as coals and the heavy fuel oils. Serious problems of hot corrosion in second and third generation energy conversion systems that use these fuels will have to be considered. Thus it is important that well focused studies be undertaken on the basic physicochemical processes that determine the nature and rates of the corrosion processes. Work is needed on the phenomena that determine the nature and extent of volatilization of cororsive agents from fuels, ash, and slags; the factors that determine how and where corrosive species can condense from the gas stream; and what roles acid/base reactions, surface forces, and electrochemical processes play in the hot corrosion of metal and ceramic parts.

Summary

The problem of hot corrosion of metals and ceramic materials is explored in terms of the current understanding of sulfation of superalloys. The corrosion process is considered to result from the dissolution and destruction of the adherent protective layer of oxide on the surface of the part because of a basic (or acidic) reaction between liquid sodium sulfate on the part and the oxide layer. It is suggested that structural defects in the oxide layer and interfacial tensions between metal and sulfate, sulfate and oxide layer, and metal and oxide may be important. It is also suggested that electrochemical processes and local cell action may accelerate the corrosion process.

Literature Cited

1. Hart, A. B., Cutler, A. J. B., Eds., "Deposition and Corrosion in Gas Turbines," Wiley, New York, 1973.
2. Schirmer, R. M., Quigg, H. T., *in* "Hot Corrosion Problems Associated with Gas Turbines," Special Technical Publication No. 421, p. 270, ASTM, 1967.
3. Johnson, H. R., Littler, D. J., Eds., "The Mechanism of Corrosion by Fuel Impurities," Butterworths, London, 1963.
4. Coward, G. W., *J. Met.* (1970) **22,** 31.
5. Grisaffe, S. J., "Superalloys," C. T. Sim, W. C. Hagel, Eds., p. 341, Wiley, New York, 1972.

6. Simons, E. L., Browning, G. V., Liebhafsky, H. A., *Corrosion* (1955) **11**, 505t.
7. Bergman, P. A., *Corrosion* (1967) **23**, 72.
8. Seybolt, A. U., *Trans. Metall. Soc. AIME* (1968) **242**, 1955.
9. Seybolt, A. U., "Na_2SO_4—Superalloy Corrosion Mechanism Studies," General Electric Research and Development Center, Report No. 70-C-189, June 1970.
10. Bornstein, N. S., DeCrescente, M. A., *Metall. Trans.* (1971) **2**, 2875.
11. Bornstein, N. S., DeCrescente, M. A., *Metall. Trans.* (1973) **4**, 1799.
12. Brown, C. T., Bornstein, N. S., DeCrescente, M. A., "High Temperature Metallic Corrosion of Sulfur and Its Compounds," Z. A. Faroulis, Ed., p. 170, Electrochem. Soc., 1970.
13. Goebel, J. A., Pettit, F. S., *Metall. Trans.* (1970) **1**, 1943.
14. Goebel, J. A., Pettit, F. S., Goward, G. W., *Metall. Trans.* (1973) **4**, 261.
15. McKee, D. W., Romeo, G., *Metall. Trans.* (1975) **6A**, 101.
16. Chatterji, D., McKee, D. W., Romeo, G., Spacil, H. S., *J. Electrochem. Soc.* (1975) **122**, 941.
17. Johnson, D. M., Whittle, D. P., Stringer, J., *Corros. Sci.* (1975) **15**, 721.
18. DeCrescente, M. A., Bornstein, N. S., *Corrosion* (1968) **24**, 127.
19. Cubicciotti, D., Keneshea, F. J., *High Temp. Sci.* (1972) **4**, 32.
20. Halstead, W. D., *Trans. Faraday Soc.* (1970) **66**, 1966.
21. Lay, K. W., *Am. Soc. Mech. Eng. Pap.* 73-WA/CD-3 (1973).
22. Lee, S. Y., Young, W. E., Vermes, G., *Am. Soc. Mech. Eng. Pap.* 73-GT-1 (1973).
23. Liang, W. W., Elliott, J. F., unpublished work, 1975.
24. Erdoes, E., Altorfer, H., *Electrochemica Acta* (1975) **20**, 937–944.
25. Flood, H., Forland, T., Motzfeldt, K., *Acta Chem. Scand.* (1952) **6**, 257–269.
26. Flood, H., Boye, N. C., *Z. Elektrochem.* (1962) **66**, 184–189.
27. Johnson, K. E., Laitinen, H. A., *J. Electrochem. Soc.* (1963) **101**, 314–318.
28. Liu, C. H., *J. Phys. Chem.* (1962) **66**, 164–166.
29. Danner, G., Rey, M., *Electrochim. Acta* (1961) **4**, 274–287.
30. Liang, W. W., Bowen, H. K., Elliott, J. F., "Metal–Slag–Gas Reactions and Processes," Z. A. Foroulis, W. W. Smeltzer, Eds., p. 608, Electrochem. Soc., 1975.
31. Liang, W., Elliott, J. F., *J. Electrochem. Soc.* (1976) **123**, 617.
32. Bornstein, N. S., DeCrescente, M. A., Roth, H. A., *Metall. Trans.*, TMS-ASM (1973) **4**, 1799–1810.
33. Shores, D. A., *Corrosion* (1975) **31**, 434.
34. Cutler, A. J. B., Grant, C. J., "Metal–Slag–Gas Reactions and Processes," Z. A. Foroulis, W. W. Smeltzer, Eds., p. 608, Electrochem. Soc., 1975.

RECEIVED July 27, 1976. Work supported by the Energy Research and Development Administration.

14

Nonstoichiometry, Order, and Disorder in Fluorite-Related Materials for Energy Conversion

LEROY EYRING

Department of Chemistry and the Center for Solid State Science,
Arizona State University, Tempe, Ariz. 85281

Ordered intermediate phases in fluorite-related model systems and in materials useful as fast ion conductors and nuclear energy sources are reviewed. The rare earth oxides are emphasized as models of extended defects and ordering in such materials. Modern high resolution (3.5 Å) electron optical methods are used to deduce unit cells, suggest structures, and reveal reaction mechanisms. The relationships between ordered structures in the ternary and binary oxides are emphasized; nevertheless, a wide range of order and disorder is observed. Electron micrographs are presented to illustrate the range of direct observations possible and to underline the importance of the technique in elucidating not only structural defects but reaction mechanisms at the unit cell level.

The structure of a material is an encyclopedia on the properties of its constituent atoms and is therefore at the root of all its chemical and physical properties. We mean, of course, the structure of the real material, which includes its defects most responsible for the reactivity and dynamics of chemical and physical change. Mechanism in reactions cannot be understood without knowledge of the defect structure. Understanding can lead to improved control of reaction properties and to the ability to design new materials. Elucidation of the structure of real materials of practical importance in the most direct way possible is therefore important, and these studies will serve as a basis for further understanding. We seek here not a shallow classification but deep insight of subtle shading.

Fluorite-related materials are among those commonly encountered in energy conversion and storage. Their varied usefulness and their limitations can be best understood in terms of their structure, which is a palpable expression of their nature. The fluorite structure may be visualized in many ways. For example, if all the tetrahedral interstices in a cubic close-packed array of atoms are filled by atoms of a different kind, the fluorite structure results. The lattice is face-centered cubic ($Fm3m$) with metal atoms (M) at (0, 0, 0) and the nonmetal atoms (X) at (¼, ¼, ¼) and (¼, ¾, ¼) and equivalent positions. This gives four formula units per unit cell. An alternative way of visualizing the structure is to consider the coordination cubes of the metal atoms (MX_8) sharing all edges in a three-dimensional chessboard network. Coordination tetrahedra of nonmetal atoms (XM_4) share all edges in a nonspace-filling three-dimensional network. Still another way of expressing this structure is to consider it as a stacking of close-packed layers of atoms along the [1 1 1] direction in the sequence α B γ β C α γ A β α B γ β C α . . . where Roman capitals represent metal atoms and Greek letters nonmetal, each in equally spaced cubic close-packed layers. Figure 1 illustrates a unit cell of fluorite where circles represent metal atom positions and triangles represent nonmetal atom positions. The octahedral interstices represented by the diamonds are all empty.

Textbook examples of the fluorite structure are usually the fluorides of Ca, Sr, and Ba and the oxides of Th and U. To this list must be added many others where the atom radius ratio of metal/nonmetal ≥ 0.73. These would include many of the hydrides and oxides of the rare earth and actinide elements. Furthermore, when we consider adding or subtracting atoms of either type to give fluorite-related defect structures which may be ordered or disordered, the possibilities boggle the mind.

Fluorite-related materials have a known compositional variation at least from M_4X_9 ($MX_{2.25}$) to MX. Although there are many other fluorite-related materials of importance in energy conversion, such as the fluorides

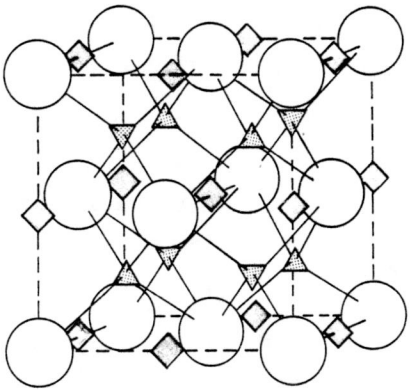

Figure 1. The fluorite structure. The large circles represent metal atom positions; the triangles represent nonmetal positions; and the diamonds represent the empty octahedral positions in the cubic close-packed structure.

and hydrides, here we shall focus attention almost exclusively on oxygen-deficient phases in the composition range $MO_x (2.0 \geq x \geq 1.5)$ where the oxygen deficiency is either ordered or disordered.

In these phases a good approximation is to consider the metal atom substructure intact as in fluorite with small displacements understood. The oxygen substructure is then characterized as possessing vacancies on normal sites which may be ordered in long or short range or disordered. This means that there will always be coherence, but there may be a regular or periodic variation in oxygen composition in the structure.

Many metals capable of tetravalency form extremely thermally stable fluorite-related oxides. Indeed, ThO_2 is the highest melting oxide known ($3300° \pm 100°C$). The congruently melting compositions of all these refractory oxides in vacuum are lower than the dioxide (even for ThO_2). CeO_2 loses oxygen in vacuum to a composition of $CeO_{1.51}$ at $2000°C$ in a tungsten cell, and Pr and TbO_2 lose oxygen to substoichiometric $MO_{1.5-\delta}$ at the melting point. The melting point of reduced substances is $> 2000°C$. The enthalpies of formation of the oxides for the elements forming fluorite-related phases are very high, ~ 210–260 kcal/g atom of metal.

A few other general statements about properties should also be made. Diffusion coefficients of metal atoms are generally very low up to temperatures in excess of one-half the melting point; in contrast, the diffusion coefficients of the oxygen atoms are relatively large. Under these same conditions the vapor in equilibrium with the solid is oxygen. Much higher temperatures must be attained before metal-containing vapor species are detectable. In short, the metal substructure is rigid and nonvolatile, while the nonmetal substructure is mobile and volatile.

Fluorite-Related Materials in Energy-Winning Roles

There is a romantic history and great literature on these anion deficient fluorite-related phases. This includes Welsbach gas mantles and Nernst glowers. This chapter is not intended to be comprehensive; rather, references are made arbitrarily to that work which highlights current efforts to clarify the structural principles behind defect fluorite-related materials.

Solid Electrolytes. ZIRCONIA- AND HAFNIA-BASED MATERIALS. When zirconia or hafnia react with the alkaline earth oxides (especially calcia) or rare earth oxides, pseudobinary fluorite-related phases with anion vacancies are formed. At high temperatures the phase fields are broad cubic solid solutions, although in some cases there may be considerable diffuse scatter in their diffraction patterns. These materials are technically important because of the rapid transport of oxide ions at moderately high temperatures without electronic or cationic conduction. This ensures a variety of high temperature electromechanical uses.

At higher temperatures cubic phases of continuously changing composition cover the field. At lower temperatures ($\leq 1500°C$) in the few cases known so far ordering occurs, probably rapidly with respect to the anion vacancies and with sufficiently long anneals, the cations may also order.

Duclot, Vicat, and Deporter (1) tabulate the investigations of calcia- and rare earth fluorite-related phases formed separately with hafnia and zirconia up to about 1970. They also give evidence of an ordered phase $Y_2Hf_7O_{17}$ and have investigated the order–disorder transition.

Carter and Roth (2) have studied the structure of calcia-stabilized zirconia using neutron diffraction. They concluded that in the ordered structure the oxygen atoms were displaced in a way similar to that found in monoclinic ZrO_2. The disorder–order transition was seen to be the differentiation of the disordered region into domains in which the distortions of the oxygen coordination polyhedra are in the same sense.

Three ordered phases (Φ, Φ_1, Φ_2) were detected in the $CaO–HfO_2$ system (3), two of which (Φ_1 and Φ_2) appear also in $CaO–ZrO_2$ (4).

Allpress, Rossell, and Scott have investigated the ordered phases (5) in the calcia-stabilized hafnia system using electron diffraction to determine the unit cells of three ordered phases: $Ca_2Hf_7O_{16}$ (Φ), rhombohedral, $a = 9.5273Å$, $\alpha = 38.801°$; $CaHf_4O_9$ (Φ), monoclinic, $a = 17.698Å$, $b = 14.500Å$, $c = 12.021Å$, $\beta = 119.47°$; and $Ca_6Hf_{19}O_{44}$ (Φ_2), rhombohedral, $a = 12.059Å$, $\alpha = 98.31°$. Rossell and Scott (6) determined the space group of Φ phase to be $R\overline{3}$ and have found cations ordered on cation sites of the fluorite structure with the calcium ions segregated into discrete layers parallel to the (1 1 1) fluorite plane. They also suggest that the two oxygen vacancies in the unit cell are paired across the body diagonal of the coordination cube of one of the Hf atoms and that the remaining six Hf atoms are near neighbors and seven-coordinated, while the calcium atoms are eight-coordinated.

Subsequently Allpress, Rossell, and Scott (7) have published the structure of $CaHf_4O_9$ (Φ_1) and $Ca_6Hf_{19}O_{44}$ (Φ_2). As in the Φ phase the Ca atoms are all eight-coordinated, while Hf may be six-, seven-, or eight-coordinated. Other common features include probable $½ <1\,1\,1>_F$ vacancy pairs with a cation between. Allpress et al. suggest that these clusters may coalesce to form finite groups in Φ_1 or extended chains in Φ_2. This information was obtained from powder diffraction data which provide only limited conclusions about vacancy ordering.

Electron diffraction patterns of calcia-stabilized hafnia and zirconia show diffuse scattering in addition to the srtong reflections from the fluorite subcell. Allpress and Rossell (8) have analyzed this diffuse scatter and have obtained good agreement with the expected diffraction from specimens with domains of the Φ_1 phase of 30-Å diameter coherently embedded in specific orientations within the cubic matrix. This is reminiscent of the domains of Carter and Roth (2).

Lefevre (9) noted ordering in ZrO_2–Sc_2O_3 caused by a phase which turned out to be rhombohedral $Zr_3Sc_4O_{12}$ (10) and two others with wider composition widths as well as higher ZrO_2 content. Bevan (11) and co-workers have determined that the composition width of these phases is less than at first thought and have assigned them the ideal formulas $Zr_3Sc_4O_{12}$, $Zr_{10}Sc_4O_{26}$, and $Zr_{48}Sc_{14}O_{117}$. $Zr_3Sc_4O_{12}$ was found (12) to be isomorphous with M_7O_{12} in the binary phases to be described later. The structural feature emphasized in these results is a string of oxygen vacancies along the $[1\,1\,1]_F$ which gives units of M_7O_{12} composed of a six-coordinated M atom surrounded with trigonal symmetry by six seven-coordinated M atoms. These units are separated by a fully oxidized M_7O_{14} group along $[1\,1\,1]_F$-like beads of two types in a string to form the structure of $Zr_{10}Sc_4O_{26}$ (space-group $R\bar{3}$). These beaded strings, all oriented parallel to the threefold axis, are edge-shared to form an interpenetrating network of the two kinds of units. The Zr and Sc atoms occupy the metal sites randomly in both structures.

In the $Zr_3Yb_4O_{12}$ phase (13), where the larger Yb is present, two modifications are observed. One is isostructural with $Zr_3Sc_4O_{12}$, and the other (a low temperature form) has some ordering of metal atoms with Zr occupying the six-coordinated metal sites and random occupancy of the seven-coordinated sites. In the other rare earth zirconia solid solutions (11), where the size discrepancy between the atoms increases, there is little evidence of ordering.

The ordered phases of similar composition observed by Komessarova and Spiridinov (14) in the HfO_2–Sc_2O_3 system and indexed as rhombohedral unit cells are probably closely related to these $ZrO_2 \cdot Sc_2O_3$ phases.

Collongues and co-workers (10, 15) as well as many others (1) have found pyrochlore phases in these ternary oxide systems. Although this phase, $A_2^{3+}B_2^{4+}O_7$, is related to fluorite, the shifts in oxygen positions are so great that the structure is best considered as interpenetrating frameworks of $[AO_2]$ and $[BO_6]$ with the A in linear coordination and B in octahedral coordination. The radius ratio, $r(A^{3+})/r(B^{4+})$, is between ca. 1.20 and 1.60.

THORIA–RARE EARTH OXIDE SYSTEMS. Solid solutions between thoria and the rare earth oxides are well known and have been used for years as oxygen-conducting solid electrolytes. In each case studied there is a wide-range solid solution having a fluorite structure up to a maximum at moderate temperatures of about 40% (M_2O_3) in ThO_2. No ordered intermediate phases in this system at low temperatures have been described.

On the M_2O_3-rich side at very high temperatures, however, Sibieude and Foex (16) have found evidence of a remarkable complexity showing phases related to the A-, B-, C-, X-, and H-type sesquioxides of variable composition as well as to numerous other hexagonal phases of apparently narrow composition. The C-type phase is, of course, fluorite-related, but no ordered intermediate narrow composition phases have been indicated in the ternary system thus far.

URANIUM DIOXIDE–RARE EARTH OXIDE SYSTEMS. Weitzel and Keller (17) have recently prepared several oxides of composition $(M_{0.5}U_{0.5})O_2$ where M is a rare earth atom; they have shown them to be strictly fluorite in spite of the unusual presence of $U(V)$ and $U(VI)$ in eight-coordination.

Many studies have shown the formation of fluorite solid solutions of wide composition range. Our attention is drawn to $3Y_2O_3 \cdot UO_3$ or UY_6O_{12} which was the first crystal of this structure determined (18). It is isostructural with the $Zr_3M_4O_{12}$ phases discussed above and with the binary oxides of composition M_7O_{12} to be discussed below as the structural end member of the homologous series in the rare earth higher oxides. These are the only known ordered fluorite-related ternary oxides in this family.

Fluorite-Related Oxides for the Nuclear Industry. Many reactor fuels and radiation power sources are fluorite-related oxides. These substances characteristically must operate in place for long periods of time, at high temperatures, in strong radiation fields, and with growing levels of impurity. Underlying any rationally based program to improve the performance characteristics of these materials must be a sound knowledge of their structure and texture, especially their defect structure, as a function of temperature, pressure of oxygen, radiation field, and level of impurity.

Attention will be focused here on the structures of oxygen-deficient fluorite-related oxides of the actinide elements. These include UO_2, ThO_2, PuO_2 $(U, Pu)O_2$ as well as the plutonium oxides.

FLUORITE $MO_{2-\delta}$. Dioxides are known for Th, Pa, U, Np, Pu, Am, Cm, Bk, and Cf. The lattice parameters are compared in Figure 2. The enormous variation of the thermodynamic stability is reflected in the oxygen activity required to maintain stoichiometry at 1000°C which varies more than 20 orders of magnitude.

ThO_2 is only slightly substoichiometric in the presence of Th even near the melting point. UO_2 begins to lose oxygen appreciably in the presence of U at $\sim$ 1500°C and reaches a monotectic at 2500°C at a composition of $\sim UO_{1.65}$. The heavier actinide dioxides typically show little deviation from stoichiometry until a temperature of a few hundred

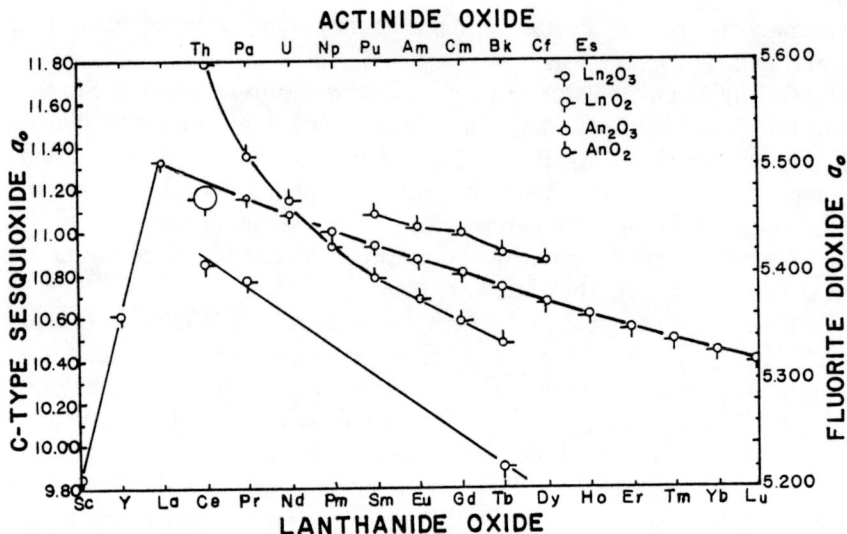

Figure 2. Lattice parameters for fluorite-related rare earth and actinide oxides

degrees is reached, and then quite suddenly they enter a phase of wide composition range of fluorite structure. An understanding of the degree and nature of short- or long-range ordering of defects in these $MO_{2-\delta}$ phases has not progressed very far. A brief review of work up to about 1970 along these lines offers phase diagrams and references (19).

Recently Sørensen (20) has undertaken measurements (and the correlation of the work of others) of the partial molar thermodynamic quantities for oxygen in nonstoichiometric plutonium and plutonium-uranium oxides. He has previously done a similar analysis on the $CeO_{2-\delta}$ system (21). In each of these systems a careful analysis of the thermodynamic data shows surprising deviations from ideal solution. In each case a derived phase diagram indicates several subregions in the $MO_{2-\delta}$ phase in which the slopes of $\Delta \overline{G}_{O_2}$ vs. log X have characteristic values. In the $CeO_{2-\delta}$ phase in addition to numerous nonstoichiometric subregions within which consistent thermodynamic properties are observed, there are superimposed indications of stability at particular stoichiometries belonging to a homologous series M_nO_{2n-2}. In the PuO_x system three such subregions (from PuO_2–$PuO_{1.82}$) and two two-phase regions are indicated. The two-phase regions are in the composition range $PuO_{1.9945}$ and $PuO_{1.9998}$. The $(U, Pu)O_{2-\delta}$ phase separates, by the same treatment, into five subregions with no indication of special stability at any narrow composition. The results cover the composition range $2.00 > x > 1.85$.

High temperature x-ray diffraction observations (20, 21) in these systems have shown superstructures of monoclinic symmetry about 900°C in the $CeO_{2-\delta}$ system. However, in $PuO_{2-\delta}$ or with an admixture of urania only fluorite structures were formed. It seems apparent that even at these high temperatures (although only up to little more than one-half the melting temperatures) there is considerable order, at least short-range order. This order exists in spite of a high mobility of oxygen atoms which at the highest temperatures must spend a good deal of time between positions of minimum energy (22).

The M_7O_{12} phase has definitely been established for Cm and Cf (23, 24), and there is clear evidence for order in intermediate regions in the $CmO_{1.82}$ and at two compositions in the $BkO_{2-\delta}$ system (25).

Table I summarizes the known intermediate phases among the actinide oxides in MO_x, $1.5 < x < 2.0$. The names of the phases are used advisedly following the nomenclature for the lanthanide oxides.

Table I. Intermediate Phases among the Actinide Oxides

M	$MO_{1.5+\delta}(\sigma)$	$MO_{1.71}(\iota)$	$MO_{1.82\pm\delta}(\delta)$	$MO_{2.00-\delta}(\alpha)$
Th				X
Pa				
U				X
Np				X
Pu	X			X
Am	X			X
Cm	X	X	X	X
Bk	X	← indications of order →		X
Cf	X	X		

In summary, many of the fluorite-related actinide oxides exhibit wide-range nonstoichiometry at high temperatures with an apparent tendency to at least short-range order which at low temperatures in the heavier members gives ordered intermediate phases of narrow composition range. Evidence for ordering in these materials is growing steadily.

A word should be said about the σ phase ($MO_{1.5+\delta}$) which is observed in several of the actinide oxide systems as well as in the rare earth–zirconia, -hafnia, or -thoria systems. This is a phase of variable composition related to the C-type rare earth sesquioxides and is also a prominent feature of the Ce, Pr, and Tb oxide systems. In all cases where this phase exists it appears to be separated at high temperatures from the α ($MO_{2-\delta}$) phase by a narrow miscibility gap. Since the metal positions in both phases are essentially those of the fluorite structure, at ~1200°C the ordering of oxygen determines phase stability.

Fluorite-Related Ordered Phases in the M_2O_3–WO_3 System. Oxides of this type are being considered as phosphors, laser hosts, and nuclear control materials. McCarthy et al. (26) have reviewed the literature on the crystal chemistry of fluorite-related rare earth–tungsten oxides. They have prepared fluorite-related compounds of the composition M_6WO_{12}, $M_{10}W_2O_{21}$, and $M_{14}W_4O_{33}$. M_6WO_{12} is isomorphous with Y_6UO_{12}. However, although the other two are thought to be similar to other members of the rare earth series, their structures have not been determined.

Structural Characteristics of the CeO_x System. As expected, the well established higher oxides of cerium possess fluorite-related structures. Many powder diffraction studies of the intermediate phases have been carried out both by quenching the specimens and by high temperature techniques. Table II presents what appears to be the best parameters presently available for the known phases.

Sørensen (21) has reported finding a phase (Ce_6O_{11}) at 900°C isostructural with Pr_6O_{11} with a monoclinic unit cell with $a = 6.781 \pm 0.006$ Å, $b = 11.893 \pm 0.009$ Å, $c = 15.823 \pm 0.015$ Å, and $\beta = 125.04$

Table II. Cell Parameter Data for the Ce_nO_{2n-2} Series (30)

Phase	Unit Cell	Fluorite-Type Pseudocell
Ce_2O_3 (θ)	Hexagonal $a_o = 3.8905 \pm 0.0003$Å $c_o = 6.0589 \pm 0.0003$Å	No pseudocell
Ce_7O_{12} (ι)	Rhombohedral $a_r = 6.784$Å $\alpha_r = 99.302°$ Hexagonal representation $a_h = 10.3410 \pm 0.0007$Å $c_h = 9.6662 \pm 0.0006$Å	 Rhombohedral $a_F = 0.5443 \pm 0.0004$Å $\alpha_F = 89.642 \pm 0.000°$
Ce_9O_{16} (ζ)		Triclinic $a_F = 5.5106 \pm 0.0012$Å $b_F = 5.5155 \pm 0.0012$Å $c_F = 5.5213 \pm 0.0012$Å $\alpha_F = 90.339 \pm 0.000°$ $\beta_F = 90.595 \pm 0.000°$ $\gamma_F = 90.204 \pm 0.000°$
$Ce_{10}O_{18}$ (ϵ)		Triclinic $a_F = 5.5076 \pm 0.0009$Å $b_F = 5.5100 \pm 0.0009$Å $c_F = 5.5112 \pm 0.0008$Å $\alpha_F = 90.360 \pm 0.000°$ $\beta_F = 90.176 \pm 0.000°$ $\gamma_F = 90.160 \pm 0.000°$
$Ce_{11}O_{18}$ (δ)		Rhombohedral $a_F = 5.5023 \pm 0.0007$Å $a_F = 89.762 \pm 0.000°$
CeO_2 (α)	Face centered cubic $a_o = 5.4109 \pm 0.0003$Å	Cubic $a_F = 5.4109 \pm 0.0003$Å

± 0.04 Å. There has been no verification of the composition, however, and the relative intensities of the subcell and supercell reflections do not appear reasonable.

Ray and Cox (27) recently determined the structure of Ce_7O_{12} by neutron diffraction measurements from powders and single crystals. They confirm the compound to be isostructural with Pr_7O_{12}. The space group is $R\bar{3}$ with hexagonal unit cell dimensions $a = 10.37$ Å and $c = 9.67$ Å (rhombohedral cell $a = 10.37$ Å and $\alpha = 99.4$ Å). The hexagonal cell contains three formula units of Ce_7O_{12}. An important observation in this study was that the crystals were extensively twinned and these twin domains had to be sorted out. The powdered substance had a greater degree of disorder than the single crystals. Considerable disorder has been seen in electron micrographs of Pr_7O_{12}; hence it may be characteristic of these materials.

Neutron and x-ray powder diffraction patterns of specimens of composition $CeO_{1.710}$, $CeO_{1.781}$, $CeO_{1.822}$, and CeO_2 have been studied by Ray, Nowick, and Cox (28). They confirm distinct structures for these phases establishing $Ce_{10}O_{18}$ ($n = 10$) for the first time. The structures of phases $n = 9$ and 10 appear to be different from the unit cells obtained by Kunzmann and Eyring (29) in the PrO_x system. The x-ray results of Height and Bevan (30) give unit cells which are also different from those of PrO_x, but it is not known whether they agree with the results of Ray and Cox (27).

The M_7O_{12} phase is isomorphous in Ce, Pr, and Tb oxides, and appears to be the lower structural end-member of the M_nO_{2n-2} series. The relationships between the structures of the other members which belong compositionally to the series is not known. Clearly there are many possible ways of accommodating a deficiency of 1/9, 1/10, or 1/11 of the oxygen atoms.

Structure and Texture in the Binary Rare Earth Oxides. HOMOLOGOUS SERIES. Structurally the binary rare earth oxides display a full range of behavior. They exhibit an homologous series of intermediate phases with the generic formula M_nO_{2n-2} where n is an integer between 4 and ∞. (These phases have a small but definite range of composition.) The ordered phases disorder at higher temperatures and high oxygen pressures to form compounds of wide composition range which are among the last embattled remnants of grossly nonstoichiometric phases. They also indicate the formation of what have been called pseudophases—regions of two-phase areas where the material behaves monophasically. Chemical hysteresis is also an almost universal structurally induced characteristic of phase transitions.

High oxygen mobility in these phases, together with modest electronic conductivity, obviates the need for cation mobility and allows the

Table III. Structural Data on

Composition	n (in R_nO_{2n-2})[a]	Symmetry	Lattice Parameters	
$RO_{1.500}$	4	$Ia3$	$a = 11.152$Å	($PrO_{1.5}$)
			$a = 10.728$Å	($TbO_{1.5}$)
$RO_{1.714}$	7	$R\bar{3}$	$a = 6.750$ Å, $\alpha = 99°23'$	($PrO_{1.714}$)
			$a = 6.509$Å, $\alpha = 99°21'$	($TbO_{1.714}$)
$PrO_{1.778}$	9	Triclinic	$a = 6.5$Å, $b = 8.4$Å, $c = 6.5$Å, $\alpha = 97.3°$, $\beta = 99.6°$, $\gamma = 75.0°$	
$PrO_{1.800}$	10	Pn	$a = 6.7$Å, $b = 19.3$Å, $c = 15.5$Å, $\beta = 125.2°$	
$TbO_{1.809}$	$10\frac{1}{3}$	Triclinic	$a = 13.8$Å, $b = 16.2$Å, $c = 12.1$Å, $\alpha = 107.4°$, $\beta = 100.1°$, $\gamma = 92.2°$	
$RO_{1.818}$	11	Triclinic	$a = 6.5$Å, $b = 9.9$Å, $c = 6.5$Å, $\alpha = 90.0°$, $\beta = 99.6°$, $\gamma = 96.3°$	
$TbO_{1.833}$	12	Pn	$a = 6.7$Å, $b = 23.2$Å, $c = 15.5$Å, $\beta = 125.2°$	
$PrO_{1.833}$	12	Pn	$a = 6.687$Å, $b = 11.602$Å, $c = 15.470$Å, $\beta = 125°15'$	
$RO_{2.000}$	∞	$Fm3m$	$a = 5.393$Å	(PrO_2)
			$a = 5.220$Å	($TbO_{1.95}$)

[a] If a phase occurs in the PrO_x as well as in the TbO_x system R is used, otherwise the specific symbol is used.

Intermediate Rare Earth Oxide Phases

Relation to Basic Structure	Relative Volume	No. of Vacancies/ Unit Cell
$a = 2a_f$	8	16
$a = a_f + 1/2 b_f - 1/2 c_f$	7/4	2
$a = a_f + 1/2 b_f - 1/2 c_f$ $b = 3/2 b_f + 1/2 c_f$ $c = 1/2 a_f - 1/2 b_f + c_f$	9/4	2
$a = a_f + 1/2 b_f - 1/2 c_f$ $b = 5/2(-b_f - c_f)$ $c = 2(-b_f + c_f)$	10	8
$a = -1/2 a_f + 5/2 c_f$ $b = -2a_f - 2b_f - c_f$ $c = 2a_f - b_f$	31/2	12
$a = a_f + 1/2 b_f - 1/2 c_f$ $b = -1/2 a_f + 3/2 b_f + c_f$ $c = 1/2 a_f - 1/2 b_f + c_f$	11/4	2
$a = a_f + 1/2 b_f - 1/2 c_f$ $b = 3(-b_f - c_f)$ $c = 2(-b_f + c_f)$	12	8
$a = a_f + 1/2 b_f - 1/2 c_f$ $b = 3/2(-b_f - c_f)$ $c = 2(-b_f + c_f)$	6	4
	1	0

formation of each phase with astonishing rapidity even at moderate temperatures ($> 300°C$).

The unit cells for several members of the homologous series M_nO_{2n-2} where $n = 7, 9, 10, 11,$ and 12 have been determined (*29*) using electron diffraction and have been added to those already known for $n = 4$ and ∞. These are shown in projection along $[2\bar{1}\bar{1}]$ in Figure 3 and are described in detail in Table III. In addition, polymorphism occurs in the zeta ($n = 9$) phase and perhaps in others.

All the phases have the same a axis. The arrangement of vacancies along this axis, which is a vector of the kind $\frac{1}{2}[2\bar{1}\bar{1}]_F$, are shown in Figure 4. This feature of the iota structure is considered the largest held in common by the members of the homologous series of binary oxides ($n = 7, 9, 10, 11, 12$). The odd members of the series also have the same c axis and hence the iota ac plane $(135)_F$ as a common feature. As Figure 3 shows, the even members have an ac plane in common also, but c is different in this case. Another element which went into this structural principle hypothesis was the high resolution structure images of the delta

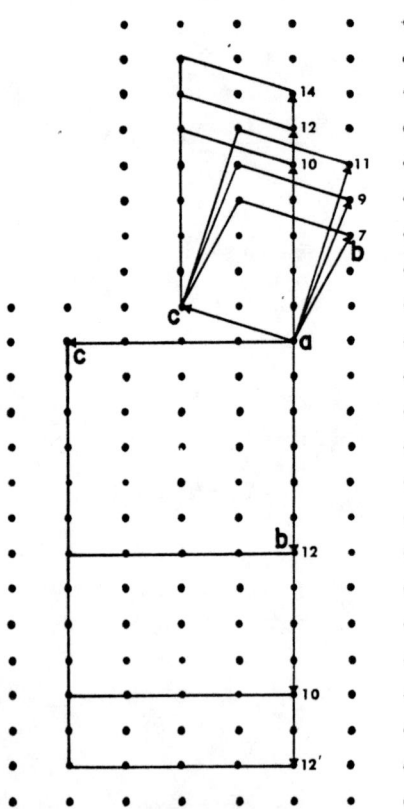

Figure 3. Projections of the unit cells of members of the homologous series along the common a axis, $[2\bar{1}\bar{1}]_F$

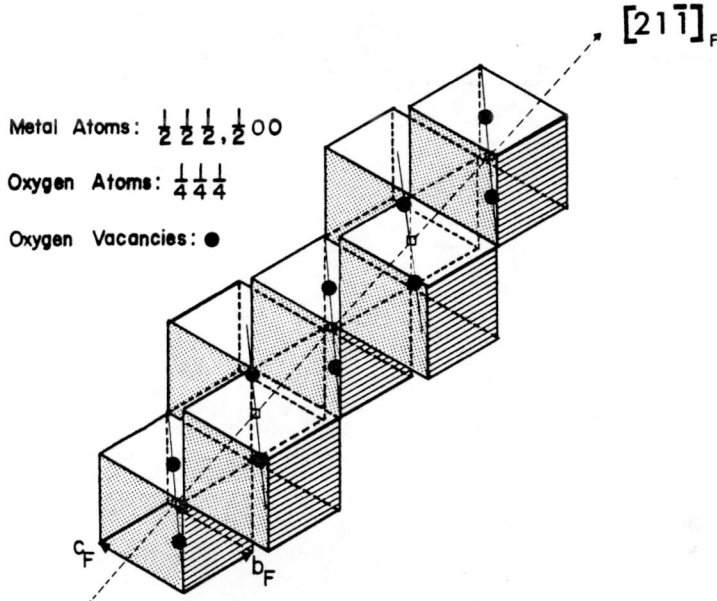

Figure 4. Arrangement of vacancy pairs along the a axis, $[21\bar{1}]_F$ common to members of the homologous series

phase ($Tb_{11}O_{20}$) which appeared to show (Figure 5) the positions of the vacancies in the unit cell as would be predicted (Figure 6).

Despite this apparent success, application of high resolution crystal structure imaging to structures such as the fluorite type had not been developed previously; hence great care must be taken in the interpretation of the images obtained.

High resolution crystal structure imaging is capable of giving a two-dimensional point-to-point image of the crystal potential to a resolution of about 3.5 Å. This is possible for crystals when there is a short axis of only one-coordination polyhedron in the direction of viewing and rather large axes in the other two directions. Such a structure exhibits large variations in potential when viewed down this short axis. In addition, the crystal must be thin (< 100 Å), oriented within $\sim 0.1°$, and viewed at the correct defocus. Under these conditions crystal structure images of structures based on the ReO_3 structure, for example, capable of intuitive interpretation have been produced (*31*).

The fluorite superstructures, on the other hand, are not likely to exhibit large potential variations in any direction. An added disadvantage is that the shortest axis for these materials is the 6.75 Å *a* axis. This strains the thin phase object approximation on which the calculations are based and hence requires that the technique be shown to apply. To add to the

Figure 5. Crystal structure image of $Tb_{11}O_{20}$. Dark field, optically enhanced.

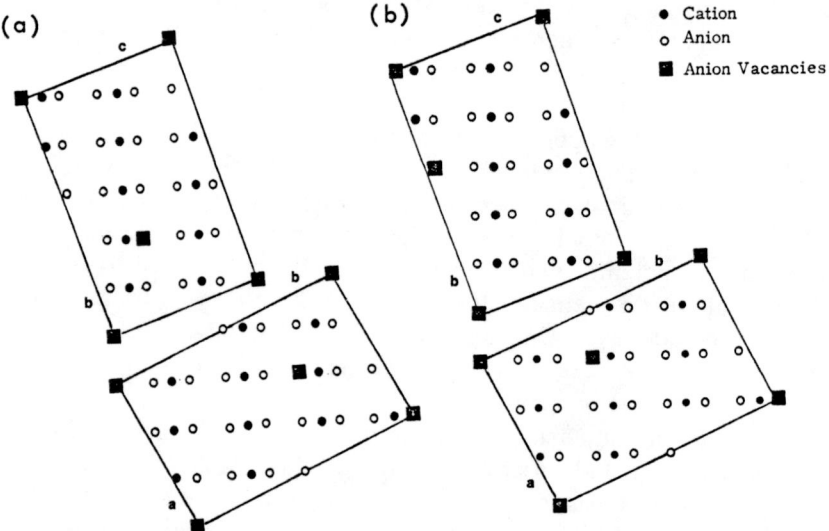

Figure 6. Possible structures of delta ($Pr_{11}O_{20}$) projected down $[211]_F$

difficulty, the reciprocal lattice axes perpendicular to the projection axis are rather long, producing few beams in a reasonable sized aperture.

The stakes in success are high, however, since this technique alone can reveal textural details directly, and the capability of treating any important class of substances such as the fluorite materials is vital. In view of such unfavorable circumstances it is imperative to compare calculated and observed images to avoid serious errors in the interpretation.

With this in mind Skarnulis, Summerville, and Eyring (*32*) have tested the applicability of the method on the prototype of the homologuos series, the iota phase (Pr_7O_{12}), the only one of the intermediate phases for which complete structural data are available (*33*). The n-beam multislice method of Cowley and Moodie (*34*) is the basis of the image contrast calculations.

Calculated Images of the Iota Phase

(Pr_7O_{12}), $[2\bar{1}\bar{1}]_F$ Zone. Figure 7 shows the projection of the ideal atoms for Pr_7O_{12} when viewed down the $\langle 111 \rangle_F$ and the $\langle 2\bar{1}\bar{1} \rangle_F$ axes. The MO_8 coordination cube is also shown to facilitate orientation.

Calculated n-beam images of the $[2\bar{1}\bar{1}]$ zone for 25-, 165-, and 235-Å thick crystals are shown in Figure 8 using ionic scattering factors. The defect of focus of the microscope is taken to be -1000 Å. The origin is placed at the top left corner with two of the axes ($a \equiv b \equiv c$) along the figure edges.

The calculated thin crystal images at 1000-Å underfocus correspond to the projection of the columns of oxygen vacancies in the structure—there being two per unit cell in this case. Many varied images are calculated when many thicknesses and defects of focus are used (*32*), but thicker crystals tend to show only one white spot per unit cell (e.g., at 235-Å thickness, 1000-Å underfocus with ionic scattering factors). This feature correlates with the strings of six-coordinated metal atoms.

The images for thicker crystals do not have a one-to-one correlation with the projected potential and hence cannot be considered true crystal structure images. They do, however, have the periodicity and symmetry of the structure and for this reason are powerful tools in identifying and correlating observed images.

$Zr_3Sc_4O_{12}$, $[2\bar{1}\bar{1}]_F$ Zone. This phase is isostructural with Pr_7O_{12}, and for this reason differences in the images may be used to infer differences in form factors of the metal atoms. The n-beam calculations for this phase using atomic scattering factors are shown in Figure 8. Some generalizations can be made. As in the case for Pr_7O_{12} the thin crystal images correlated well with the projected potential (e.g., 24–48-Å thickness, 900–1000-Å underfocus, atomic scattering factors) (*32*). The same

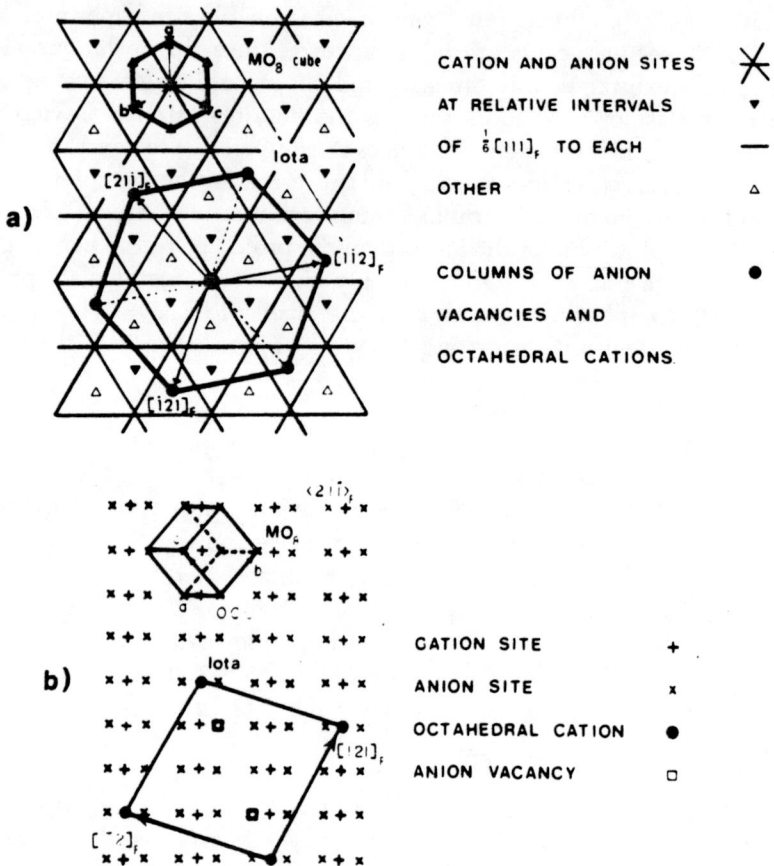

Figure 7. Projections of the ideal atom positions of Pr_7O_{12} down (a) the $[111]_F$ axis where the line intersections, filled triangles, and empty triangles represent columns of metal and nonmetal sites at relative intervals of $1/12[111]_F$ to each other. The columns of nonmetal vacancies and six-coordinated metal atoms are indicated by filled circles and (b) the $[211]_F$ axis where metal atom sites are marked with a plus sign; the nonmetal sites are marked by a multiplication sign; nonmetal vacancies are marked with a square; and filled circles again indicate columns of six-coordinated metal atoms. MO_8 cubes are also outlined to facilitate interpretation.

is true except for a shift in origin for crystals about 165 Å thick as observed for Pr_7O_{12}. The images derived using ionic scattering factors have more fringe, but images showing two spots per unit cell occur at 120–144-Å and 1000-Å underfocus (32). In both cases the thickest crystals show one spot per unit cell

Calculated Images of Pr_7O_{12}, $[111]_F$ Zone. These images showing an hexagonal array of spots at all thicknesses and defects of focus almost

14. EYRING Nonstoichiometry, Order, and Disorder 257

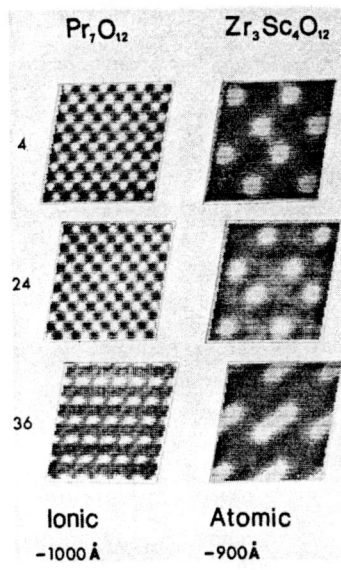

Figure 8. Calculated images of Pr_7O_{12} and $Zr_3Sc_4O_{12}$. The type of scattering factors and the defect of focus are shown at the bottom, while the number of slices (~ 6.75Å per slice) is shown at the left.

uniformally indicate one spot per unit cell, although there is an origin shift in some cases.

Observed Images of the Iota Phase

Images of R_7O_{12} Phases, $[21\bar{1}]$ Zone. Figure 9(a) and (b) show typical images of Pr_7O_{12} with one spot per unit cell. Such images are comparable to some of those calculated for thick crystals. Figure 9(c) shows an image of $Zr_3Sc_4O_{12}$ in which two spots per unit cell appear in

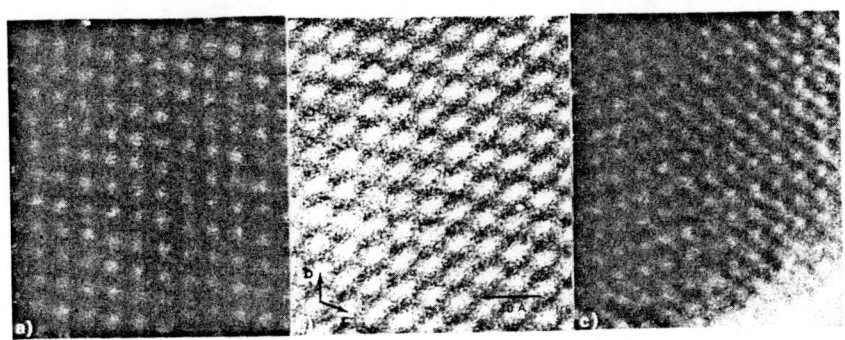

Figure 9. Observed $\langle 100 \rangle_7$ crystal structure images of Pr_7O_{12} and $Zr_3Sc_4A_{12}$: (a) typical image of Pr_7O_{12}; (b) thick-crystal image of Pr_7O_{12}, and (c) thin-crystal image of $Zr_3Sc_4O_{12}$ showing vacancy arrangement

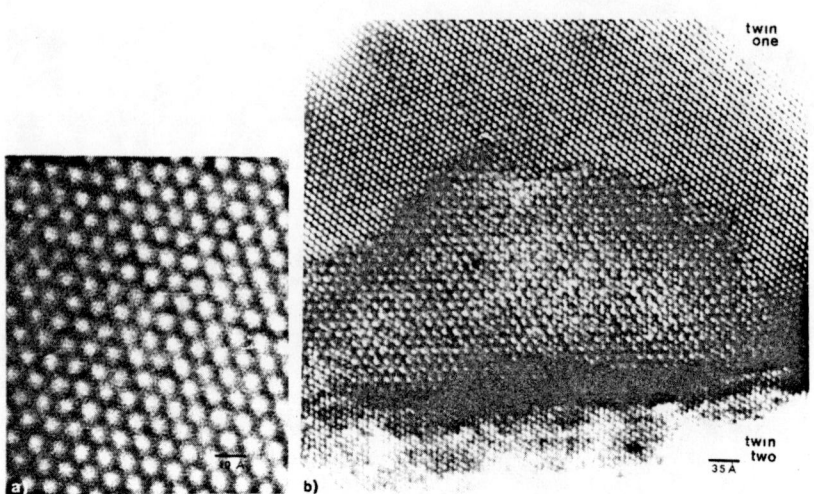

Figure 10. Observed $\langle 111 \rangle_7$ images of Pr_7O_{12} and $Zr_3Sc_4O_{12}$: (a) typical image, (b) two twin orientations of $Zr_3Sc_4O_{12}$ with a region of C-type oxide between

the upper right side. This could be a reasonably thin crystal, e.g., 120–144 Å to correlate with the calculated images. In general the observed images of the iota phase show variations as do the calculated images; however, the correlations are sometimes ambiguous.

Images of Pr_7O_{12} Phases, $\langle 111 \rangle$ Zone. There is good agreement between calculated and observed images in this zone; Figure 10(a) shows the observed image. The spots can be correlated with columns of six-coordinated cations in the direction of the beam. In Figure 10(b) one may observe two regions of $Zr_3Sc_4O_{12}$ in twin orientation with an overlay of M_2O_3 (C-type) at the twin boundary. The $[111]_F$ zone axis is common to all three regions.

Discussion of the Iota Phase Images. Despite its limitations this technique shows great promise in correlating calculated and observed images even if the point-to-point correspondence with the projected potential does not exist for the thick crystals imaged so far. It is difficult in these oxygen labile materials to observe thin crystals without composition change in the vacuum of the microscope under the heating of the electron beam. At the least an unambiguous identification can be made of each phase, and the image, while not intuitively interpretable, can give some help in sorting out the structure since images calculated from wrong structures do not agree well. In the case of passive crystals, thin enough samples could be used to observe images with contrast agreeing with the projected potential.

Images of Beta, Epsilon, and Zeta Phases

Observations. Summerville and Eyring (*35*) have continued high resolution crystal structure imaging of other ordered intermediate phases in an attempt to clarify the structural principle which would elucidate the homologous series in these fluorite-related materials.

The existence of a bifurcated, even-odd, homologous series in the rare earth oxides was clearly indicated in earlier work (*29*). It was important to image members of the even group. Some images of the beta phase ($Pr_{24}O_{44}$) are shown in Figure 11. The strong spots at the corners of a

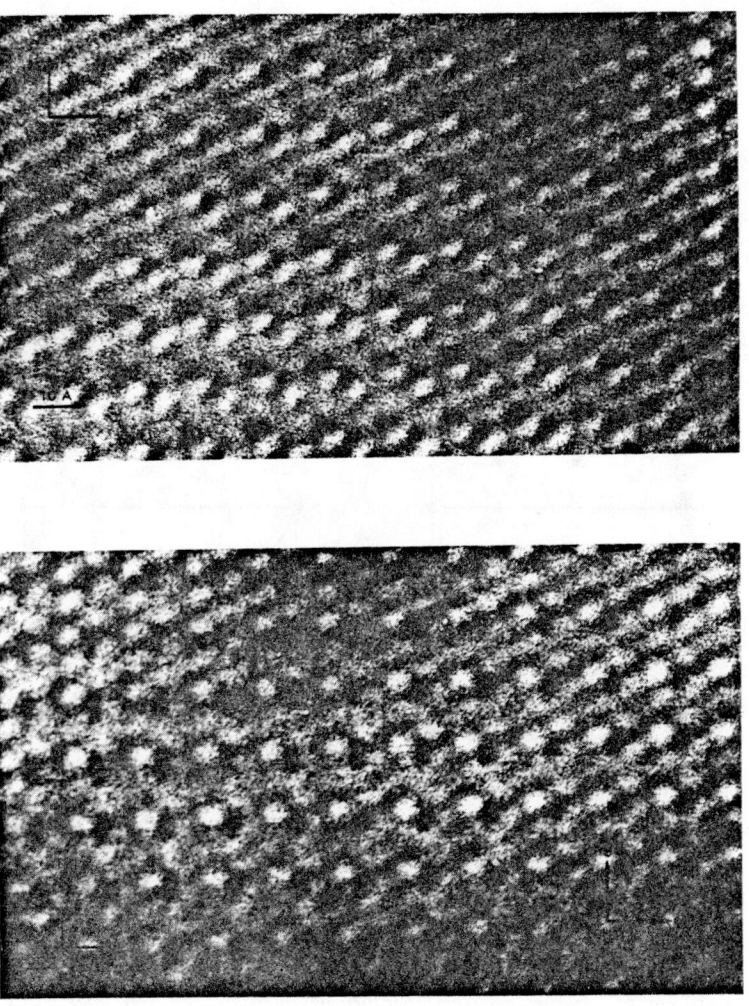

Figure 11. $\langle 100 \rangle_{12}$ *Crystal structure images of beta*

rectangle (nearly a square) correspond to one unit cell in this [$21\bar{1}$] projection (Figure 3). There is considerable contrast within the unit cell.

If the triclinic unit cell with $n = 12$ shown in projection in Figure 3 were twinned along $(110)_F$, the unit cell actually observed would result. Such a unit cell with expected vacancies is shown in Figure 12 in projection.

Images calculated on the basis of this structure, except that the metal atoms about the vacancies were allowed to relax just as they do in iota phase (32), are shown in Figure 13 for three thicknesses and a −900-Å defect of focus. The oxygen atoms are not shifted in this calculation except to remove those corresponding to the vacancies.

The calculated images for thin crystals at −900-Å defocus correspond to the projected anion vacancies of the model as they did in the case of iota phase; however, images for thicker crystals are quite different. The observed images of Figure 11 correlate well with those calculated at a defocus of −900-Å and at 162- and 243-Å thickness respectively as shown. All things considered, the agreement is better than should be expected. At least the *P1* structure of Figure 12 is compatible with the observations.

The thermal decomposition temperature of members of the homologous series decreases from iota to beta, which is the last of the series observed in PrO_x except for PrO_2. This decrease of thermal stability

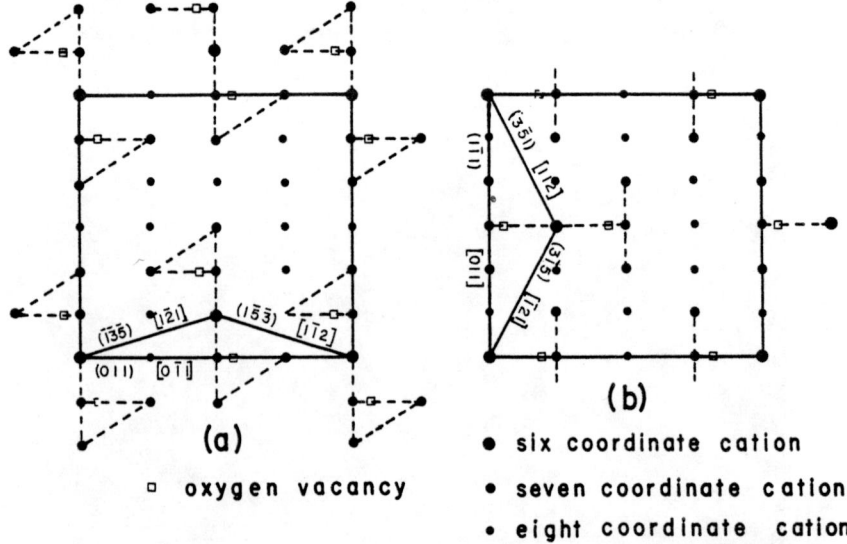

Figure 12. Diagrammatic representations of the proposed structures of the beta phase. (a) Model with symmetry P1. (b) Model with symmetry Pm. Indices refer to fluorite subcell.

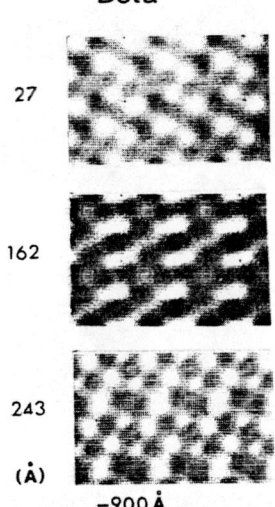

Beta

27

162

243

(Å)

−900 Å

Figure 13. Calculated images of beta phase with the defect of focus and the thickness of the crystal indicated

suggests the possibility of disorder in these phases. Figures 14 and 15 illustrate such disorder. In one case there is a shift in the intensity of certain spots along a definite boundary, and in the other there is a loss of the regular pattern over one unit cell.

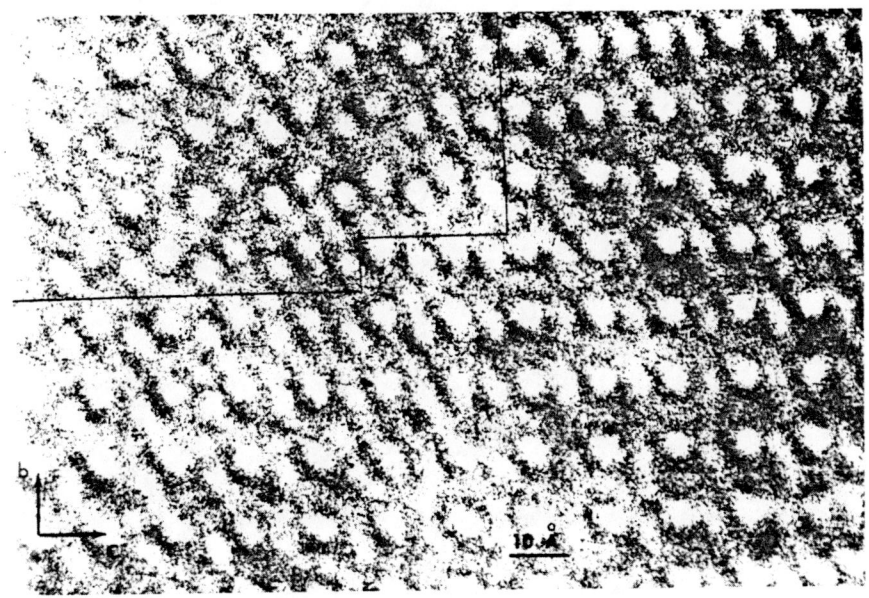

Figure 14. Crystal structure image from $\langle 100 \rangle_{12}$, showing image variations which may correspond to domains of each of the proposed polymorphs of beta

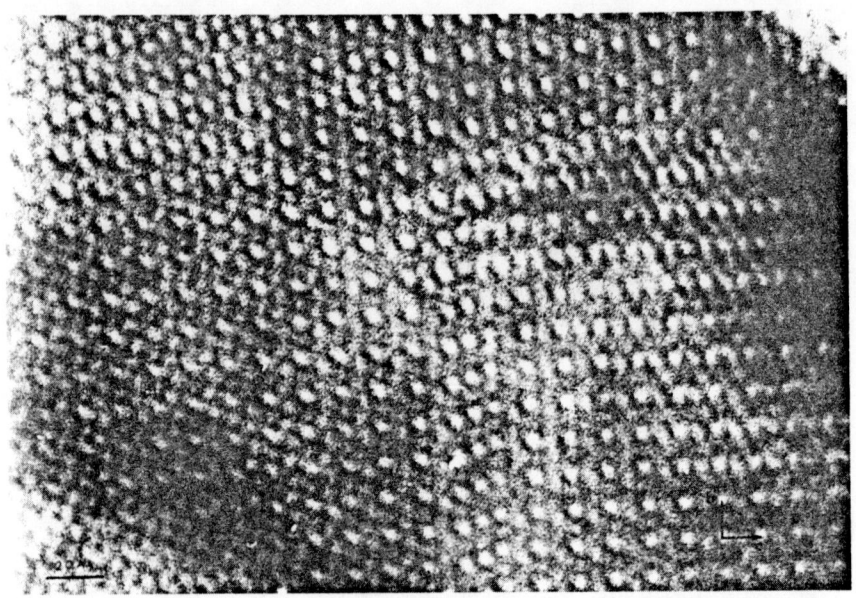

Figure 15.   Crystal structure image of beta from a $\langle 100 \rangle_{12}$ zone showing apparent stacking faults

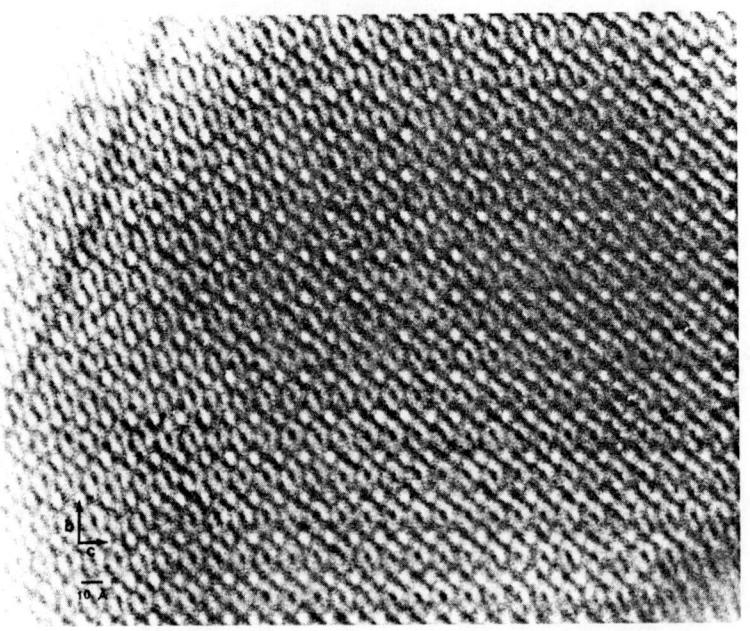

Figure 16.   Crystal structure image from a $\langle 100 \rangle_{10}$ zone of epsilon

An image of the epsilon phase ($Pr_{10}O_{18}$) is shown in Figure 16. The corners of the unit cell are sharp spots with considerable detail in the contrast within the unit cell. Images for this phase are presently being calculated, but the results are not far enough along to merit discussion.

Figure 17 depicts images of zeta phase. At the top of the figure a projection of the bc plane is observed to be fairly regularly spaced but with obvious fluctuation of the intensities. An intergrowth of iota and zeta with a repeat distance of $b_7 + b_9$ is clearly discerned at the bottom but in this case with the ab plane projected. The zeta unit cells have two spots per unit cell, whereas they are not resolved in iota. Calculations have not been made for zeta, but since two spots per unit cell are seen in the images of the delta phase (Figure 5), it may not be surprising to see them in zeta.

Discussion of the Observed Results. The interpretations as set forward here (32) extend the structural model of Kunzmann and Eyring

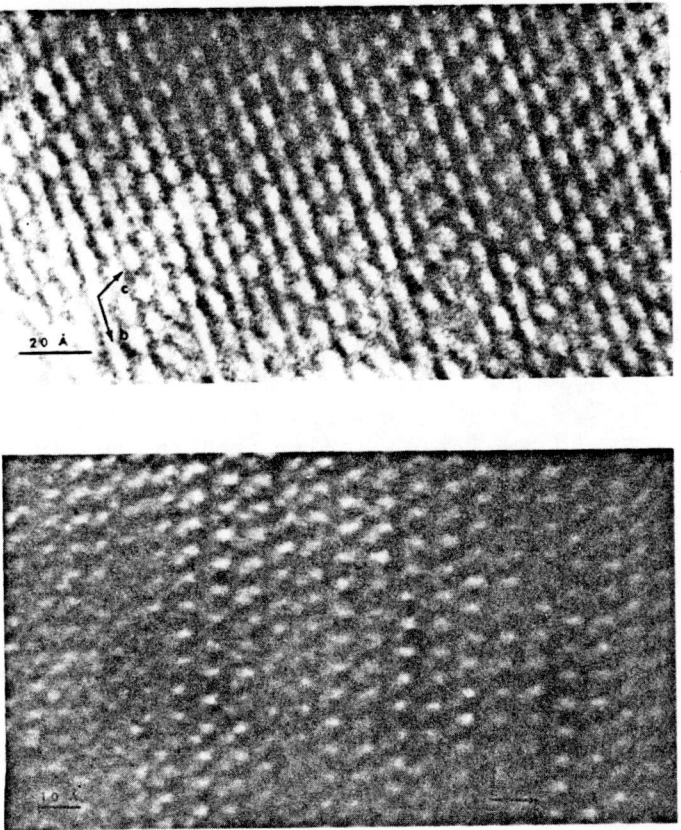

Figure 17. $\langle 211 \rangle_F$ crystal structure image of the zeta phase

(29) and are consistent with the view that in the odd members of the series every nth $(\overline{1}35)_F$ plane of cations is six-coordinated, with those in between being all seven- or eight-coordinated. The oxygen vacancies on each six-coordinated cation occur along the body diagonal [111] of the coordination cube at an angle of 73° to the $(\overline{1}35)_F$ plane.

In the even members twinning at the unit cell level along $(110)_F$ produces puckered $\{135\}_F$ planes such that their average direction is parallel to $(110)_F$. In the case of beta, for example, there are alternate segments of $(135)_F$ and $(53\overline{1})_F$ planes of six-coordinated cations which make up the b-face of the monoclinic unit cell. These are $(101)_F$ on average.

Images of Phase Transformations and Intergrowth in Fluorite-Related Binary Oxide Systems

When the rare earth oxides which have appreciable oxygen dissociation pressures at moderate temperatures are studied in the vacuum of the microscope under the heating of the electron beam, phase reactions are not only possible, they are inevitable. Therefore, while images of the ordered phases were observed, specimens in the act of transformation were recorded. Summerville and Eyring (32) have observed numerous cases, a few of which will be rehearsed here.

Diffraction Patterns. Streaking along b^* is always observed in the diffraction pattern when disorder is observed in the image. Before the streaking becomes extensive, it is replaced by the diffraction spots of the new phase. It is very common to find diffraction patterns containing diffraction spots from two or more phases in the same zone indicating the topotactic intergrowth of these phases. These observations suggest that during reaction there is disorder along b which is soon replaced by order in this direction but by more than one phase. Then finally the new phase occurs alone. These changes in the homologous series occur with an advancing front parallel to $\{135\}_F$ or to the puckered planes if the reaction occurs between even members.

Images of Systems in Phase Reactions. Conventional domains are rarely seen in images of these materials. As mentioned above and as shown in Figure 18(a), among the homologous series a planar reaction front seems most typical. In this case the zeta phase is decomposing to iota, and alternating layers one unit cell wide suggest the mechanism of reaction. On the other hand, in the Pr_7O_{12}–Pr_2O_3 phase region where the structures are not so closely related, coherent intergrowth results from nucleation and growth in a more conventional way as illustrated in Figures 18(b) and 19. The latter is of a $Zr_xSc_yO_z$ composition of $\sim$ $MO_{1.64}$. The edge of the conventional domains are seen to be quite clean

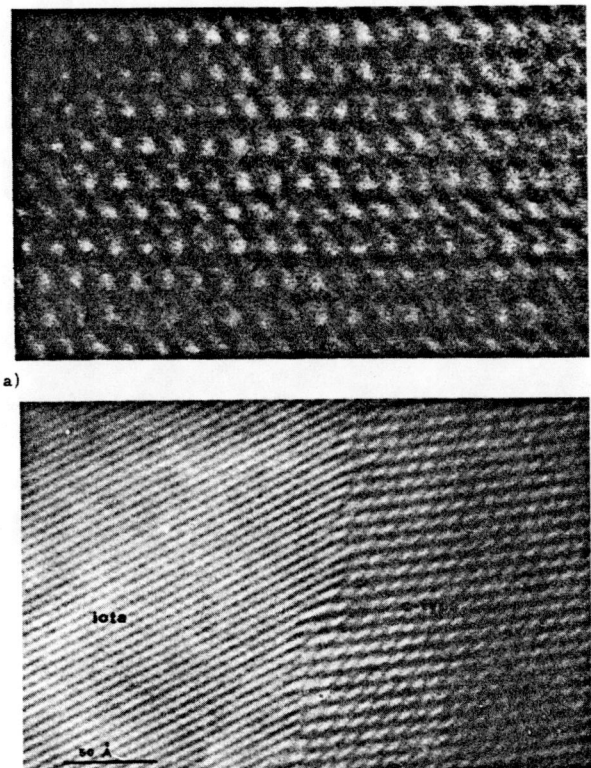

Figure 18. (a) Crystal structure image from $\langle 100 \rangle_9$ showing intergrowth with iota. (b) $\langle 111 \rangle_F$ Crystal structure image showing a domain of Pr_7O_{12} in a crystal which is largely sigma phase.

with little disturbance in the semicoherent interface. In Figure 19 perfect register between the [111] direction of both phases is observed between the $Zr_3Sc_4O_{12}$ substrate and the Sc_2O_3 inclusions of 70Å in diameter.

Disorder in Images. The occurrence of fine scale disorder in images is commonly observed. At equilibrium the Zr^{4+} and Sc^{3+} in $Zr_3Sc_4O_{12}$ are random and the oxygen lattice is regular. Figure 20 shows disorder which may arise from incomplete randomization of the Zr^{4+} and Sc^{3+} in the $Zr_3Sc_4O_{12}$ specimen which leads to disorder in the oxygen substructure. In other cases disorder is apparent on a larger scale, giving almost a superstructure repeat pattern. Figure 21 shows not only this kind of disorder in the Pr_7O_{12} matrix but also epsilon and beta (and perhaps Φ') in a crystal of iota. The periodicity characteristics of the ordered inclusions change with time. Figure 22 shows the relative orientation of the intergrown regions in the specimen of Figure 21.

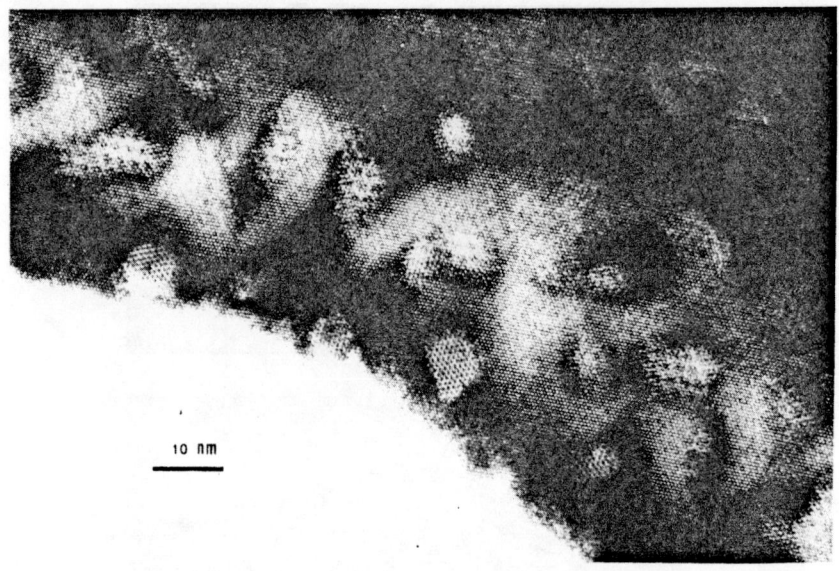

Figure 19. Domains of Sc_2O_3 in a matrix of $Zr_3Sc_4O_{12}$ as seen in $\langle 111 \rangle_F$ images

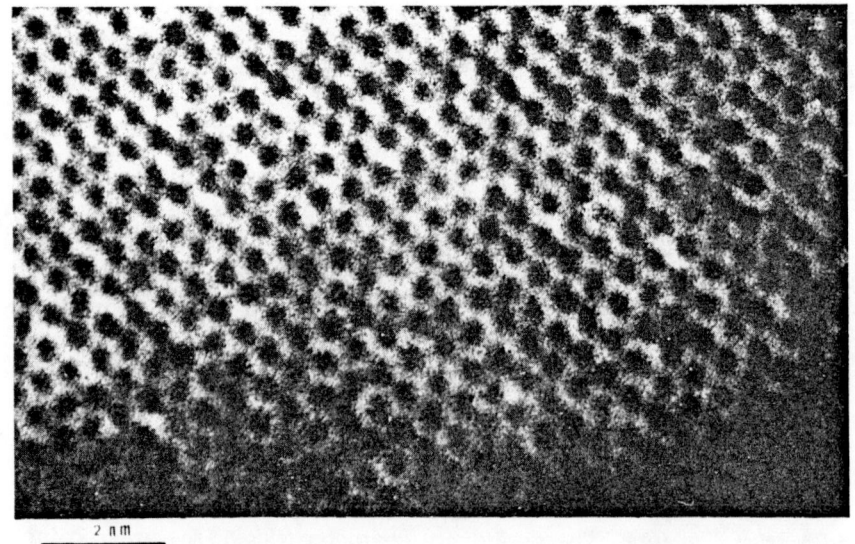

Figure 20. Disorder evident in $\langle 111 \rangle_7$ images of $Zr_3Sc_4O_{12}$

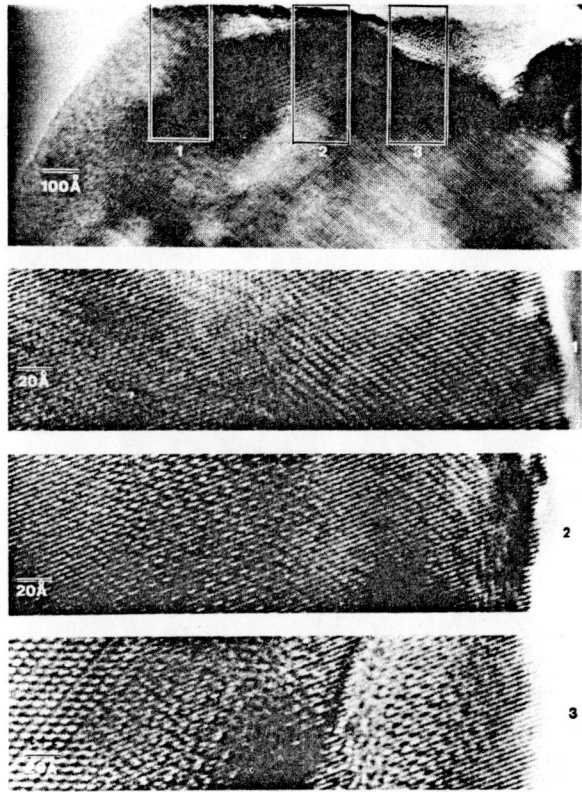

Figure 21. Intergrowth in Pr_nO_{2n-2} $[21\bar{1}]_F$ zone

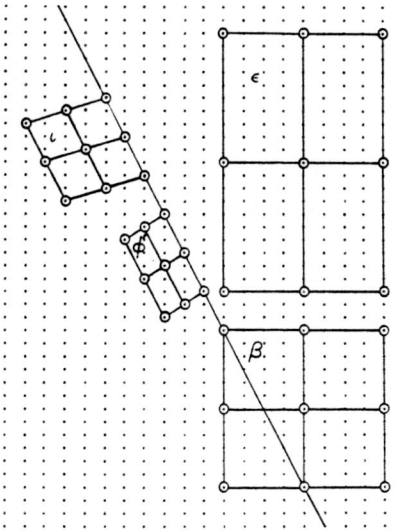

Figure 22. Schematic representation of the intergrowth phases shown in Figure 21

Comments on the Results of Structure Imaging of Fluorite-Related Rare Earth Oxides

(a) Crystal structure images of good resolution have been obtained from crystals of several phases in the homologous series of fluorite-related oxides.

(b) The calculated images of thin crystals correspond to the projected potential of the known structure.

(c) The observed images correlate well with those calculated for thick crystals in which the structure is known.

(d) The thick crystal images do not provide an intuitively interpretable picture, but the correlation of periodicity and symmetry affords unequivocal identification of the phases observed.

(e) Phase reactions can be followed in substantial detail; however, more must be done to model them accurately.

(f) It will be necessary to continue using calculated images to minimize errors in interpretation.

(g) In systems in which thin crystals may be observed intuitive interpretation should be possible even to the point of locating oxygen vacancies in fluorite-related crystals.

General Comments

Fluorite-related materials are important to new energy conversion and storage procedures. Some of these are binary, others ternary, or even more complex. They involve many metal atoms and usually oxygen, fluorine, or hydrogen. Significant improvement in existing materials and the creation of new materials are required as fossil fuels, which have been the conventional sources of energy, are displaced.

Central to this material's evolution is a knowledge of structural details at the unit cell level. Such information is needed to understand their function and failure and to direct the process of generating alternatives.

Although the fluorite structure itself is ubiquitous and simple, almost all usable materials having structures related to it have extended defects. A review of these ordered and disordered phases has been sketched, revealing a multitude of phases of fine scale compositional variability with undoubted short-range order and, in addition, a multitude of ordered intermediate phases which are often compositionally members of an homologous series M_nO_{2n-2}.

Many details of the homologous series in the rare earth oxide system have been discussed—principally those aspects which have been discovered by means of high resolution electron optical techniques. This has included the determination of the unit cells of the homologues. The use of calculations to give confidence to image interpretation has been demonstrated. The images can then be used to show reaction patterns and disorder in phase reactions.

Finally recent results on the Ce_nO_{2n-2} series were presented to illustrate that even in this closely related oxide system not all the homologues have the same structure as those of PrO_x or TbO_x.

Much remains to be done before a clear understanding of the manifold of structures which are called fluorite-related is evoked, but the quest will be worth the effort.

Acknowledgment

The current support of the U.S. Energy Research and Development Administration (and earlier support of the U.S. Atomic Energy Commission) for that part of the work described here and attributed to the author and his co-workers is gratefully acknowledged.

Literature Cited

1. Duclot, M., Vicat, J., Deporter, C., *J. Solid State Chem.* (1970) **2**, 236.
2. Carter, R. E., Roth, W. L., in "Electrochemical Force Measurements in High Temperature Systems," C. B. Alcock, Ed., pp. 125–144, Institution of Mining and Metallurgy, London, 1968.
3. Delamarre, C., *Rev. Int. Hautes Temp. Refract.* (1972) **9**, 209.
4. Michel, D., *Mater. Res. Bull.* (1972) **8**, 943.
5. Allpress, J. G., Rossell, H. J., Scott, H. G., *Mater. Res. Bull.* (1974) **9**, 455.
6. Rossell, H. J., Scott, H. G., *J. Solid State Chem.* (1975) **13**, 345.
7. Allpress, J. G., Rossell, H. J., Scott, H. G., *J. Solid State Chem.* (1975) **14**, 264.
8. Allpress, J. G., Rossell, H. J., *J. Solid State Chem.* (1975) **15**, 68.
9. Lefevre, J., *Ann. Chem.* (1963) **8**, 135.
10. Collongues, R., Queipoux, F., Perez y Jorba, M., Gilles, J. C., *Bull. Soc. Chim. Fr.* (1965) 1141.
11. Thornber, M. R., Bevan, D. J. M., Summerville, E., *J. Solid State Chem.* (1970) **1**, 545.
12. Thornber, M. R., Bevan, D. J. M., Graham, J., *Acta Cryst.* (1968) **B24**, 1183.
13. Thornber, M. R., Bevan, D. J. M., *J. Solid State Chem.* (1970) **1**, 536.
14. Komissarova, L. N., Spiridinov, F. M., *Dokl. Akad. Nauk SSSR* (1968) **182**, 834.
15. Collongues, R., Perez y Yorba, M., Lefevre, J., *Bull. Soc. Chim. Fr.* (1961) 70.
16. Sibieude, F., Foëx, M., *J. Nucl. Mater.* (1975) **56**, 229.
17. Weitzel, H., Keller, C., *J. Solid State Chem.* (1975) **13**, 136.
18. Bartram, S. F., *Inorg. Chem.* (1965) **5**, 749.
19. Eyring, L., in "Solid State Chemistry," C. N. R. Rao, Ed., pp. 565–634, Marcel Dekker, New York, 1974.
20. Sørensen, *Proc. Int. Conf. Plutonium and Other Actinides*, 5th, Sept. 1975, Baden-Baden, Germany
21. Sørensen, O. T., *J. Solid State Chem.* (1976) **18**, 217.
22. O'Keeffe, M., in "Fast Ion Transport," W. van Gool, Ed., pp. 233–247, North Holland, 1973.

23. Chikalla, T. D., Turcotte, R. P., *Natl. Bur. Stand. (U.S.) Spec. Publ.* **364** (1972) 319.
24. Turcotte, R. P., Haire, R. G., private communication.
25. Turcotte, R. P., Chikalla, T. D., Eyring, L., *J. Inorg. Nucl. Chem.* (1971) **33**, 3749.
26. McCarthy, G. J., Fischer, R. D., Johnson, G. G., Jr., Gooden, C. E., *in Natl. Bur. Stand. Spec. Publ.* **364** (1972) 397.
27. Ray, S. P., Cox, D. E., *J. Solid State Chem.* (1975) **15**, 333.
28. Ray, S. P., Nowick, A. S., Cox, D. E., *J. Solid State Chem.* (1975) **15**, 344.
29. Kunzmann, P., Eyring, L., *J. Solid State Chem.* (1975) **14**, 229.
30. Height, T. M., Bevan, D. J. M., private communication.
31. Iijima, S., *Acta Cryst.* (1973) **A29**, 18.
32. Skarnulis, A. J., Summerville, E., Eyring, L., to be published.
33. Von Dreele, R. B., Eyring, L., Bowman, A. L., Yarnell, J. L., *Acta Cryst.* (1975) **B31**, 971.
34. Cowley, J. M., Moodie, A. F., *Acta Cryst.* (1957) **10**, 609.
35. Summerville, E., Eyring, L., to be published.

RECEIVED July 27, 1976.

15

Solid Metal Hydrides: Properties Relating to Their Application in Solar Heating and Cooling

G. G. LIBOWITZ and Z. BLANK[1]

Materials Research Center, Allied Chemical Corp., Morristown, N. J. 07960

Concepts for using solid metal hydrides for solar heating and cooling are described. In solar heating the enthalpy of formation of the metal hydride provides a means of storing solar thermal energy, while in cooling the endothermic dissociation of the hydride is used. The properties of metal hydrides required for these applications are reviewed, the most important properties being large enthalpies of formation (but relatively low thermal stabilities) and high hydrogen-to-metal ratios. There are two approaches to developing new hydrides to meet these requirements: (a) modifying the properties of known hydrides—examples based on the thermodynamics of solids are discussed in some detail; and (b) synthesizing new intermetallic-compound hydrides.

The metal hydrides under consideration form by direct combination of a transition metal or alloy with hydrogen as follows:

$$M + \frac{x}{2} H_2 \rightleftarrows MH_x \qquad (1)$$

The formation of hydride MH_x is usually a spontaneous exothermic reaction which can be reversed easily by applying heat. The hydrogen densities in these metal hydrides are extremely high (e.g., the number of hydrogen atoms per cm^3 is greater than in liquid hydrogen) (1). For this reason, and also because of the ease of reversibility of Equation 1,

---

[1] Current address: Corporate R&D Laboratories, The Singer Co., Fairfield, N. J. 07006.

these materials are being widely investigated (2) as a storage medium for hydrogen in its possible use as a fuel (3). However, the applications of metal hydrides proposed in this paper use primarily the relatively high enthalpies of formation of metal hydrides rather than their hydrogen storage capability.

Thermal Storage

Figure 1a illustrates the concept proposed for solar heating (4, 5), whereby metal hydrides are used for storage of thermal energy. A metal hydride contained in a reservoir in the basement of a house is heated by solar energy via a heat transfer medium such as water or air. The hydrogen released by the reverse of Equation 1 is transferred to a large storage reservoir. Because many metal hydrides have hydrogen dissociation pressures in the range of tens of atmospheres (at temperatures obtainable from solar heat), the hydrogen gas can be compressed in the storage tank without employing auxiliary compressors. Heat is recovered by allowing the stored hydrogen to flow back to the de-hydrided metal (which is now under low pressure because it is not being heated). The heat evolved by Equation 1 (forward reaction) is used for hot water and space heating.

An alternative, illustrated in Figure 1b, would be to store the hydrogen in a secondary less stable hydride. This configuration has the advan-

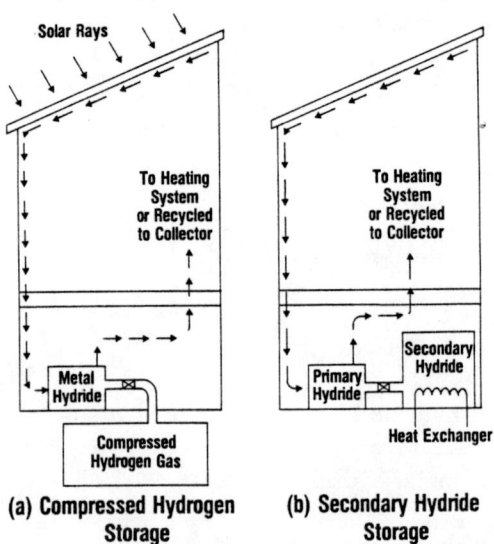

Figure 1. Systems for storing solar thermal energy using metal hydrides

Table I. Thermodynamic Properties of Hydrides for Solar Thermal Storage

Hydride System	Enthalpy (kJ/mol H_2)	Dissociation Pressure (atm) 367°K	Dissociation Pressure (atm) 294°K	Thermal Storage Capacity (J/g)
$VH_{0.95} \rightleftarrows VH_2$	−40	41	1.5	397
$FeTiH_{0.1} \rightleftarrows FeTiH_{1.0}$	−28	34	3.5	120

tage of requiring considerably less volume to store the hydrogen. Also, because of the endothermic nature of the dissociation reaction, less hydrogen would be evolved if a leak developed, thus increasing the safety factor.

With a secondary storage hydride the equilibrium dissociation pressure of the primary hydride at elevated temperatures (during solar heating) should be higher than that of the secondary hydride at room temperature to permit spontaneous flow of hydrogen gas from the primary to the secondary hydride. Conversely, for the same reason it would be desirable for the equilibrium dissociation pressure of the secondary hydride to be higher than that of the primary hydride at room temperature. This is illustrated in Table I which lists the properties of two metal hydrides that can be used for thermal storage of solar energy. Vanadium dihydride which has the higher thermal storage capacity and enthalpy of formation could be used as the primary hydride, while iron–titanium hydride, which is less stable, can be used as the secondary storage hydride. The solar heated VH_2, which dissociates to the monohydride VH, has a much higher dissociation pressure (6) (39 atm at 367°K) than iron–titanium hydride (7) (3.5 atm) at room temperature (294°K). However, the dissociation pressure of VH_2 at room temperature (1.5 atm) is less than that of FeTiH (3.5 atm). The relative pressure–temperature relationships will be discussed further in the section on solar cooling.

Solar Cooling

In the application of metal hydrides for solar cooling the endothermic dissociation of the hydride (reverse of Equation 1) is used. The concept is illustrated in Figure 2. Warm ($\geq 27°C$) air to be cooled is passed over a heat exchanger containing a hydride (Hydride I) which has a relatively high dissociation pressure ($\sim 10–20$ atm) at about 27°C. The dissociating hydride removes heat from the air, and the cooled air is ejected into the house. The hydrogen evolved from Hydride I is absorbed by an alloy which forms a more stable hydride (Hydride

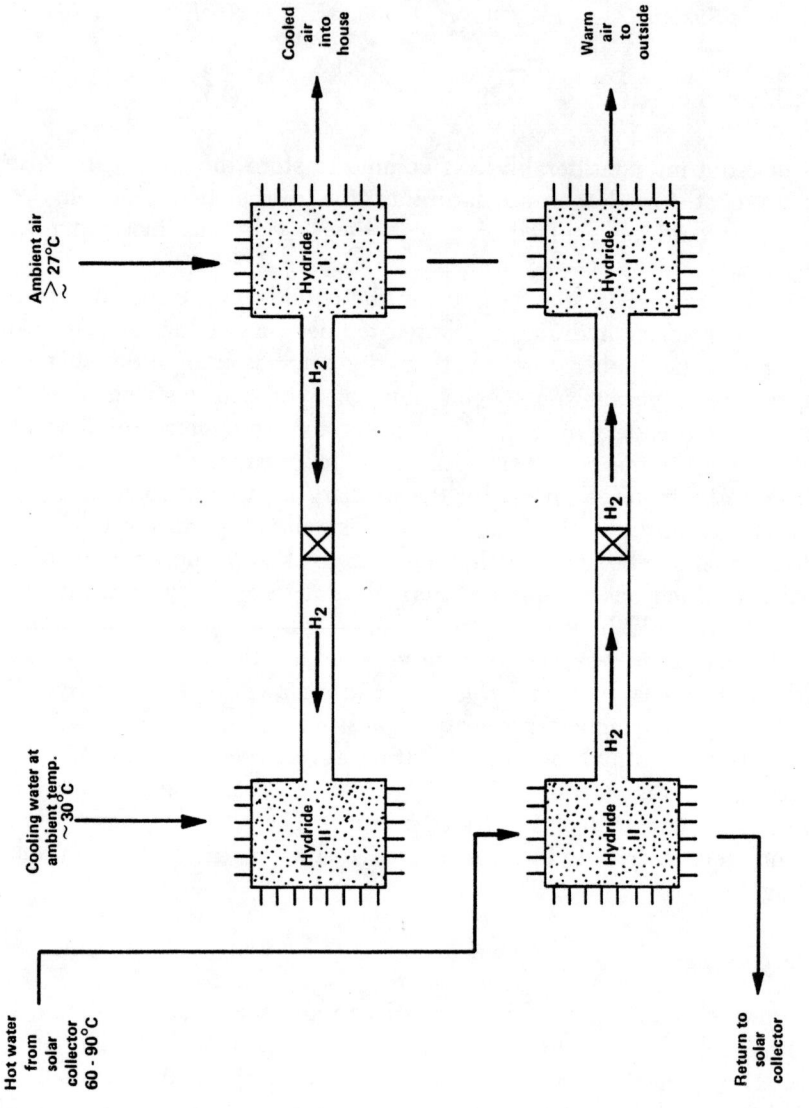

Figure 2. Solar cooling with metal hydrides

II) which is cooled by water at ambient temperature. The rate of cooling is determined by the rate of dissociation of Hydride I which is controlled by the valve shown.

Two such units operate simultaneously. While one is in a cooling cycle (described above), the second unit is in a charging cycle, whereby Hydride II is heated by solar energy so that its dissociation pressure becomes higher than that of Hydride I. The hydrogen is then re-absorbed by the metal or alloy of Hydride I, and the heat of this reaction can be ejected to the outside.

The general pressure–temperature relationships of the two hydrides are illustrated in Figure 3. Point A represents the temperature and pressure (P_I) of operation of the cooling hydride (I), and point B shows the corresponding pressure (P_{II}) of Hydride II at the same temperature. Since $P_I > P_{II}$, the secondary alloy absorbs the hydrogen evolved from Hydride I to form Hydride II. The heat of reaction tends to raise the temperature of Hydride II, although it is cooled by water at the ambient temperature. However, as seen in Figure 3, the temperature may attain a value corresponding to point B' before hydrogen absorption ceases because $P_{II} = P_I$.

In the re-charge cycle Hydride II is heated to a temperature corresponding to point C (60°–90°C), and hydrogen flows from Hydride II back to the metal of Hydride I, provided Hydride I does not reach a temperature corresponding to point D. An efficient heat exchanger

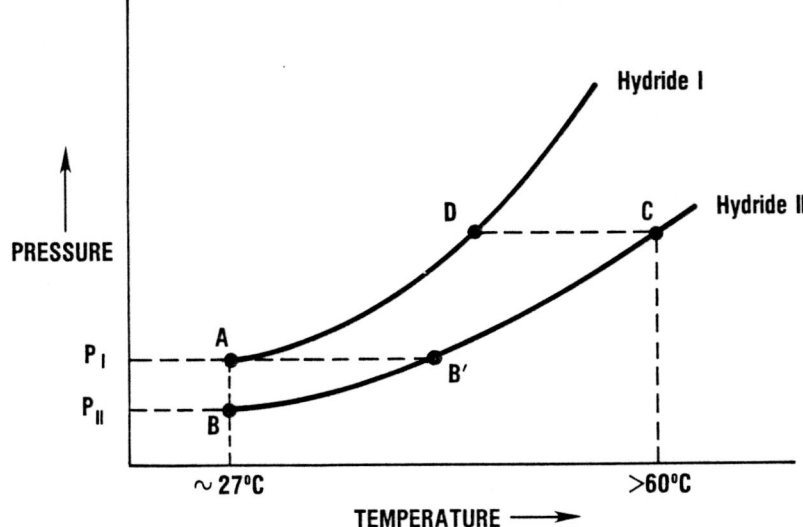

Figure 3. Pressure–temperature relationships for metal hydrides used in solar cooling

which is sufficiently cooled by the ambient air will prevent this from happening.

In many metal-hydrogen systems there is an hysteresis effect (8) in the absorption and desorption of hydrogen. In such cases the absorption pressure is always higher than the desorption pressure. Figure 3 is drawn under the assumption that there is no hysteresis effect in either hydride. However, if a particular hydride exhibits hysteresis, two pressure–temperature curves must be drawn, one representing absorption and the other desorption, four curves in all if both hydrides exhibit hysteresis. Points B (B′) and D would then be on the absorption curves, and points A and C on the desorption curves.

Since the slopes of the pressure–temperature curves are a function of the enthalpy of formation ΔH_f of the hydride, the curves must cross at some temperature (unless the ΔH_f values of the two hydrides are identical). In the case of FeTiH and VH_2 they cross at about 79°C as illustrated in Figure 4. In this case FeTiH is the primary hydride, Hydride I, and VH_2 the secondary hydride, Hydride II. At 27°C the dissociation pressure of FeTiH is about twice that of VH_2 (points A and

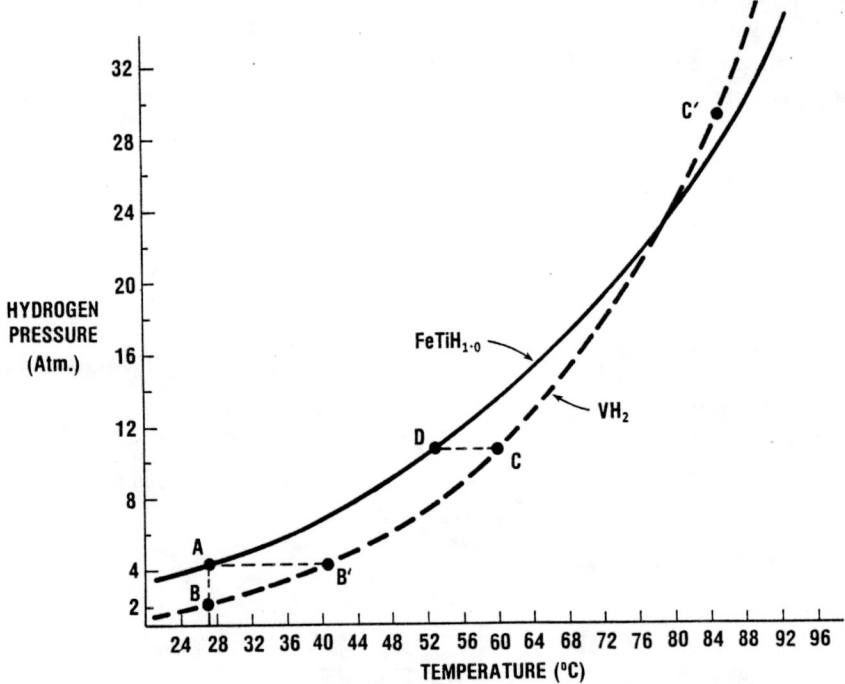

Figure 4. Pressure–temperature relationships for vanadium dihydride and iron titanium hydride under conditions used in solar cooling

B). As the vanadium absorbs hydrogen, its temperature can reach a value of about 41°C (point B') before it would stop absorbing hydrogen.

In the recharge cycle, if the VH_2 is heated to 60°C by solar energy (point C), the iron-titanium hydride must be kept below 53°C (point D) for recharging to occur. If the VH_2 is heated to 85°C (point C'), because of the cross-over of the curves, the iron-titanium hydride can reach an even higher temperature (88°C) before recharging ceases.

The curves in Figure 4 are desorption curves. However, because of hysteresis in the FeTi-H system (7), point D should actually be on the absorption curve (9) of this system; this would shift point D to the left in Figure 4 to a value of 31°C rather than 53°C. Consequently the vanadium dihydride should be heated to a higher temperature, e.g., for point C' at 85°C absorption of hydrogen in FeTiH (point D) would occur at 59°C.

Properties of Metal Hydrides

Although two specific hydrides were mentioned above, at the present time there are no known hydrides whose properties are ideally suited for solar heating and cooling. To find new hydride systems it is necessary to understand the fundamental solid state chemistry of transition metal hydrides, including the nature of the chemical bonding, the electronic and crystal structures, and thermodynamic and transport properties.

Table II. Properties of Metal Hydrides for Solar Heating and Cooling

1. Enthalpy of formation
2. High hydrogen-to-metal ratio
3. Good thermal conductivity
4. Rapid rates of formation and dissociation
5. Stability towards oxygen and moisture

Table II summarizes some of the properties which would be important in utilizing metal hydrides in the systems discussed above for solar heating and cooling. Since it is the enthalpy of formation (or dissociation) ΔH_f of metal hydrides which will be primarily utilized in this application, this property must be considered the most important. A thermal figure of merit, M_{th}, may be defined (5) as follows:

$$M_{th} = \frac{x}{2} \frac{\Delta H_f}{MW}$$

where x is the hydrogen-to-metal ratio of the hydride as shown in Equation 1, and MW is the molecular weight. Thus a high value of x is as desirable as a high absolute value of ΔH_f. For the case of the secondary or storage hydride, the hydrogen-to-metal ratio is most important, while it is desirable to have a low absolute value of ΔH_f.

For efficient heat transfer the thermal conductivities of the metal hydrides should be high. This is generally true for transition metal hydrides because of their metallic bonding (10).

The kinetics of hydride formation and dissociation must be considered because the rate of heat recovery or cooling will depend upon these factors. Therefore, the diffusion rates of hydrogen in the metal (or alloy), and in some cases in the hydride phase, are important. However, the available surface area of metal or hydride may be of greater importance. Because the transition metals expand on hydride formation, the corresponding cracking and spalling which occur increase the rate of hydride formation. On the other hand, most alkali and alkaline earth metals contract upon forming the hydride, and this has a detrimental effect on the rate of reaction (11).

Although the hydrides would be contained in a closed system as illustrated in Figures 1 and 2, it is desirable that they be stable with respect to oxygen and moisture because of the possibility of leaks. Most metal hydrides oxidize easily, particularly at elevated temperatures. In some cases, however, oxidation is not complete. For example, in the case of FeTi, the surface of the alloy becomes coated with a thin oxygen-rich film (12) which blocks hydrogen absorption. However, the alloy can be re-activated by heating in hydrogen gas. It has also been reported that some alloys will absorb hydrogen in the presence of oxygen or H_2O (13).

Another requirement of metal hydrides for these applications is that the cost of the corresponding metal or alloy be relatively low. This is the major disadvantage of vanadium hydride. The disadvantage of high cost can be partially overcome by finding alloy hydrides which have high values of ΔH_f and high hydrogen-to-metal ratios, so that M_{th} is increased and less alloy is required.

New Alloy Hydrides

There are two approaches which can be taken in the development of new hydrides: (a) the properties of known hydrides can be modified by appropriate alloying, and (b) new intermetallic compounds which form hydrides with required properties can be synthesized.

Property Modification by Alloying. An example of the first approach is the modification of thermodynamic properties. Only the properties of the primary hydrides in the applications proposed above are considered since the properties required of the secondary or storage hydride are the same as those needed when storing hydrogn as a fuel, and this latter application has been amply described elsewhere (2, 14).

As mentioned above, for solar heating and cooling the primary hydride must have a high absolute enthalpy of formation, which is an indication of the bond strength in a compound. Consequently, if a particular alloy hydride has satisfactory properties in other respects, it may be possible to increase ΔH_f by appropriate alloying. An example of this is the recent work of Van Mal et al. (*13*) on the intermetallic-compound hydride $LaNi_5H_7$. These authors have proposed a "Rule of Reversed Stability" which can be stated as follows: the less stable an intermetallic compound, the greater the stability of the corresponding hydride. Since the dissociation pressure of a hydride is a measure of its stability, a decrease in dissociation pressure will indicate a more stable hydride and therefore an increase in the absolute value of the enthalpy of formation as shown by the van't Hoff relation for Equation 1:

$$\ln P_H = \frac{2}{x} [(\Delta H_f/RT) - \Delta S/R] \qquad (2)$$

Since Equation 1 is an exothermic reaction, ΔH_f is negative, so that an increase in absolute value of ΔH_f will result in a decrease in hydrogen dissociation pressure P_H, according to Equation 2.

Van Mal and co-workers (*15*) added the alloying elements cobalt, iron, and chromium to the intermetallic compound $LaNi_5$ by substituting them for 20% of the nickel present. According to the enthalpies of formation ΔH_{IC} of the intermetallic compounds shown in Table III, the substitution of Co, Fe, and Cr for Ni in $LaNi_5$ should decrease the stability of the intermetallic compound $LaNi_5$ in the order shown, i.e., Cr should have the greatest effect since $LaCr_5$ is the least stable (high positive ΔH_{IC}) of these compounds, with Fe and Co having a lesser effect. Since the stability of the intermetallic compound is decreased by alloying, the stabilities of the corresponding hydrides should be increased. This would be indicated by a decrease in dissociation pressure P_H. The results of the dissociation pressure measurements (*15*) are shown by the 40°C pressure–composition isotherms in Figure 5. The constant pressure

Table III. Calculated Enthalpies of Formation of Some Intermetallic Compounds (*17*)

Intermetallic Compound	ΔH_{IC} (kcal/mol)
$LaNi_5$	−40
$LaCo_5$	−17
$LaFe_5$	+ 4
$LaCr_5$	+12

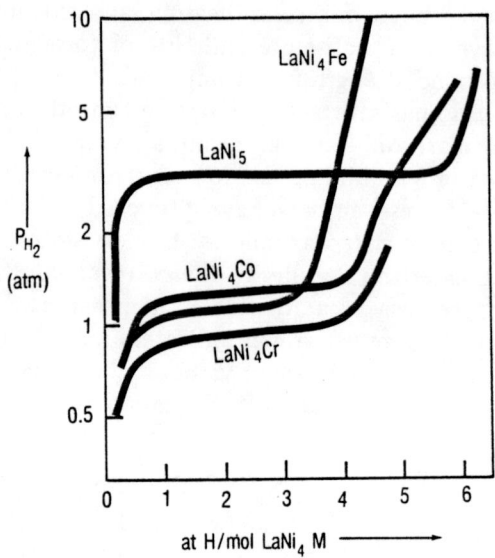

Figure 5. Effect of alloying on the stability of lanthanum pentanickel hydride (adapted from Ref. 15)

plateaus (two phase regions of the phase diagram) represent the dissociation pressures of the hydride (16). The hydrides of the Co, Fe, and Cr substituted alloys are more stable than the hydride of $LaNi_5$ in the order expected.

Although it is desirable to increase the absolute value of ΔH_f as much as possible for the primary hydride, as pointed out above, the dissociation pressure also is decreased, and this may result in an equilibrium hydrogen pressure which is less than that of the secondary hydride under the conditions of operation. Furthermore, if the hydrogen pressure is too low, the rate of mass transfer of hydrogen from one part of the cooling or heating system to another becomes too low for effective operation.

It can be seen from Equation 2 that the detrimental effect of large values of ΔH_f on hydrogen pressure may be partially offset by decreasing the entropy change of the hydride formation reaction. This can be accomplished by decreasing the vibrational entropy of the hydride phase. A lower vibrational entropy is associated with a higher vibrational frequency and stronger bonding. Hence, alloying elements which increase ΔH_f by strengthening the chemical bonds will also tend to have a favorable effect on the entropy change.

Another possibility for decreasing ΔS of Equation 1 is to increase the entropy of the alloy phase by forming a random alloy. A possible illustration of this effect can be seen in an investigation of the hydrides of Pd$_3$Fe by Flanagan and co-workers (18) in which it was observed that at 25°C the hydrogen pressure in equilibrium with the disordered alloy was orders of magnitude higher than the pressure in equilibrium with the ordered alloy (cubic Cu$_3$Au structure). The authors explained the greater stability of the ordered hydride by the existence of a larger number of Pd–H bonds than in the disordered phase (assuming the hydrogen atoms enter the body-centered site of the unit cell). However, another contributing factor to the higher equilibrium pressure of the disordered phase hydride may have been the larger configurational entropy of the disordered alloy.

Synthesis of New Intermetallic Compounds. The second approach to developing new hydrides for solar heating and cooling applications is to synthesize new intermetallic compounds which would form hydrides meeting the property requirements listed in Table II. In general, intermetallic-compound hydrides appear to bear little or no resemblance to the component metal hydrides. This is illustrated by the properties of the hydrides of the intermetallic compound ZrNi listed in Table IV. Zirconium forms a dihydride, and nickel normally does not form a hydride except under unusual circumstances (10). However, the hydrogen-to-metal ratio in ZrNiH$_3$ is higher than would be expected on the basis of the constituent metals Zr and Ni. The structures (19) of the two hydrides are different. The dissociation pressure (20) of ZrNiH$_3$ is almost nine orders of magnitude higher than that of ZrH$_2$, although the Zr–H distance in ZrNiH$_3$ is less, which usually indicates stronger bonding.

The Rule of Reversed Stability (15) has been proposed (15, 21) as a method of predicting the hydride-forming tendencies of intermetallic compounds. Although this rule appears to be of value in predicting the effect of alloying elements on the properties of known hydrides as discussed above, for reasons presented elsewhere (14) it appears to have very limited applicability in finding new intermetallic compound hydrides. At the present time there is no obvious way of reliably predicting the properties of a hydride of an intermetallic compound from a knowledge of the properties of the constituent metal hydrides. Consequently, an intermetallic-compound hydride should be viewed as a pseudo-binary

Table IV. Comparison of Properties of ZrH$_2$ and ZrNiH$_3$

	ZrH$_2$	ZrNiH$_3$
Structure	Tetragonal (distorted fluorite)	Orthorhombic
Dissoc. press. at 250°C (torr)	4×10^{-9}	200
Zr–H distance (Å)	2.09	1.96

metal hydride with the intermetallic compound being considered a new metal. The behavior of the intermetallic compound towards hydrogen then depends upon its particular crystal structure and electronic structure (22, 23). Thus, to develop new intermetallic-compound hydrides, basic alloy theory must be used, as well as a fundamental knowledge of the nature of the metal–hydrogen bonds in metal hydrides.

Conclusion

A review of the known metal and alloy hydrides reveals that none meets all the requirements listed in Table II at a cost low enough to make the solar heating and cooling concepts described in this paper economically feasible. Therefore, for these concepts to be utilized new alloy hydrides must be discovered. In the discussion on modification of hydride properties by alloying, the variation of thermodynamic properties is considered. However, alloying of known hydrides also may be employed to increase hydrogen-to-metal ratios by changing crystal structures and increasing lattice parameters (14) or by modification of the electronic band structures (22, 23). Furthermore, electronic structure modifications will have an effect on the thermal conductivity of a hydride. Rates of hydride formation and dissociation also have been varied by alloying (24).

Because metal hydride systems permit indefinite heat storage (5) as opposed to other thermal storage materials, they offer a distinct advantage to solar heating and cooling if alloy hydrides can be found which would make this concept economically competitive.

Literature Cited

1. Libowitz, G. G., "The Solid State Chemistry of Binary Metal Hydrides," p. 47, Benjamin, New York, 1965.
2. *World Hydrogen Energy Conf. Proc., 1st*, Univ. of Miami, 1976, Session 8B.
3. Hagenmuller, P., ADVAN. CHEM. SER. (1977) **163**, 1.
4. Libowitz, G. G., *Intersoc. Energy Convers. Eng. Conf., 9th*, 1974, 322–325.
5. Libowitz, G. G., Blank Z., *Intersoc. Energy Convers. Eng. Conf., 11th*, 1976, 673–680.
6. Reilly, J. J., Wiswall, R. H., *Inorg. Chem.* (1970) **9**, 1678.
7. Reilly, J. J., Wiswall, R. H., *Inorg. Chem.* (1974) **13**, 218.
8. Libowitz, G. G., "The Solid State Chemistry of Binary Metal Hydrides," pp. 83–86, Benjamin, New York, 1965.
9. Reilly, J. J., Johnson, J. R., *World Hydrogen Energy Conf. Proc., 1st*, 1976, 8B3–8B26.
10. Mueller, W. M., Blackledge, J. P., Libowitz, G. G., "Metal Hydrides," Academic, New York, 1968.
11. Libowitz, G. G., "The Solid State Chemistry of Binary Metal Hydrides," p. 13, Benjamin, New York, 1965.
12. Sandrock, G. D., *Intersoc. Energy Convers. Eng. Conf. Proc., 11th*, 1976, 967–971.

13. Wiswall, R. H., Reilly, J. J., *Intersoc. Energy Convers. Eng. Conf. Proc.*, 7th, 1972, 1342–1348.
14. Libowitz, G. G., in "Critical Materials Problems in Energy Production," C. Stein, Ed., Academic, 1976.
15. Van Mal, H. H., Buschow, K. H. J., Miedema, A. R., *J. Less-Common Metals* (1974) **35**, 65.
16. Libowitz, G. G., "The Solid State Chemistry of Binary Metal Hydrides," pp. 50–55, Benjamin, New York, 1976.
17. Miedema, A. R., *J. Less-Common Metals* (1973) **32**, 117.
18. Flanagan, T. B., Majchrzak, S., Baranowski, B., *Philos. Mag.* (1972) **25**, 257.
19. Peterson, S. W., Sadana, V. N., Korst, W. L., *J. Phys. (Paris)* (1964) **25**, 451.
20. Libowitz, G. G., Hayes, H. F., Gibb, T. R. P., *J. Phys. Chem.* (1958) **62**, 76.
21. Buschow, K. H. J., Van Mal, H. H., Miedema, A. R., *J. Less-Common Metals* (1975) **42**, 163.
22. Switendick, A. C., *Solid State Commun.* (1970) **8**, 1463.
23. Switendick, A. C., *Int. J. Quant. Chem.* (1971) **5**, 459.
24. Douglass, D. L., *Met. Trans.* (1975) **6A**, 2179.

RECEIVED July 27, 1976.

16

Storage of Hydrogen Isotopes in Intermetallic Compounds

S. A. STEWARD, J. F. LAKNER, and F. URIBE

Lawrence Livermore Laboratory, University of California, Livermore, Calif. 94550

Reaction of $LaCo_5$ under high pressure has produced a hydride with the $LaCo_5H_9$ composition, which is the expected maximum stoichiometry. A comparison of hydrogen solubility in $ErCo_5$ with solubilities of previous studies in $PrCo_5$, $PrCo_3$, and $ErCo_3$ show that hydride stability decreases with lanthanide atomic number and with increasing atom ratio of transition metal to lanthanide metal. Empirical methods for estimating ternary hydride enthalpies and free energies are evaluated and are found inadequate for calculating approximate hydrogen plateau pressures.

The oil embargo imposed in 1973 by the Organization of Petroleum Exporting Countries (OPEC) quickly impressed upon the industrialized countries their dependency on cheap, abundant sources of energy, principally fossil fuels, particularly oil. Although a pre-embargo indifference has settled upon the world again, a large segment of the technical community remains acutely aware of our inefficiencies and lack of flexibility in the amounts and types of fuels consumed. Consequently, the options available over the next several decades have been examined. Important areas for consideration have been conservation, greater efficiency in production and utilization of existing fuels, alternative fuels, and the new technologies needed for their development. This book focuses on advanced energy research such as solid state batteries, catalysis, and the subject of this presentation, hydrogen.

While hydrogen is often considered an alternative fuel, it is a secondary source. It is abundant, clean, and produces water as a combustion product. The hydrogen isotopes will also be used as fuel for fusion reactors, which are expected to be in operation by the end of this century.

Although hydrogen and deuterium are usually produced by electrolysis, thermochemical cycles are also being considered. For nuclear applications tritium is generated by neutron irradiation of ^{6}Li. The generation of large quantities of the gases creates a storage problem, and until recently, high pressure or cryogenic methods of storage were the only solutions. Both methods are costly; cryogenic storage uses considerable energy for liquification, with substantial evaporative losses.

The use of metal hydrides as storage materials has been of increasing interest. Many have greater hydrogen densities (mol H_2/cm^3 material) than liquid hydrogen (1). To be a good storage medium, a hydride should: (a) have in its absorption curve a reasonable two-phase plateau pressure (1–10 atm) at room temperature, (b) reversibly absorb and desorb gas, (c) have a high hydrogen density, and (d) be relatively insensitive to gaseous impurities such as carbon, nitrogen, and oxygen. The hydride should be economical. The most interesting regions of the absorption curves are the plateaus, where the hydrides absorb or release quantities of gas over a considerable composition range at constant pressure.

Prior to 1968 most of the hydrides available were those of the metallic elements. The saline hydrides are quite stable and in some cases very difficult or impossible to prepare by direct combination of the elements. Transition metals of Groups III and IV also form quite stable hydrides (2). Group V metals (vanadium, niobium, and tantalum) dissolve large quantities of hydrogen. Their two-phase (mono- and dihydride) plateau regions are in the 1–10 atm ranges near room temperature. Impurities, especially oxygen, can increase the plateau pressure considerably and inhibit hydride formation at low pressures and compositions.

Except for palladium, reactions of other transition metals with hydrogen are endothermic, with little or no gas dissolved at low temperatures. The lighter rare earth elements have very stable isostructural dihydrides and trihydrides. However, with the exception of europium and ytterbium, the dihydrides of samarium through lutetium change structure as the MH_3 composition is approached. In every case the equilibrium hydrogen pressures are below one atm at temperatures near 500°C. The vanadium, niobium, and tantalum hydrides are the only binary metal hydrides with suitable absorption pressures at temperatures of interest. Their cost and sensitivity to impurities, however, inhibit their use as storage materials.

Intermetallic Compounds

Studies of hydrogen absorption by intermetallic compounds and alloys were reported only intermittently until the early 1960s. Most of the early work dealt with alloys of palladium with platinum, nickel, copper, and especially silver. An alloy of palladium and 30% silver is used as a diffusion membrane in hydrogen purifiers. This alloy is more resistant to deformation than pure palladium when exposed to hydrogen (3). As the silver concentration increases, the hydrogen solubility decreases, but the diffusion constant of the Pd/30% Ag alloy is still about half the value of pure palladium (3).

In 1961–62 the Denver Research Institute published hydrogen absorption measurements for over 300 intermetallic compounds (4). Although many of these measurements have since been refuted and additional compounds having high hydrogen solubility have been found, these studies were the first large-scale investigation of hydrogen interaction with intermetallic compounds.

A literature review of metallic ternary and quaternary hydrides published by Newkirk (5) covers a wide range of alloys and intermetallic compounds that have been examined recently. Several investigators have reported the formation of alkali and alkaline earth ternary hydrides. These are listed in Table I with references and are generally stable in air.

Since 1968, research in this area has concentrated on the family of compounds with AB_5 stoichiometry, where A is a lanthanide and B is a transition metal, usually nickel, cobalt, or iron. The discovery of high hydrogen absorption by such compounds is a classic tale of scientific serendipity (10). Some of these compounds had been investigated for their magnetic properties, especially as permanent magnets. At the Philips Laboratories in The Netherlands, acid etching was used in polishing crystals to decrease the surface effects on coercivity. The investigators postulated that hydrogen produced by the etching process might influence the magnetic properties. The coercivity of $SmCo_5$ was therefore measured while the sample was exposed to hydrogen. Surprisingly, the $SmCo_5$ absorbed large quantities of the gas. This finding initiated several studies of hydrogen absorption by similar AB_5 compounds (11, 12, 13).

Table I. Alkali and Alkaline Earth Ternary Hydrides

Ternary Hydride	Ref.
$LiSrH_3$	6
$LiEuH_3$	7
Ca_2IrH_5	8
Sr_2IrH_5	8
Ca_2RhH_5	8
Sr_2RhH_5	8
Ca_2RuH_6	8
Sr_2RuH_6	8
Eu_2RuH_6	9

In addition to ternary hydrides to be discussed later, recent work has shown hydrogen absorption in several related intermetallic compounds: $TbFe_3$, $DyFe_3$, $HoFe_3$, $ErFe_3$, Tb_2Co_7 (14), and La_2Mg_{17} (15).

Structural Relationships. Intermetallic compounds of lanthanide and transition metals form a very interesting class of structures. The AB_5 series crystallizes in the hexagonal $CaCu_5$ (P6/mmm) type of structure (11) shown in Figure 1. Generally, radius ratios (r_A/r_B) greater than 1.30 form the $CaCu_5$ type configuration, whereas compounds with r_A/r_B less than 1.30 prefer the cubic UNi_5 structure. As compounds are formed by rare earths to the right of lanthanum in the periodic chart, the so-called lanthanide contraction results in a decrease in AB_5 unit cell volume, caused chiefly by a contraction in the basal plane. The AB_5 phase is generally stable over the composition range ($AB_{4.8-5.5}$).

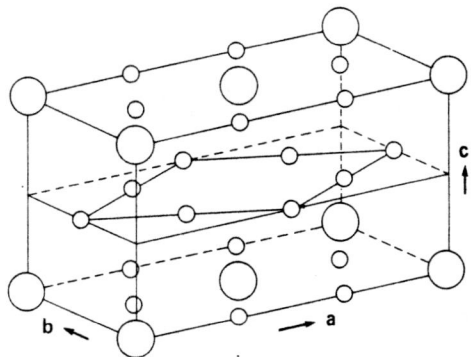

Figure 1. The $CaCu_5$ (AB_5) structure. The large circles represent the Ca (or A) atoms. The small circles represent the Cu (or B) atoms.

These metal groups, including some of the actinides that consist of structures of sequential blocks of $CaCu_5$ and Laves phase-type layers, form other phases (16). Atomic substitution can lead to other structures. Representative structures and their relationships to the Laves and $CaCu_5$ arrangements include:

$ThMn_{12}$ Half of the calcium atoms are replaced by manganese pairs.

Th_2Zn_{17} One third of the calcium atoms are replaced by zinc pairs. Th_2Fe_{17} and Th_2Co_{17} have this structure.

Th_2Ni_{17} Similar to Th_2Zn_{17}, but the location of the replaced calcium atoms is different.

Er_2Co_7 Two double layers of $CaCu_5$ packing and a four-layer Laves sequence form an eight-layer unit

Ce_2Ni_7	Same as Er_2Co_7, but the eight-layer units have the AB hexagonal format.
$NbBe_3(AB_3)$	A group of $CaCu_5$ layers and one of a Laves type form a six-layer unit in an ACB sequence.
$CeNi_3$	Same as $NbBe_3$, but the unit stacking is BC.

stacked in the cubic ABC manner. Many rare earths form this structure.

The hexagonal AB_5 compounds form orthorhombic hydrides. The changes in structure are undoubtedly a result of expansion of the basal plane caused by hydrogen atom occupation of the interstitial sites that lie in the metal layers (17). These asymmetric sites are tetrahedra formed by two lanthanide and two cobalt atoms, and octahedra formed by two lanthanides and four cobalt atoms. There are nine sites per AB_5 formula with two formulas per unit cell. Kuijpers and Loopstra found by neutron diffraction that the deuterium atoms in $PrCo_5D_4$ were ordered on certain available octahedral and tetrahedral sites. The possible interstitial positions in the $CaCu_5$ type structure are shown in Figure 2.

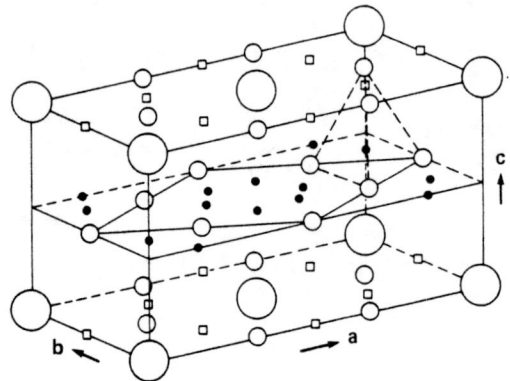

Figure 2. The $CaCu_5$ structure including the asymmetric tetrahedral (●) and octahedral (□) sites

Estimation of Hydride Stability. An empirical method by which the enthalpies of formation of alloys may be estimated quantitatively has been formulated (18, 19, 20). The approach assumes that the driving force for reactions between metals is a function of two factors: a negative one arising from the difference in chemical potential, $\Delta\phi^*$, of electrons associated with each metal atom, and a positive one that is the difference in the electron density, Δn_{ws}, at the boundaries of Wigner–Seitz type cells surrounding each atom. Values of ϕ^* for the metals are approximated by the electronic work functions; n_{ws} is estimated from compressibility data.

The atomic concentrations in the alloy must be included in the calculation. The most recent formulas for performing these calculations (20) are:

$$\Delta H = N_0 f(C_A{}^s, C_B{}^s) g(C_A, C_B) [-Pe(\Delta\phi^*)^2 + Q_0(\Delta n_{ws})^{2/3}] \quad (1)$$

$$f(C_A{}^s, C_B{}^s) = C_A{}^s C_B{}^s [1 + 8(C_A{}^s C_B{}^s)^2] \quad (2)$$

$$g(C_A, C_B) = 2(C_A V_A{}^{2/3} + C_B V_B{}^{2/3})/(V_A{}^{2/3} + V_B{}^{2/3}) \quad (3)$$

$$C_A{}^s = C_A V_A{}^{2/3}/(C_A V_A{}^{2/3} + C_B V_B{}^{2/3}) \quad (4)$$

$$C_B{}^s = C_B V_B{}^{2/3}/(C_A V_A{}^{2/3} + C_B V_B{}^{2/3}), \quad (5)$$

where $C_A{}^s, C_B{}^s$ = surface concentrations of each component, C_A, C_B = atomic concentrations for each component, V_A, V_B = molar volumes, P, Q_0 = constants that vary between systems of alloys; i.e., liquid alloys, transition-p-metal alloys, etc., e = electronic charge, and N_0 = Avogadro's number.

Values for ϕ^*, n_{ws}, and V_m are given elsewhere (20), as are the constants P and Q_0 for each of the systems studied. When an earlier form of the equation was used, a qualitative analysis of the sign of enthalpies of formation for binary hydrides was made (19). Values for ternary hydrides can also be predicted (21) since

$$\Delta H(AB_x H_y) = \Delta H(AH_{y-a}) + \Delta H(BH_a) - \Delta H(AB_x) \quad (6)$$

This relationship would be strictly true if the exact composition of the binary hydrides forming the ternary hydride were known. In practice, they are unknown, and selection of the appropriate binary hydride composition is somewhat arbitrary. These calculations require values for ϕ^* and n_{ws} for hydrogen. So that the model may be applied to binary hydrides, properties of metallic hydrogen were estimated.

Application of Equation 6 illustrates the rule of "reversed stability," i.e., the greater the intermetallic compound stability, the lower the ternary hydride stability, which results in a larger hydrogen overpressure. In a qualitative sense this rule is correct, but, as we shall show, quantitative calculations using Equations 1 and 6 are very unreliable.

The exact nature of the calculated ΔH is not clear. In metal–hydrogen systems three ΔH's are present: (a) heat of solution of hydrogen in metal, ΔH_s, (b) the enthalpy for the phase change in the plateau region (22), ΔH_p, and (c) a second heat of solution for hydrogen in the hydride phase, ΔH_h. These should be additive, yielding the enthalpy of formation of the stoichiometric hydride. When the plateau region is very wide, i.e., small solution regions, enthalpy for the derived phase change approximates the enthalpy of formation of the hydride. For this reason it has been assumed previously, as well as in this paper, that the empirically calculated enthalpies represent the phase changes and are, therefore, related to the equilibrium hydrogen pressures in the plateau regions. The values are usually listed in kilocalories per mole of hydrogen gas.

The development of some means of predicting the potential stability of ternary hydrides is of interest. We have, therefore, utilized the calculated ΔH's discussed previously to estimate plateau pressures of several compounds for which data are available.

Experimental enthalpy and entropy values in the plateau regions may be derived from van't Hoff plots using

$$\ln P_{eq} = \frac{\Delta H}{RT} - \frac{\Delta S}{R} \qquad (7)$$

Tabulations of entropies for most binary hydrides allowed us to estimate entropy values according to hydrogen-to-metal ratio. These are given in Table II.

Since heats of formation of most binary hydrides have been determined experimentally, these should be used in Equation 6 with calculations for intermetallic compounds. Once a semiempirical value for the

Table III. Calculated Free Energies of Formation Comparison with

Hydride	ΔG/mol Hydride, Exposed at Room Temperature (kcal)	Pressure (atm)	ΔH/mol H_2, Calculated (kcal)
ScH_2	−37.6	10^{-28}	−38.3
YH_2	−43.9	10^{-32}	−31.8
YH_3	−18.1	10^{-14}	−24.8
TiH_2	−23.1	10^{-17}	−32.5
ZrH_2	−13.6	10^{-10}	−42.8
$VH_{0.5}$	−2.4	0.2	−12.4
$NbH_{0.5}$	−3.0	6×10^{-3}	−19.2
ThH_2	(−26)	10^{-19}	−35.5
UH_3	−17.5	10^{-13}	−14.5
LiH	−14.0	10^{-11}	−20.0
NaH	−7.7	10^{-6}	−29.0
CsH	−7.4	10^{-6}	−31.4
BeH_2	(+5)	(4×10^3)	−15.5
CaH_2	−34.0	10^{-25}	−25.0

*Values in parentheses are less certain.

Table II. Estimated Entropies of Binary Hydrides

Hydride	ΔS cal/K(mol H_2)
$MH_{0.5}$	−20
MH	−40
MH_2	−35
MH_3	−30

enthalpy is calculated, free energy values at room temperature may be derived using the appropriate entropy values from Table II. The expected plateau pressure can then be determined.

Calculations of binary hydride stabilities were attempted first, since Miedema (19) had previously applied his concept to these compounds. We selected ϕ^* at 5.1, n_{ws} as 4.66, and V_m as 7.1 for the hydrogen parameters. Table III presents data selected from over 25 binary hydrides studied. Only six, or a quarter of the total, have calculated pressures within two orders of magnitude of the experimental values. Also listed in the same table are the ϕ^* values of hydrogen that would yield the correct ΔH. This is the least certain parameter for hydrogen. The value varies 20% or more from the 5.1 used in the calculation. A small error is greatly magnified when the term is squared. This variation produces an

and Plateau Pressures of Binary Hydrides and
Experimental Results[a]

ΔG/mol Hydride, Calculated (kcal)	Pressure, Calculated (atm)	Pressure Ratio Calculated/ Experimental	ϕ_H^*, Yields Experimental ΔH
−27.8	10^{-21}	10^7	5.31
−21.3	10^{-16}	10^{16}	5.38
−23.7	10^{-18}	10^4	5.39
−22.0	10^{-16}	10	5.03
−32.3	10^{-24}	10^{14}	5.05
−1.6	0.07	0.35	5.22
−3.3	4×10^{-3}	0.67	5.1
−25.0	10^{-19}	1	5.09
−8.3	10^{-6}	10^7	5.27
−4.0	10^{-3}	10^8	5.25
+8.5	10^6	10^{12}	5.48
+9.7	10^7	10^{13}	5.68
−10.0	10^{-8}	10^{-11}(?)	4.72
−14.5	10^{-11}	10^{14}	5.26

Table IV. Comparison of Calculated and Experimental Plateau

Ternary Hydride	ΔH Calculated (kcal/mol H_2)	ΔG Calculated (kcal/mol hydride)
$LaNi_5H_6$	−14.4	−7.2
$LaNi_5H_4$	−16.5	−15
$LaCo_5H_6$	−16	−12
$LaCo_5H_4$	−19	−20
$LaPd_5H_6$	+12	+72
$LaFe_5H_6$	−25	−39
YNi_5H_4		
$Th_2Ni_{13}H_{28}$	−24.0	−189
$ThCo_5H_{4.6}$	−14.0	−8.0
$FeTiH_{2.5}$	−11.1	−0.7
$Th_2Co_7H_5$	−14.5	−10.0
$Th_2Fe_7H_{4.6}$	−15.6	−11.7
$ZrNiH_3$	−37	−40

even greater error in the pressure calculations since they are related exponentially to the derived free energies. An error of ∼ 0.7 kcal in ΔG will yield an order of magnitude error in the pressure.

Returning to the ternary hydrides, Equation 1 was used to calculate the heats of formation of several intermetallic compounds for which hydrogen absorption has been measured. These values and the experimental enthalpies for binary hydrides were substituted into Equation 6 to obtain the heat of formation of the ternary hydrides. Table IV shows the results of some of these calculations. Hydrogen absorption data are not available for some of the compounds. Listing of hydrogen pressures less than 0.01 atm for the thorium compounds (23) indicates that plateau pressures were lower but apparently could not be measured with the manometers used.

Although measured hydrogen pressures for $Th_7Ni_{13}H_{28}$, $Th_7Co_3H_{30}$, and $Th_7Fe_3H_{30}$ were below 0.01 atm, the calculated 10^{-50} atm is unlikely. Comparisons of the available measured and calculated pressures show, with the exception of $FeTiH_{2.5}$, differences of several orders of magnitude as do the binary hydrides.

Our estimated entropies should be reasonably good. In any case, an error here would not produce the magnitudes of error discussed above. Since experimental, rather than calculated, binary hydride enthalpy data were used, the error must lie in the calculated enthalpies for the intermetallic compounds. The empirical theory was restricted to compounds where the A metal, or lanthanide, is the minority metal. However, our evaluation shows that this requirement makes little practical difference in pressure estimates. The present level of sophistication given by these

Pressures for Ternary Hydrides at Room Temperature

Pressure, Calculated (atm)	Pressure, Experimental (atm)	$P_{calc.}/P_{exper.}$
6×10^{-6}	2.5	10^{-6}
10^{-11}	2.5	10^{-12}
10^{-9}	160	10^{-11}
10^{-15}	0.05	10^{-14}
$> 10^{50}$	(high)	
10^{-29}	$\gg 1$	
10^{7}	$> 10^{2}$	$< 10^{5}$
$< 10^{-100}$	< 0.01	10^{-7}
10^{-6}	20	0.07
0.3	4	19^{-8}
10^{-8}	0.25	10^{-9}
10^{-9}	0.35	10^{-23}
10^{-30}	10^{-7}	

empirical formulas does not allow reasonable accuracies in calculating ternary hydride stabilities in terms of their plateau pressures.

Experimental

Two experiments were conducted to determine the ultimate solubility of hydrogen in AB_5 compounds and to observe the effect of phase and lanthanide change on the room temperature equilibrium pressures.

High Pressure Studies. Since the $CaCu_5$ structure has nine available interstitial sites equivalent to those occupied by the deuterium atoms in orthorhombic $PrCo_5D_4$ structure, it was of interest to determine whether the other sites could be filled at higher pressures. An apparatus capable of measurement to 10,000 atm was used to measure hydrogen equilibrium pressures. This system is described in another report (24).

The compound $LaCo_5$ was chosen because it had the lowest plateau pressure, contained only four hydrogen atoms vs. six in $LaNi_5$, and had been well studied. Figure 3 shows the room-temperature isotherm. Kuijpers' results (25) are included for comparison. The material absorbed over eight hydrogen atoms per formula unit at 1300 atm. This result is consistent with the availability of the nine asymmetric octahedral and tetrahedral interstitial sites per formula unit in the $CaCu_5$ structure.

The presence of a third plateau indicates the presence of another hydride phase. It may be assumed that this is the stoichiometric phase $LaCo_5H_9$. Studies are being extended to higher pressures and different temperatures, using both hydrogen and deuterium. Such high hydrogen composition may result in a greatly expanded and altered superlattice. A high pressure x-ray camera is being constructed to measure lattice parameters under these extreme conditions. A large isotope effect might also be expected at such high compression.

Structure and Atomic Size Effects. Previous studies have shown the relationship between decreasing lanthanide atomic radius and either

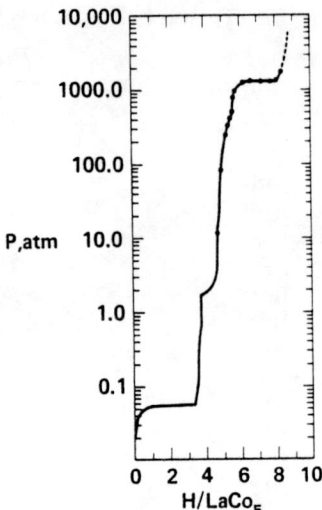

Figure 3. Room temperature absorption isotherm (21°C) of $LaCo_5H_x$ up to 2000 atm. The low pressure data coincide with those in Ref. 25.

increasing plateau pressures or hydride instability. Erbium is to the far right of the lanthanide series and forms $ErCo_5$ with the $CaCu_5$ structure. One author reported the preparation of $ErCo_6$ ($Er_{0.86}Co_{5.14}$) instead of $ErCo_5$ (26). This may be in error, since other investigators (27) had previously found $ErCo_5$. AB_5 phase generally has a wide composition range and the x-ray data for our sample corresponded to both patterns ($ErCo_5$ and $ErCo_6$) in powder files, indicating they are one in the same compound. The existence of $ErCo_6$ is also doubtful since AB_6 is not reported in other lanthanide-transition metal systems and there is no single crystal data available to support an assumption of ordered vs. random substitution of cobalt for erbium.

Surprisingly, to the best of our knowledge, the reaction of $ErCo_5$ with hydrogen has not been reported. Hydrogen absorption in several praesodymium-cobalt phases (28) and in $ErCo_3$ (29) has been reported recently. The experimental completion of the simple $PrCo_5$, $PrCo_3$, $ErCo_5$, and $ErCo_3$ matrix would allow an experimental determination of the

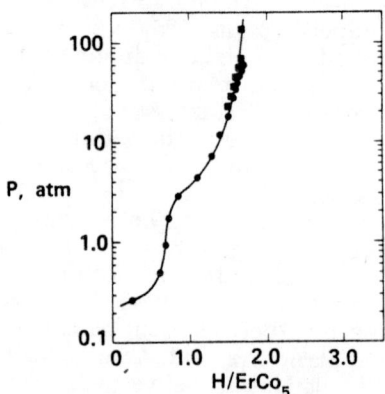

Figure 4. Hydrogen absorption isotherm for $ErCo_5$ at 25°C

effect of lanthanide contraction and phase change on the hydrogen plateau pressures. The previous data were extrapolated to 25°C by the use of Equation 7 and compared with the present results. The $ErCo_5$ room-temperature isotherm is presented in Figure 4. The critical temperature seems to be somewhat below 25°C, indicating that $ErCo_5H_x$ is less stable than the $ErCo_3H_x$.

The room-temperature plateau pressures are shown in Table V. These values indicate that the stability of the hydrides in these compounds decreases with an increase in lanthanide atomic number and also with the cobalt-to-lanthanide ratio. The stability change with phase is seen in Clinton's experiments on $PrCo_x$ (28), except that $PrCo_2$ and $PrCo_3$ stabilities are reversed. They are very similar, however.

Table V. Comparison of Equilibrium Pressures at Room Temperature

Ternary Hydride	Equilibrium Pressure (atm)
$PrCo_5H_{0.5}$	0.9
$PrCo_3H$	10^{-4}
$ErCo_5H$	~ 4 [a]
$ErCo_3H_3$	0.09

[a] There is no plateau, indicating that the critical temperature is below room temperature. The composition and pressure are inferred from the inflection point.

Both observations seem reasonable. As the lanthanide radius decreases, the interstitial sites will become smaller, and thus less void space is available for the hydrogen atoms. As the transition metal-to-lanthanide atomic ratio increases, there is less atom-atom contact between hydrogen and the lanthanide, which forms the more stable binary hydride.

Conclusions

The use of previously published empirical theories does not provide reliable estimates of equilibrium pressures of ternary hydrides. Either a substantial modification of these theories or another approach is required.

High pressure experiments have shown the existence of $LaCo_5H_9$. This hydrogen composition is twice the value previously measured and greater than any compound with the $CaCu_5$ structure, i.e., $LaNi_5H_{6.7}$.

Comparisons of current $ErCo_5H_x$ data with previously published results for $PrCo_5$, $PrCo_3$, and $ErCo_3$ show that hydride stability decreases with lanthanide atomic number and with increasing atom ratio of transition metal-to-lanthanide metal.

Acknowledgments

The authors wish to thank Herman Leider, who performed many of the calculations and participated in important discussions of subjects in this paper.

Literature Cited

1. Libowitz, G. G., *J. Nucl. Mater.* (1960) **2**, 1.
2. Mueller, W. M., Blackledge, J. P., Libowitz, G. G., Eds., "Metal Hydrides," Academic, New York, 1968.
3. Lewis, F. A., "The Palladium Hydrogen System," Academic, New York, 1967.
4. "Investigation of Hydriding Characteristics of Intermetallic Compounds," Denver Research Institute, University of Denver: Report **LAR-55**, Nov. 15, 1961; **DRI-2059**, Oct. 15, 1962.
5. Newkirk, H. W., "A Literature Study of Metallic Ternary and Quaternary Hydrides," **UCRL-52110**, Aug. 2, 1976, Lawrence Livermore Laboratory, Livermore, CA.
6. Messer, C. E., Eastman, J. C., Hers, R. G., Maeland, A. J., *Inorg. Chem.* (1964) **3**, 776.
7. Messer, C. E., Hardcastle, K., *Inorg. Chem.* (1964) **3**, 1327.
8. Moyer, R. O., Jr., Stanitski, C., Tanaka, J., Kay, M., Kleinberg, R., *J. Solid State Chem.* (1971) **3**, 541.
9. Thompson, J. S., Moyer, R. O., Jr., Lindsay, R., *Inorg. Chem.* (1975) **14**, 1866.
10. Zijlstra, H., *Chem. Technol.* (1972) **2**, 280.
11. van Vucht, J. H. N., Kuijpers, F. A., Bruning, H. C. A. M., *Philips Res. Rep.* (1970) **25**, 133.
12. Kuijpers, F. A., van Mal, H. H., *J. Less-Common Met.* (1971) **23**, 395.
13. van Mal, H. H., Buschow, K. H. J., Kujpers, F. A., *J. Less-Common Met.* (1973) **32**, 289.
14. Wallace, W. E., Rao, V. U. S., "Thermal, Structural and Magnetic Studies of Metals and Intermetallic Compounds," Annual Report to ERDA, Contract E(11-1)-3429, June 1, 1975.
15. Toma, H., "Rare-Earth Info. Ctr. News," **X**, No. 1, Mar. 1, 1975, Iowa State Univ., Ames, Iowa.
16. Pearson, W. B., "The Crystal Chemistry and Physics of Metals and Alloys," Wiley-Interscience, New York, 1972.
17. Kuijpers, F. A., Loopstra, B. O., *J. Phys. Chem. Solids* (1974) **35**, 301.
18. Miedema, A. R., de Boer, F. R., de Chatel, P. F., *J. Phys. F.* (1973) **3**, 1558.
19. Miedema, A. R., *J. Less-Common Met.* (1973) **32**, 117.
20. Miedema, A. R., Boom, R., de Boer, F. R., *J. Less-Common Met.* (1975) **41**, 283.
21. van Mal, H. H., Buschow, K. H. J., Miedema, A. R., *J. Less-Common Met.* (1974) **35**, 65.
22. Flanagan, T. B., Oates, W. A., *Ber. Bunsenges. Phys. Chem.* (1972) **76**, 706.
23. Buschow, K. H. J., van Mal, H. H., Miedema, A. R., *J. Less-Common Met.* (1975) **42**, 163.
24. Lakner, J. F., Steward, S. A., Uribe, F., "High Pressure Hydrogen Apparatus for PCT Studies to 700 MPa at 200°C. Preliminary Results on $LaCo_5$ Hydride at 21°C," **UCRL-52039**, Feb. 27, 1976, Lawrence Livermore Laboratory, Livermore, CA.
25. Kuijpers, F. A., *Philips Res. Rep. Suppl.*, 1973, No. 2.
26. Buschow, K. H. J., *Z. Metallkd.* (1966) **57**, 728.
27. Wernick, J. H., Geller, S., *Acta Crystallogr.* (1959) **12**, 662.
28. Clinton, J., Bittner, H., Oesterreicher, H., *J. Less-Common Met.* (1975) **41**, 187.
29. Takeshita, T., Wallace, W. E., Craig, R. S., *Inorg. Chem.* (1974) **13**, 2282.

RECEIVED August 16, 1976. This work was performed under the auspices of the U.S. Energy Research & Development Administration under contract No. W-7405-Eng-48.

This report was prepared as an account of work sponsored by the United States Government. Neither the United States nor the United States Energy Research & Development Administration, nor any of their employees, nor any of their contractors, subcontractors, or their employees, makes any warranty, express or implied, or assumes any legal liability or responsibility for the accuracy, completeness or usefulness of any information, apparatus, product or process disclosed, or represents that its use would not infringe privately-owned rights.

# 17

Chemical Conversion Using Sheet-Silicate Intercalates

JOHN M. THOMAS, JOHN M. ADAMS, SAMUEL H. GRAHAM, and D. TILAK B. TENNAKOON

Edward Davies Chemical Laboratories, University College of Wales, Aberystwyth, SY23 1NE, U.K.

> *The use of layered silicates as matrices within and upon which novel chemical reactions may be carried out are summarized. The feasibility of controlled variation in the nature and siting of certain transition-metal and other cations, the magnitude of the interlamellar spacing, and orientation as well as two-dimensional ordering of intercalated organic molecules is demonstrated. Specific examples of structures based on one-dimensional Fourier plots derived from x-ray and neutron diffraction data are cited, and one three-dimensional crystal structure for a rather special intercalate (dickite:formamide) is reported. The selective generation of a variety of aromatic products by thermostimulation is summarized, and the particular ability of copper montmorillonite to activate, preferentially, olefinic double bonds (in oligomerizations) is illustrated by reference to the thermal reactivity of intercalated indene and* trans-*stilbene.*

Because of the mystery that still surrounds the phenomenon of catalysis, it is not yet generally possible to design, ab initio, new chemical agents that can serve as catalysts for the synthesis of desired structurally or stereochemically specified products. Some progress along specific directions has been made, however, a fact borne out both by the success of Zeigler–Natta catalysts and the existence of an established procedure for the "solid-state" syntheses of polypeptides. Catalyst design, however, is still in its infancy, notwithstanding the significant advances that have been made recently using zeolitic solids and transition-metal complexes.

Heterogeneous catalysts cannot as yet be designed with the same facility, precision, and variation now achievable with the syntheses of many classes of organic molecules—from organic fluorescers and photochromic materials to synthetic steroids—or with the same success in engineering super-ionic inorganic conductors, such as β-alumina, or continuously variable electronic band-gaps in III–V ternary or quaternary semiconductors.

One particularly productive route to the syntheses of new and interesting molecular catalysts relies on the use of metal vapors which are usually co-condensed with a hydrocarbon reactant. In essence, this approach, whose lineage may be traced to Pimentel's matrix isolation technique, circumvents the necessity to provide thermal activation during reaction—as would be required by processes involving solid metals—thereby rendering feasible the formation of a number of novel molecular species such as dibenzene titanium or polybutadiene, cyclododecatrienes, and vinylcyclohexenes. The metal–vapor synthesis, which has been exploited by Timms (*1*), Green (*2*), Skell (*3*) and others (*4, 5*), yields the desired product under solvent-free conditions.

Another successful route, also utilized to advantage by Green, capitalizes upon the fact that when central metal atoms are situated in ligand environments which make the metal atoms highly electron rich the resulting complexes, which now possess orbitals of strong metal character and high energy, are likely to be reactive toward relatively inert species such as N_2, N_2O, and CH_4 or inert bonds such as C–H and C–C.

The route which we ourselves have chosen to investigate (*6–15*) entails the use of certain sheet silicate structures within which ion-exchange may first be performed. To date, as is described below, our efforts have been concentrated principally on structural studies of the parent silicates and their deliberately modified derivatives. It is to be noted that: (a) wide variation is possible in regard to the nature of the particular cations that can be inserted between the infinite, two-dimensional anions; (b) the interlayer spacing, which obviously governs the ease of diffusion of intercalated reactants and products, is, to a degree, adjustable depending, inter alia, upon factors such as the humidity and the nature of the organic molecules present in the system; and (c) in certain circumstances, highly specific reactions among the intercalated species may be stimulated, and the products released, under solvent-free conditions. In one sense a new type of "homogeneous" catalysis is involved, the phase in question being the two-dimensional intercalate. Moreover, there is evidence that with certain types of silicates it may be possible to lay out an ordered two-dimensional sheet of one reactant in which the intermolecular spacing parallel to the sheets may also be adjustable.

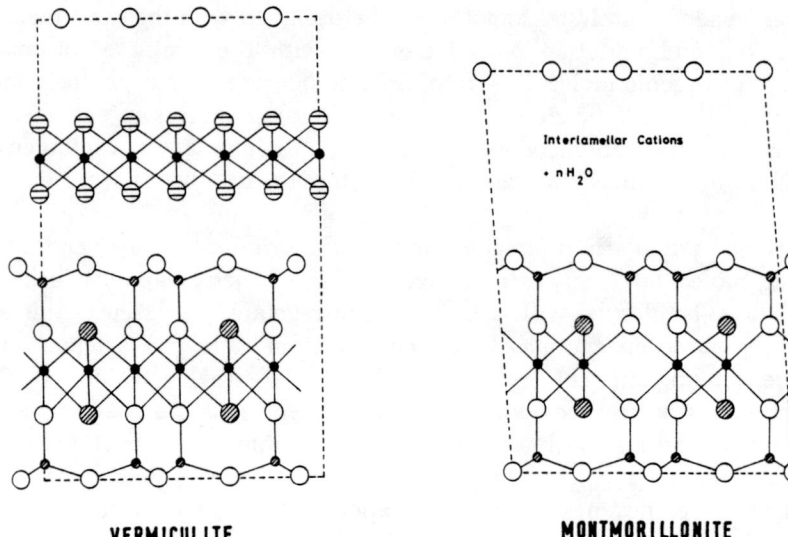

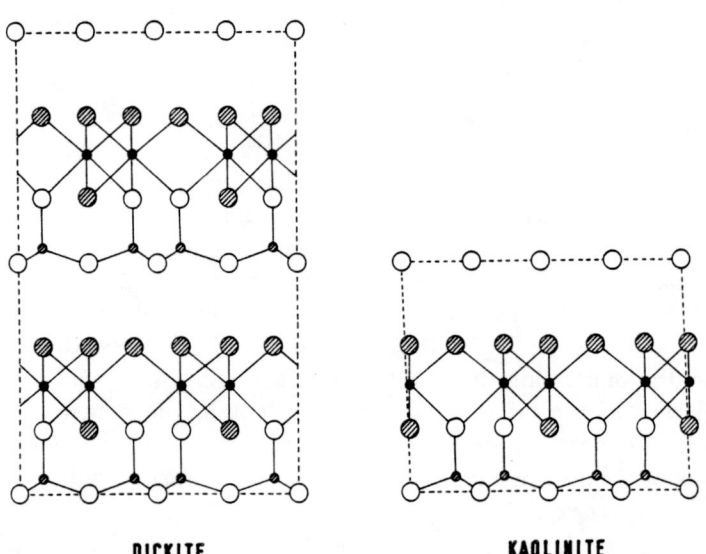

Figure 1. Projections of the structures of the sheet silicates mentioned in the text: (a) montmorillonite, (b) vermiculite, (c) kaolinite, (d) dickite. (•) octahedral cations, (⊘) tetrahedral cations, (○) oxygen, (⊘) hydroxyl, (⊖) water molecules.

Resumé of Structural Characteristics of Sheet Silicates

In previous publications (8–15) from these laboratories the various relevant attributes of the particular sheet silicates of interest in the context of intercalation have been summarized. We must recall here the salient features of montmorillonite, vermiculite, kaolinite, and dickite only (Figure 1). In the well known tetrahedral–octahedral–tetrahedral (TOT) framework characteristic of montmorillonite, the layers [idealized formula $Al_4Si_8O_{20}(OH)_4$] are negatively charged because of replacement of Al^{3+} in octahedral sites by Fe^{2+} or Mg^{2+} and of Si^{4+} in tetrahedral sites by Al^{3+}. The cation-exchange capacity (c.e.c.) is governed by the extent of these replacements, and the distribution of charge in turn depends on the location of the replaced ions. Clearly much variation is possible here, so that inhomogeneities of charge distribution in two dimensions, and consequently of stacking in three, may arise from this and other sources. Montmorillonite is almost invariably microcrystalline and possesses surface areas generally well in excess of several hundred m^2g^{-1}, and c.e.c. of 0.5–0.7 e per formula unit. Vermiculite may, for convenience, be regarded as a sheet silicate in which water has been intercalated in a more or less ordered fashion. These sandwiched water molecules may be progressively driven out with heat treatment.

In kaolinite [idealized formula $Al_4Si_4O_{10}(OH)_8$] the TO framework which extends two-dimensionally is weakly interconnected in a third direction via Al–OH . . . O–Si hydrogen bonds. Dickite, a two-layer monoclinic modification of kaolinite, is also shown in projection in Figure 1. We have found that this particular silicate forms a three-dimensionally ordered intercalate (16), which is discussed briefly below.

Formation and Structural Aspects of Sheet Silicate Intercalates

A vast amount of work (admirably summarized by Brindley (17) up to 1970 and by Theng (18) up to 1974) has already been published on the conditions under which a wide range of organic molecules may be assimilated by various sheet silicates. Here we refer to only those systems relevant to the ultimate question of reactivity or structural assessment. Many of the organic molecules or exchangeable cations chosen by us for initial study were so selected because it was felt that they would facilitate the interpretation of x-ray and neutron diffraction studies. Thus we see from Figure 2, which in turn has been derived from neutron diffraction studies, precisely where along the z-axis (perpendicular to the basal plane) the Ni^{2+} ion resides in a vermiculite, the cations of which had been exchanged for Ni^{2+}. Likewise, Figure 3 shows how replacing Na^+ with Sr^{2+}, which has approximately twice its charge density, as the interlamellar cation significantly modifies the orientation of tetrahydropyran. The oxygen-ring atom lies closest to the silicate layer when the cation is Na^+, and a carbon-ring atom lies closest when the cation is Sr^{2+}.

It is evident that with the montmorillonites both neutral and charged organic species may be assimilated into the interlamellar spaces. Neutral species may be introduced either from solution or from the vapor, the latter approach being advantageous if traces of intercalated solvent, notably water, are to be avoided. Though there is a paucity of the relevant thermodynamic information, it is clear that the binding energies of these intercalated species may vary widely. It seems possible for organic molecules to be attached both directly to an associated cation—via σ- or

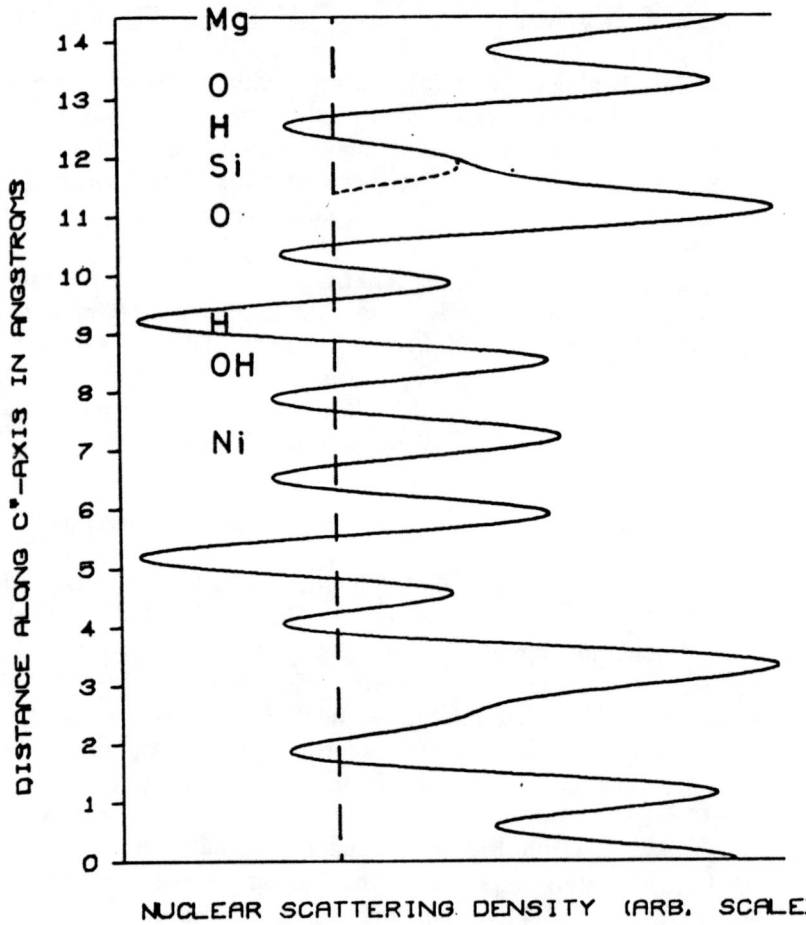

Figure 2. One-dimensional Fourier map of nuclear scattering density for Ni^{2+}-vermiculite

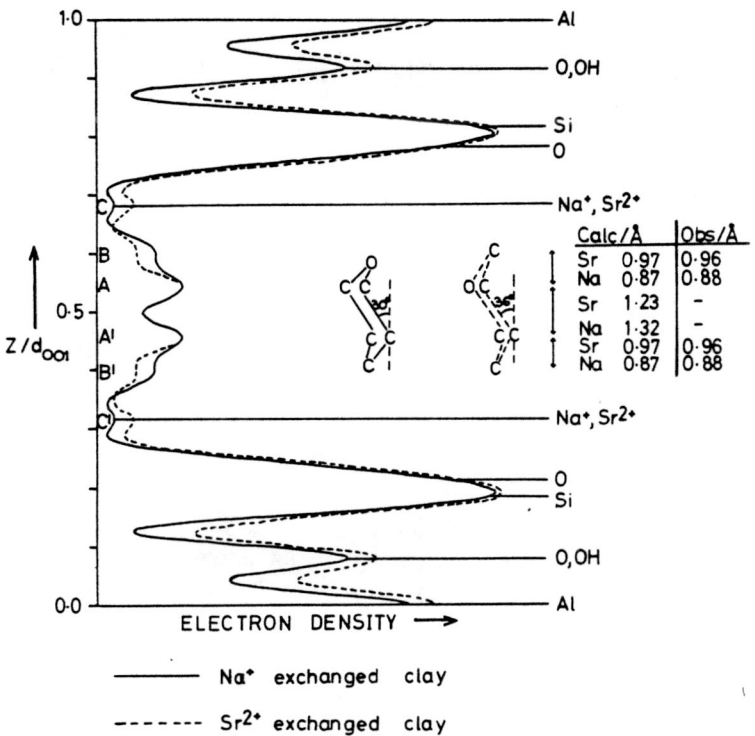

Figure 3. Electron density projections of Na^+- and Sr^{2+}-montmorillonite–tetrahydropyran intercalates

π-bonding, the relative proportion of which is expected, on the basis of the principles of general organometallic chemistry, to vary as a function of temperature and chemical environment—or loosely as in a secondary "solvation" shell or physically adsorbed state. Some organic entities (such as amines) which are capable of forming cations may displace the interlamellar cations originally present in the sheet silicate. Others show a greater propensity to be adsorbed at the external rather than the interlamellar surfaces of the sorbent. Another relevant phenomenon is the sequential self-conversion that certain organic intercalates of montmorillonite undergo. The pyridine intercalate of Na^+-exchanged montmorillonite is gradually converted at room temperature from a larger to a smaller interlamellar spacing (23.3 Å to 14.8 Å):

$$M(py)_4 \cdot 2H_2O \xrightarrow[+H_2O]{-py} M(py)_2 \cdot 4H_2O$$

where $M = 2[(Al_{3.5}Mg_{0.5})Si_8O_{20}(OH)_4Na_{0.5}]$; and the γ-butyrolactone/Sr^{2+}-exchanged montmorillonite shows a double conversion in which the interlamellar spacing changes from 23.1 Å when the intercalate is first formed to 18.3Å and then to the stable 13.2-Å intercalate. The 1D-Fourier maps (Figure 4) of these three forms of the butyrolactone-Sr^{2+} complexes, imply that the sequential changes involve a conversion of the parent intercalate, which appears to incorporate three layers of the organic species between contiguous montmorillonite sheets, first to a two-layer form and, finally, to the more stable, single-layer intercalate. (For further details on these systems see Ref. 16).

More work must be done on the structural aspects of these various types of intercalates of montmorillonite, but progress is severely hampered both by the unavailability of this mineral in other than microcrystalline forms and by the fact that the intercalates of montmorillonite do not seem to take up three-dimensionally ordered structures, a situation which perhaps is not unexpected in view of the irregularities of the charge distribution and stacking sequences in the c^* direction of the parent mineral (20). X-ray diffraction is not likely to reveal the necessary, desired information. However, combination with neutron-diffraction data on well oriented samples of the intercalates leads to a greater understanding of the delicate variations in structure that are displayed by, for example, the transition-metal exchanged montmorillonites. Figure 5, which shows Fourier maps derived from neutron-diffraction data (21), clearly indicates that whereas the Ni^{2+} ion in the tetrahydrofuran intercalate of Ni^{2+}-exchanged montmorillonite, $(Al_{3.5}Mg_{0.5})Si_8O_{20}(OH)_4$-$Ni_{0.25}(C_4H_8O)_{2.3}$, is situated centro-symmetrically in the interlamellar region, the Co^{2+}-ion in a closely similar intercalate, $(Al_{3.5}Mg_{0.5})Si_8O_{20}$-$(OH)_4Co_{0.25}(C_4H_8O)_{2.2}$, is off-center.

Kaolinite and dickite also form intercalates which, in general, do not appear to be three-dimensionally ordered (22). However, one such ordered intercalate has been discovered recently (16): it is a 1:1 complex of dickite and formamide, $Al_4Si_4O_{10}(OH)_8(HCONH_2)_2$. A projection of its structure is shown in Figure 6. Rather remarkably no two-dimensional superlattice exists in this structure. Contrast the two-dimensionally and three-dimensionally ordered intercalates of graphite (23) and the transition-metal chalcogenides (24, 25, 26). This fact arises probably because formamide, being such a small molecule, may fit snugly within the unit mesh parallel to (001) of the dickite structure. Of great interest here is the occurrence of relatively rigidly clamped intercalated species. It is known from Weiss' work (22) that the rates of intercalation of kaolinites are relatively sluggish and that at room temperature both formamide and urea, after an induction period of some one or two days, are each gradually incorporated into the host matrix, the final part of the formation curve being reached asymptotically after an intervening

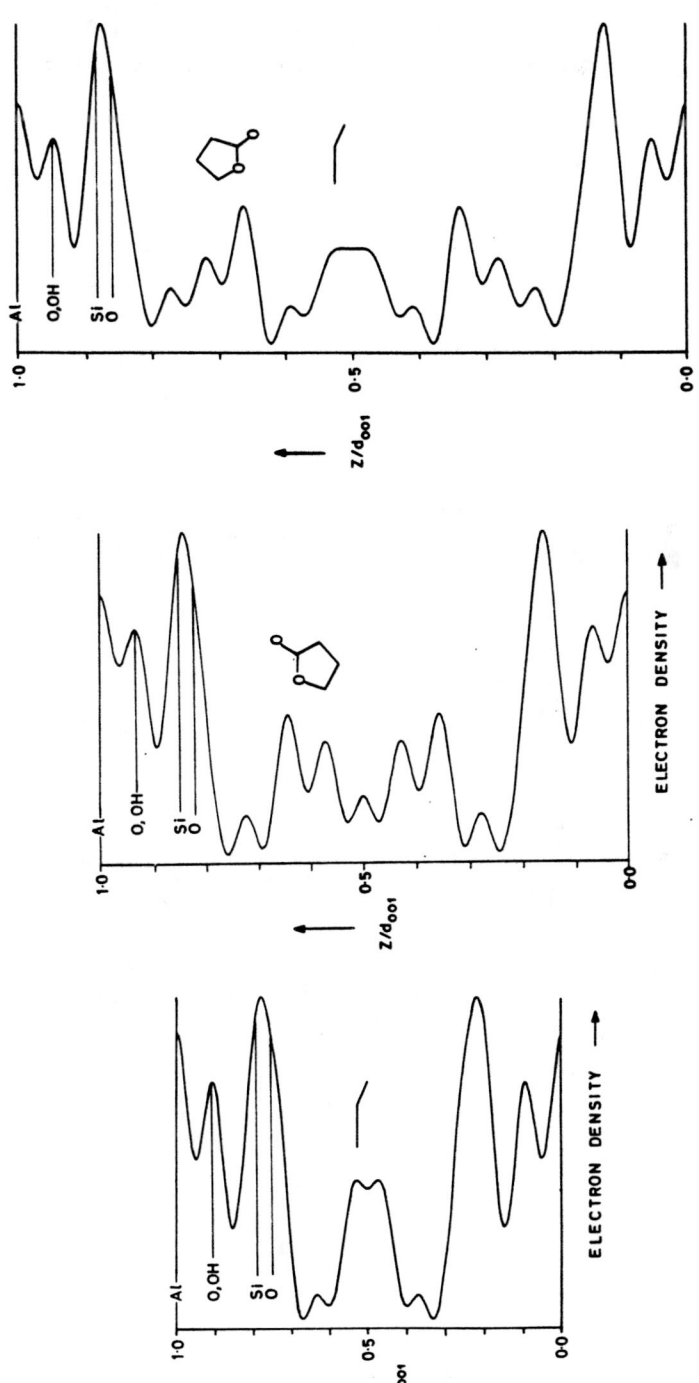

Figure 4. Electron density projections for the three distinct "phases" of the Sr^{2+}-montmorillonite-γ-butyrolactone intercalate (1.6, 3.2, and 4.8 γ-butyrolactone/formula unit)

306 SOLID STATE CHEMISTRY

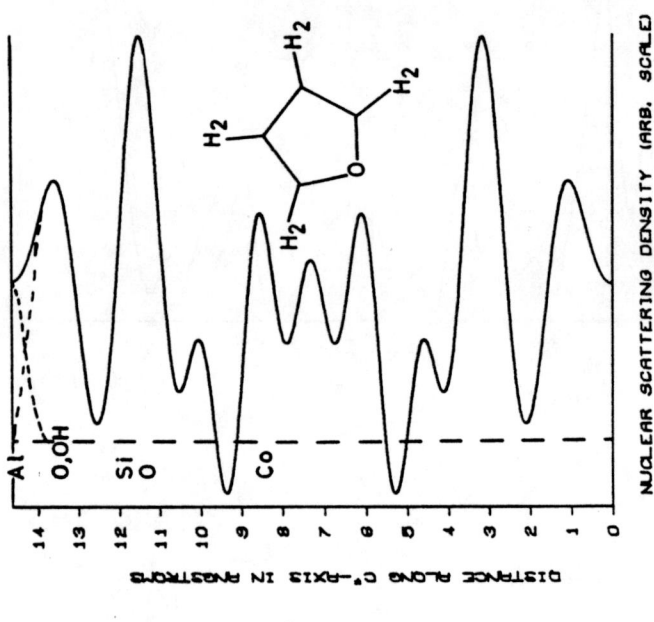

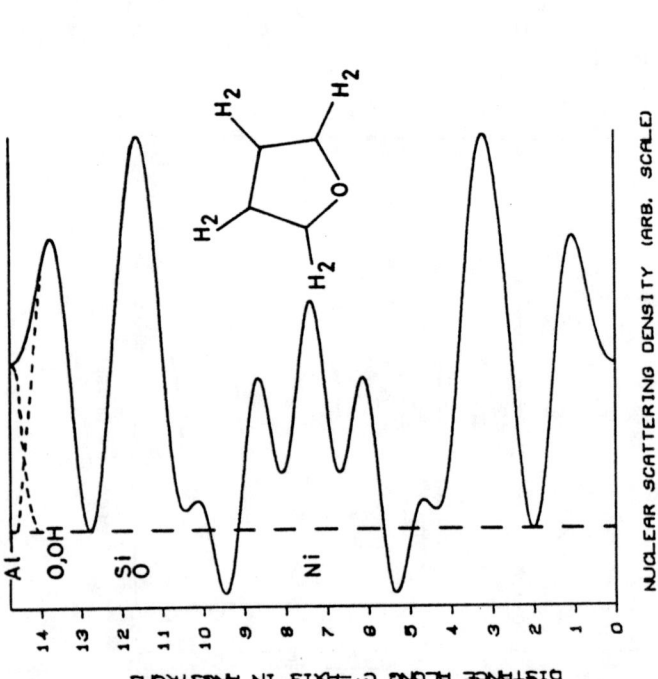

Figure 5. Projections of nuclear scattering density for (a) Ni^{2+}-montmorillonite–tetrahydrofuran (2.3 THF/formula unit) and (b) Co^{2+}-montmorillonite–tetrahydrofuran (2.2 THF/formula unit)

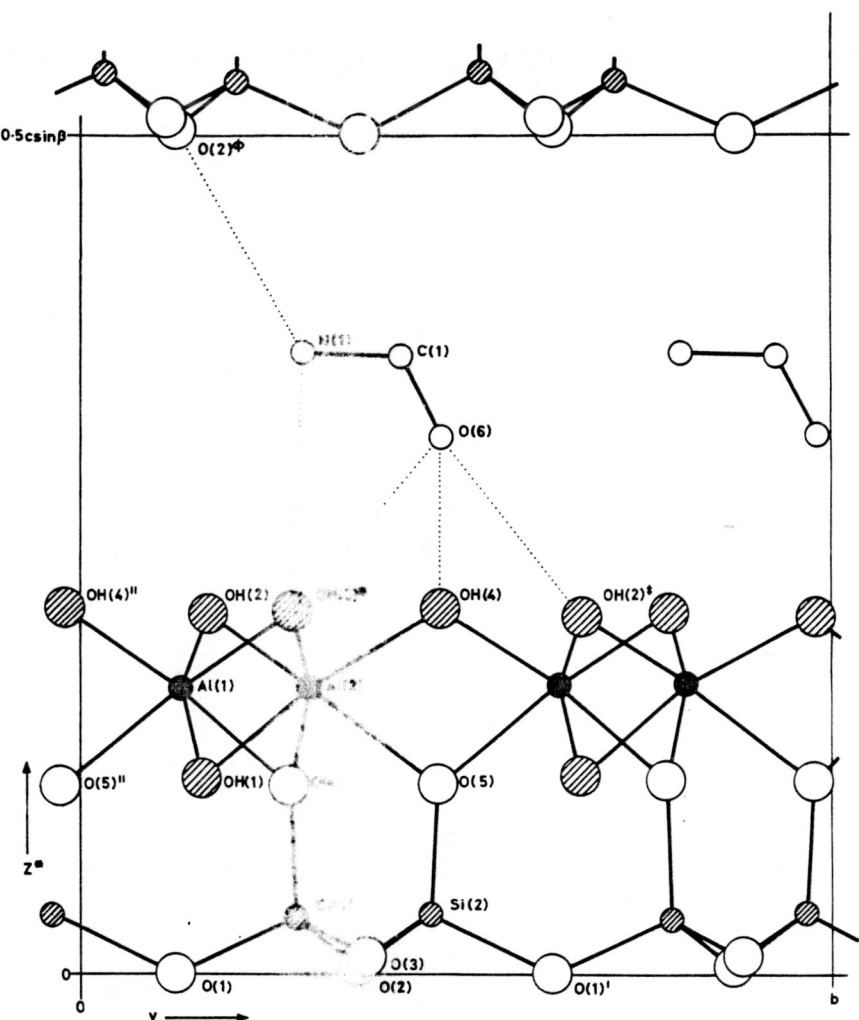

Figure 6a. Projection of the structure of the dickite–formamide intercalate along the a axis

linear rate of uptake. Clearly it would be of interest, in the context of catalysis, to introduce a second reactant into the system when the first two-dimensionally ordered reactant (in this case formamide) only partially completes the sites available for its occupation. Co-adsorption studies have been well characterized with montmorillonite, and Lailach and Brindley (27) found that thymine, which is not on its own incorporated into montmorillonite from aqueous solution, is readily taken up in the presence of adenine or hypoxanthine, each of which is also adsorbed

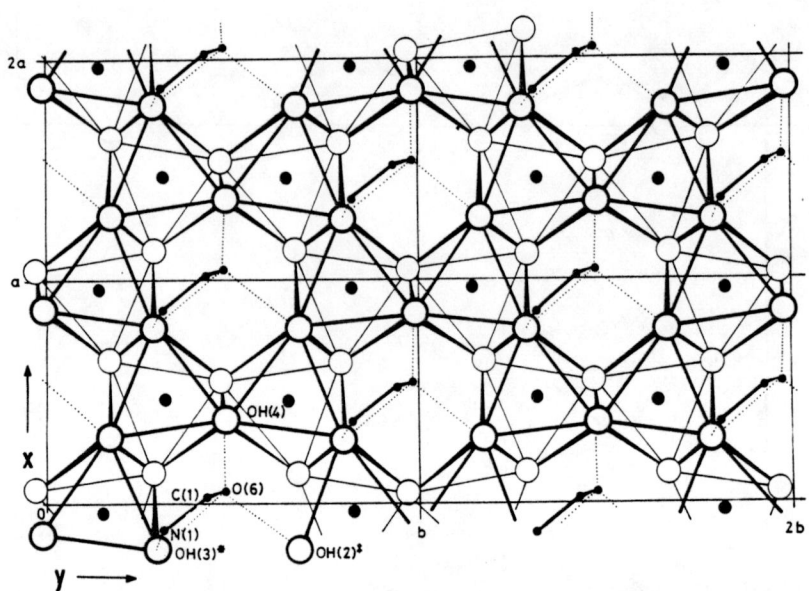

Figure 6b. Projection of the AlO_6 octahedra of the dickite–formamide intercalate onto the ab plane. The hydrogen bonding of the formamide molecules is also shown.

in isolation. It is believed (*17*) that adenine and thymine form the well-known purine–pyrimidine base-pairs in the interlamellar region.

Extensive compilations of simple crystallographic information relating to the numerous types of organic molecules that may be intercalated by sheet silicate minerals are given in the recent monograph by Theng (*18*), who also reviews the work of many of the earlier (MacEwan, Walker, Hofmann, Bodenheimer, Brindley, and Weiss) and more recent investigators (Mortland, Matsunaga, Blumstein, Fripiat, Farmer, and others). Table I summarizes a few facts relating to some of the simple structural characteristics of the intercalates discussed in this section and studied by us.

Some Specific Chemical Conversions

Ideally one would wish to have available information which relates the reactivity and associated stereo- or regio-specificity of thermally induced reactions of various organic intercalates of the sheet silicates on the one hand with the stoichiometry and detailed structural properties of the starting material on the other. Unfortunately, such information is, at present, almost totally unavailable. Where the reaction is efficient, interesting, or well-identified, the accompanying structural information has turned out to be simply inaccessible because of the lack of crystalline order, which in turn yields the necessary well-defined x-ray data from which the one-dimensional Fourier plots are extracted. Conversely, for those intercalates where it has been possible to process the x-ray or neutron-diffraction data to yield more-or-less accurate structural information, the degree of reactivity or novelty of the reaction is itself low.

One particularly efficient reaction is the thermally induced conversion to aniline of the intercalate (14.5 Å spacing) consisting of diprotonated, 4,4'-diamino-*trans*-stilbene and montmorillonite (approximate stoichiometry: $(Al_{3.5}Mg_{0.5})Si_8O_{20}(OH)_4(H_3NC_6H_4CH=CHC_6N_4NH_3)_{0.25}$. This reaction, which proceeds rapidly at ca. 270°C, yields aniline (close to 45% of the parent diamine) as the sole gaseous product. Another efficient reaction that occurs on the surfaces of montmorillonite is the almost quantitative conversion in the temperature range 40°–150°C of triphenylamine to N,N,N',N'-tetraphenylbenzidine, when the complex formed by the exposure of Na⁺-montmorillonite to alcoholic solutions of triphenylamine is heated. It seems very likely that radical cations (which certainly form with ease at Lewis acid sites in the silicate when benzidine and other similar molecules are exposed to montmorillonite (*see* the Mössbauer and spectroscopic studies of Tricker et al. (*8, 13, 28*)) are first formed and that these proceed along either of two possible pathways: (a) dimerization followed by deprotonation and the benzidine rearrangement

$$2(Ph_3N)^+ \rightarrow (Ph_3NNPh_3)^{2+}$$

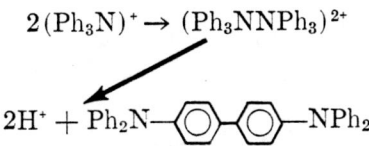

$$2H^+ + Ph_2N---NPh_2$$

or (b) a direct coupling between para positions of the benzene rings of the two triphenylamine radical cations, again followed by the elimination of two protons. It must be noted that the interaction of montmorillonite and triphenylamine may be restricted solely to the exterior surface of the clay mineral: the interlayer spacing (admittedly a crude criterion for the occurrence of intercalation) remains essentially unchanged prior to and following uptake of the tertiary amine.

When diphenylethylene is heated to reflux for 30 min in contact with Cu(II)-exchanged montmorillonite, the indan dimer (specifically 1-methyl-1,3,3-triphenyl-indan) is readily obtained in good (30%) yield. The product crystallizes out as small prisms, m.p. 142°C (lit. 142–143) from acetic acid (*12*). This new procedure seems to provide the most convenient preparation of this dimer.

Table I. A Summary of the Crystallographic Information

System[a]

1. Na^+-mont (dried)
2. Na^+-mont (hydrated)
3. Cu^{2+}-mont (hydrated)
4. Ag^+-mont (hydrated)
5. Sr^{2+}-mont (hydrated)
6. Ni^{2+}-mont (hydrated)
7. Co^{2+}-mont (hydrated)
8. Sr^{2+}-mont–γ-butyrolactone

9. Na^+-mont–tetrahydropyran
10. Sr^{2+}-mont–tetrahydropyran
11. Ni^{2+}-mont–tetrahydrofuran
12. Co^{2+}-mont–tetrahydrofuran
13. Na^+-mont–pyridine

14. Ag^+-mont–azobenzene
15. Cu^{2+}-mont–azobenzene
16. Ni^{2+}-vermiculite
17. Co^{2+}-vermiculite
18. Dickite–formamide

[a] mont = montmorillonite.

It has recently been established that transition-metal-exchanged samples of montmorillonite readily form complexes with a wide range of aromatic hydrocarbons and their simple derivatives such as chlorobenzene, anisole, and azobenzene. It seems that metal-arene complexes may be stabilized inside, or at the surfaces of, montmorillonite. There is clearly a vast territory of synthetic organic chemistry here which awaits exploration. A few investigations have already been initiated in these laboratories, and we now present a brief account of the behavior of dehydrated Cu^{2+}-montmorillonite (hereafter designated Cu–M) complexes following exposure to a number of selected aromatic hydrocarbons.

Spectroscopic and other studies (28–33) indicate that Cu–M may bind aromatic molecules, such as benzene, toluene, xylene, etc., in three distinct forms: the simple, loosely bound (physically adsorbed) state; the more strongly attached state in which the bond-strengths within the arene ligand are slightly perturbed (the so-called type I complexes in the classification of Pinnavaia and Mortland (31)); and the type-II complexes which are usually colored differently from the type I complexes, where the aromaticity and point symmetry of the arene is destroyed. It

Relating to the Intercalates Discussed in the Text

Approx. Formula[b]	Basal Spacing (Å)	No. of Orders of Diffraction Observed
M Na$_{0.5}$	9.6	3
M Na$_{0.5}$ · 3.5H$_2$O	12.4	6
M Cu$_{0.25}$ · 3.5H$_2$O	12.5	6
M Ag$_{0.5}$ · 3.5H$_2$O	12.5	6
M Sr$_{0.25}$ · 7.0H$_2$O	15.2	5
M Ni$_{0.25}$ · 7.0H$_2$O	15.1	5
M Co$_{0.25}$ · 7.0H$_2$O	15.2	5
M Sr$_{0.25}$ · (C$_4$H$_6$O$_2$)$_{1.6}$	13.2	13
M Sr$_{0.25}$ · (C$_4$H$_6$O$_2$)$_{3.2}$	18.3	14
M Sr$_{0.25}$ · (C$_4$H$_6$O$_2$)$_{4.8}$	23.1	18
M Na$_{0.5}$(C$_5$H$_{10}$O)$_{1.3}$	14.99	13
M Sr$_{0.25}$(C$_5$H$_{10}$O)$_{1.0}$	14.81	13
M Ni$_{0.25}$(C$_4$H$_8$O)$_{2.3}$	14.50	12
M Co$_{0.25}$(C$_4$H$_8$O)$_{2.2}$	14.58	12
M Na$_{0.5}$(C$_5$H$_5$N)$_{2.0}$(H$_2$O)$_{1.0}$	23.3	16
M Na$_{0.5}$(C$_5$H$_5$N)$_{1.0}$(H$_2$O)$_{2.0}$	14.8	13
M Ag$_{0.5}$(C$_{12}$H$_{10}$N$_2$)$_{0.9}$	22.3	6
M Cu$_{0.25}$(C$_{12}$H$_{10}$N$_2$)$_{0.8}$	20.5	11
V Ni$_{0.75}$	14.4	12
V Co$_{0.75}$	14.4	12
Al$_4$Si$_4$O$_{10}$(OH)$_8$ · (HCONH$_2$)$_2$	20.19	22

[b] M = (Al$_{3.5}$Mg$_{0.5}$)Si$_8$O$_{20}$(OH)$_4$; V = Mg$_6$(Al$_{1.5}$Si$_{6.5}$)O$_{20}$(OH)$_4$.

is thought that in type I complexes the arene is edge-π-bonded to the copper, rather similar to the bonding that exists in C$_6$H$_6$CuAlCl$_4$ (see Turner and Amma (34)). Mortland (32), Rupert (33), and Tennakoon (12, 28) have shown that only the symmetrical arenes such as benzene, biphenyl, naphthalene, and anthracene form type-II complexes with Cu–M; anisole is apparently an exception (32). Type-I complexes are formed by benzene and all the alkyl benzenes and symmetrical arenes studied to date, and dehydration of the Cu–M complexes of the symmetrical arenes usually results in the (reversible) conversion of the type-I to the type-II state. Ligands in type-I complexes appear to retain their aromaticity; and spectroscopic evidence also suggests that in both type-I and type-II complexes the organic moieties form radical cations. Type-II species can also be formed with Fe^{3+} and VO^{2+} ions, and it has been suggested that these species are pairs of associated radical cations rather than intercalated organometallic species (35).

The thermal reactivity of these complexes reveals interesting trends. Thus the room-temperature stable, type-II Cu–M:benzene complex (this complex is stable in a vacuum of ca. 10^{-7} torr at room temperature) will, on heating, break down to yield numerous fragments all possessing molecular masses less than that of the benzene. Type-I Cu–M complexes with toluene and with the various isomeric xylenes will, upon gentle heating, yield volatile products which show that condensation, with hydrogen elimination, of the arenes has occurred. Thus mass peaks (m/e) of 272 and 182, corresponding to three toluene units minus four hydrogen atoms and two toluenes minus two hydrogen atoms, respectively, are observed. The mass-spectrometric fragmentation patterns of the volatile products are consonant with the occurrence of the four products shown in Figure 7. (The identity of these products has not, however, been independently confirmed). All the evidence points to the fact that type-I arene complexes of Cu–M readily yield radicals or radical ions when heated, which accounts for the nature of the reaction products.

When the aromatic molecule forming the complex also contains an ethylenic linkage, fundamental differences then arise in the pattern of thermal reactivity. In short both *trans*-stilbene and indene (each introduced separately from the vapor into the dehydrated Cu–M) yield oligomeric products in which no loss of hydrogen atoms has occurred. There appears to be preferential bonding to or activation of the ethylenic link,

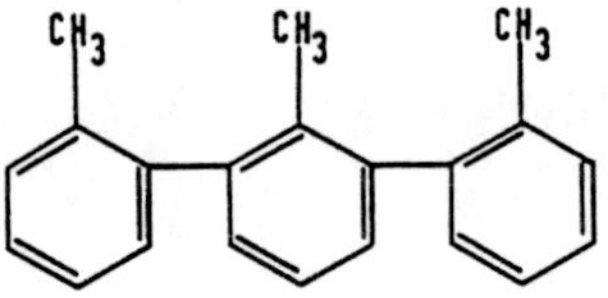

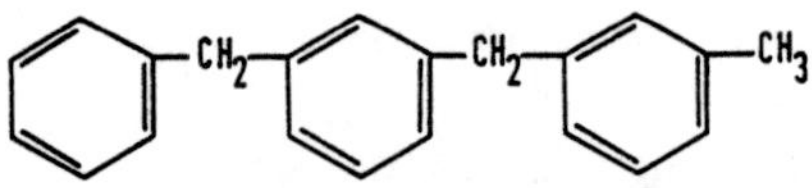

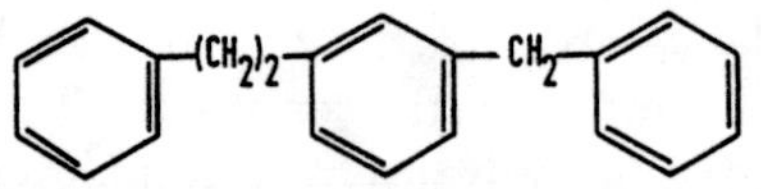

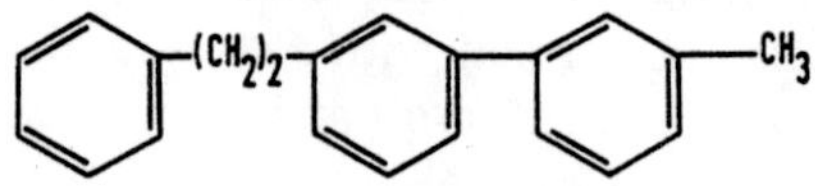

Figure 7. Polymeric material produced upon heating the Cu^{2+}-montmorillonite–toluene system

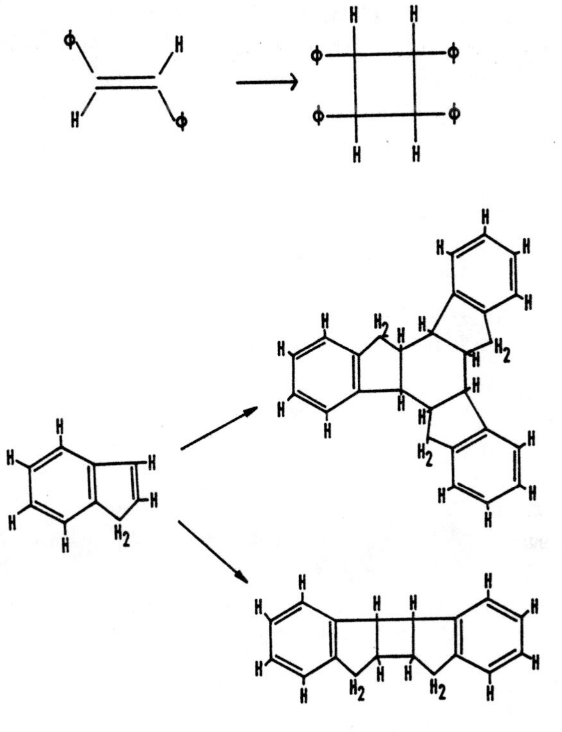

Figure 8. Volatile products formed on heating (a) Cu^{2+}-montmorillonite–trans-stilbene and (b) Cu^{2+}-montmorillonite–indene

compared with the benzene ring, by the copper sites. Reasonable amounts of the trimers and dimers of indene and of the cyclobutane product (dimer of *trans*-stilbene) are obtained on heating the Cu–M:indene and Cu–M:*trans*-stilbene complexes, respectively, in vacuo at 50°–250°C (*see* Figure 8). There is, therefore, strong evidence here that a solid matrix (the montmorillonite), which functions both as a framework to which the reactant is attached and also as a support or promoter of the catalyst (Cu^{2+} ions), can discriminate between two types of unsaturated carbon–carbon bonds. This phenomenon clearly merits further study.

In no case studied to date is there evidence that copper atoms are carried away with the volatile products. However, it is not difficult to design sheet silicate systems in which organometallic products, such as metallated benzenes, could be produced to order using organic intercalates of sheet silicates. Some of these possibilities are currently under investigation.

In this article we have restricted our attention entirely to thermally stimulated chemical conversions. It must not be thought that radiation-induced reactions are either impossible or hardly worth investigating. Indeed some elegant studies of γ-ray induced polymerizations of intercalated monomers of acrylonitrile (AN) and methacrylonitrile (in Na$^+$-montmorillonite) have already been reported (36). The insertion polymers so formed were found to be extensively cyclized, an occurrence which is interpretable in terms of the specific orientation of the intercalated monomer species. It is often possible to insert two or even three layers of organic molecules between the aluminosilicate sheets, and the resulting two-dimensional organization is then akin to that which prevails in a smectic mesophase. Again, it is obvious that much scope for further study exists with such intercalates.

Acknowledgments

We are grateful for the numerous stimulating discussions we have had with our colleagues M. J. Tricker, J. O. Williams, Stephen Evans, and J. S. Anderson. We are also indebted to P. I. Reid and M. J. Walters for their helpful contributions. J. M. Thomas also thanks the Petroleum Research Fund of the American Chemical Society for financial assistance in attending the New York Meeting of the Society.

Literature Cited

1. Timms, P. L., *J. Chem. Soc., Chem. Comm.* (1969) 1033.
2. Benfield, F. W. S., Green, M. L. H., Ogden, J. S., Young, D., *J. Chem. Soc., Chem. Comm.* (1973) 866.
3. Williams-Smith, D. L., Wolf, L. R., Skell, P. S., *J. Am. Chem. Soc.* (1972) 94, 4042.
4. Ozin, G. A., Voet, A. V., *Acc. Chem. Res.* (1973) 9, 313.
5. Ogden, J. S., Turner, J. J., *Chem. Br.* (1971) 7, 186.
6. Tennakoon, D. T. B., Thomas, J. M., Tricker, M. J., Graham, S. H., *J. Chem. Soc., Chem. Comm.* (1974) 124.
7. Tennakoon, D. T. B., Thomas, J. M., Tricker, M. J., *J. Chem. Soc., Dalton Trans.* (1974) 2207.
8. Tennakoon, D. T. B., Thomas, J. M., Tricker, M. J., *J. Chem. Soc., Dalton Trans.* (1974) 2211.
9. Adams, J. M., *J. Chem. Soc., Dalton Trans.* (1974) 2286.
10. Tricker, M. J., Tennakoon, D. T. B., Thomas, J. M., Heald, J., *Clays Clay Miner.* (1975) 23, 77.
11. Adams, J. M., Thomas, J. M., Walters, M. J., *J. Chem. Soc., Dalton Trans.* (1975) 1459.
12. Tricker, M. J., Tennakoon, D. T. B., Thomas, J. M., Graham, S. H., *Nature* (1975) 253, 110.
13. Tennakoon, D. T. B., Tricker, M. J., *J. Chem. Soc., Dalton Trans.* (1975) 1802.
14. Adams, J. M., Thomas, J. M., Walters, M. J., *J. Chem. Soc., Dalton Trans.* (1976) 1975.
15. Adams, J. M., Graham, S. H., Reid, P. I., Thomas, J. M., *J. Chem. Soc., Chem. Comm.* (1977) 67.
16. Adams, J. M., Jefferson, D. A., *Acta Crystallogr.* (1976) B32, 1180.

17. Brindley, G. W., *Reun. Hisp.-Belga Miner. Arcilla, An.* (1970) 55.
18. Theng, B. K. G., "The Chemistry of Clay-Organic Reactions," Adam Hilger, London, 1974.
19. Adams, J. M., Lukawski, K., Reid, P. I., Thomas, J. M., Walters, M. J., *J. Chem. Res.* (1977) 301.
20. Weiss, A., *in* "Organic Geochemistry," G. Eglington and M. T. J. Murphy, Eds., Springer-Verlag, Berlin, 1969.
21. Adams, J. M., Thomas, J. M., Walters, M. J., *J. Chem. Soc., Dalton Trans.* (1976) 112.
22. Weiss, A., *Angew. Chem. Int. Ed. Engl.* (1963) **2**, 697.
23. Evans, E. L., Thomas, J. M., *J. Solid State Chem.* (1975) **14**, 99.
24. Yoffe, A. D., in *Festkoerperprobleme*, XIII, H. J. Queisser, Ed., p. 1, Pergamon, London, 1973.
25. Parry, G. S., Scruby, C. B., Williams, P. M., *Philos. Mag.* (1974) **29**, 601.
26. Thomas, J. M., *Philos. Trans. R. Soc. London* (1974) **277**, 251.
27. Lailach, G. E., Brindley, G. W., *Clays Clay Miner.* (1969) **17**, 95.
28. Tennakoon, D. T. B., Ph.D. thesis, University College of Wales, Aberystwyth, 1974.
29. Donor, H. E., Mortland, M. M., *Science* (1969) **166**, 1406.
30. Mortland, M. M., Pinnavaia, T. J., *Nature (London) Phys. Sci.* (1971) **229**, 75.
31. Pinnavaia, T. J., Mortland, M. M., *J. Phys. Chem.* (1971) **75**, 3957.
32. Fenn, D. B., Mortland, M. M., Pinnavaia, T. J., *Clays Clay Miner.* (1973) **21**, 315.
33. Rupert, J. P., *J. Phys. Chem.* (1973) **77**, 784.
34. Turner, R. W., Amma, E. L., *J. Am. Chem. Soc.* (1966) **88**, 1877.
35. Pinnavaia, T. J., Hall, P. L., Cody, S. S., Mortland, M. M., *J. Phys. Chem.* (1974) **78**, 994.
36. Blumstein, R., Blumstein, A., Parkikh, K. K., *Appl. Polym. Symp.* (1974) **25**, 81.

RECEIVED July 27, 1976.

18

High-Temperature Electrolysis/Fuel Cells: Materials Problems

H. OBAYASHI and T. KUDO

Central Research Laboratory, Hitachi, Ltd., Higashi-Koigakubo, Kokubunji, Tokyo, 185, Japan

The motivation for electrolysis/fuel cell technology is summarized, and problems regarding the development of high-temperature electrolysis cells and medium-temperature fuel cells are discussed: (a) a suitable solid electrolyte, (b) a cathode for operating in a highly oxidizing atmosphere mechanically compatible with the electrolyte, (c) a cell design that minimizes material and heat transfer problems, and (d) an electronic conductor stable in a wide range of oxygen partial pressures (1–10^{-20} atm P_{O_2}) for series connection. Materials research addressed to solving these problems is reviewed.

Fuel cells convert chemical energy into electrical energy with the dissipation of heat; electrolysis cells convert heat and electrical energy into chemical energy. If the enthalpy of the chemical reaction is

$$\Delta H = \Delta G + T\Delta S,$$

the electrical energy derived from a fuel cell (or consumed in an electrolysis cell) is ΔG and the heat dissipated (or consumed) is $T\Delta S$. In the absence of overvoltages and other circuit losses an engine consisting of a high-temperature electrolysis cell and a low-temperature fuel cell would convert heat to electricity and/or fuel with an efficiency approaching the Carnot factor. In practice, losses at the electrodes are comparable with the net energy derived from such a cycle, so such heat engines have not been developed. However, as the supply of cheap fossil fuels becomes exhausted, there is renewed incentive to develop cells with reduced electrode loss, for the versatility of an engine that can convert heat to either electricity or fuel would have tremendous advantages in a nuclear-energy economy. Two approaches to this problem are (a) the development of better catalytic electrodes and (b) operation at higher temperatures. Because liquid electrolytes become too corrosive at high temperatures, solid electrolytes offer great advantages for use with liquid and gaseous fuels.

Two factors have inhibited the use of solid electrolytes: (a) adequate cell design and (b) adequate ionic mobility. Because ions move through solids via a diffusion process, the ionic mobility contains an exponential factor $\exp(-E_a/kT)$, where E_a is referred to as the activation energy. Unless E_a is unusually small (0.1–0.3 eV), the ionic mobility gives a joule loss in the electrolyte that is tolerable only at high temperatures. This situation is acceptable for a high-temperature electrolysis cell, and the reasons such cells have not been developed are, essentially, (a) inadequate cell design and (b) insufficient incentive given an available supply of cheap fossil fuels. We should anticipate better designs of high-temperature electrolysis cells as the incentive for developing synthetic fuels becomes more intense.

To date, engineering design has concentrated on fuel cells, since they offer the possibility of converting the chemical energy of an available fuel into electrical and mechanical energy without the customary restriction of the Carnot efficiency factor. Many types have been proposed (1, 2, 3, 4, 5), and each has advantages for a particular application. However, fuel cells for extensive terrestrial application are only now becoming commercially available from a single supplier (6). They use a liquid electrolyte and can accept a fairly wide range of fuels. Their efficiency is limited by the temperature of operation, which is restricted because of the corrosive character of the electrolyte. Cells that employ either solid or molten-salt electrolytes for operation above 300°C are called high-temperature (or medium-temperature) fuel cells (7, 8, 9). Since molten-salt electrolytes are corrosive, attention is given in this article to cells utilizing solid electrolytes.

The high-temperature operation made possible by solid electrolytes has several advantages. (a) Improved kinetics at the electrodes allow acceptably high current densities with methane, propane, and other hydrocarbon fuels that do not react smoothly at ambient temperatures. (b) In the absence of a liquid phase in the system, maintaining the "ternary phase boundary region" to avoid wetting of the electrode need not be considered. (c) Fabrication, operation, and maintenance of the system is made easier by reduced corrosion and sealing problems. (d) Control of the water/fuel balance is easier because the composition of the electrolyte is invariant and independent of the composition of the fuel gas. (e) A reduced polarization of the system permits operation at higher current densities, and hence higher power densities.

On the other hand, working at elevated temperatures has disadvantages: a lower output voltage and problems with mismatch of the thermal expansion of electrode and electrolyte materials. Although the latter problem can be engineered around by appropriate cell design, both problems make it desirable to operate a fuel cell at as low a temperature as possible, compatible with the power requirements. A medium-temperature range (200°–400°C) appears most desirable were it possible to find a solid electrolyte having acceptable ionic mobilities in this temperature range. The mobile ionic species of the solid electrolyte would be either O^{2-} or H^+ ions (most fuels offering H^+ as the mobile component). Because no known proton conductors with acceptable H^+-ion mobilities are stable above 200°C, the evaluation of solid electrolytes in fuel cells has been confined to O^{2-}-ion conductors. Unfortunately, acceptable O^{2-}-ion mobilities occur only above 800°C, a temperature considerably above the optimum for fuel-cell operation. Nevertheless, the materials considerations for a high-temperature fuel cell are similar to those for a high-temperature electrolysis cell, and the O^{2-}-ion conductors considered here are excellent candidates for this latter application. Clearly the discovery of a solid electrolyte capable of transporting O^{2-} or H^+ ions with acceptable mobilities in the temperature range 200°–400°C would revolutionize the design and use of fuel cells.

Development of a commercial electrolysis/fuel cell using a solid electrolyte requires the solution of several problems: (a) selection of a suitable material for the solid electrolyte; (b) identification of a chemically and mechanically compatible cathode that remains a good electronic conductor in a highly oxidizing atmosphere, is thermodynamically stable, and contains abundant, inexpensive elements; (c) simple cell design that allows series connection of individual units and makes acceptable any mismatch in thermal expansion of the electrolyte and the electrodes; and (d) identification of an inexpensive electronic conductor stable in a wide range of oxygen partial pressures ($1-10^{-20}$ atm p_{O_2}) for series connection. Of these, the most important immediate problem is the solid electrolyte. At an operating temperature near 350°C the solid O^{2-}-ion electrolyte should have a high O^{2-}-ion mobility and a high O^{2-}-ion transference number (ratio of O^{2-}-ion component to total electrical conductivity) over a wide range of oxygen partial pressures. It must be chemically stable over a wide temperature range, made of inexpensive materials, and easily fabricated into dense ceramic membranes of arbitrary shape.

This paper reviews the present status of materials research into O^{2-}-ion solid electrolytes, electrodes, and cell interconnectors.

O^{2-}-Ion Solid Electrolytes

General Considerations. The discovery and utilization of oxygen-ion conduction in a ceramic oxide dates back to before 1900, when Nernst used yttria-stabilized zirconia, $xY_2O_3 \cdot (1-2x)ZrO_2$ crystallizing in the fluorite structure, as a glower element, known as the Nernst mass (*10*). Many workers have contributed to this field since then (*11*). Although Nernst's glower element was not a commercial success, oxides conducting O^{2-} ions have been investigated for a variety of other applications, including the now commercially available oxygen sensors used for the study of the thermodynamic properties of oxides at elevated temperatures (*8, 12*). The use of ceramic O^{2-}-ion conductors as solid electrolytes in fuel cells has been studied intensively at several institutes (*13, 14, 15, 16*), especially during the late 1960s. Although the cell designs were inadequate and the high-temperature application is better suited to electrolysis, these studies have provided important information about the performance of O^{2-}-ion conductors under operating conditions and some of the problems to be encountered because of materials mismatch between electrodes and electrolyte. The O^{2-}-ion conductors most extensively studied are the ZrO_2-based systems (*13*). Other systems with the fluorite structure, for example those based on ThO_2 and CeO_2, have also been investigated.

The O^{2-}-ion conductors that have been studied most intensively generally exhibit a negligible cationic conductivity (*17, 18, 19*). A general model for such an O^{2-}-ion conductor consists of a fixed cation subarray within which the anions move by jumping from one energetically equivalent site to another.

The mobility μ_O of the ionic motion is related to the diffusion coefficient D_O through the Nernst–Einstein relation,

$$\mu_O = 2eD_O/kT \tag{1}$$

where $-2e$ is the charge on the mobile O^{2-} ion and the sign of μ_O is chosen for convenience. From random-walk theory,

$$D_O = (2n_{DO})^{-1}[V_O'']z_O a_O^2 \nu_O \tag{2}$$

where $n_{DO} = 1, 2,$ or 3 is the dimensionality of the ionic motion, $[V_O'']$ is the concentration of doubly ionized oxygen vacancies (each of which is capable of trapping two electrons), z_O is the number of equivalent nearest-neighbor sites a jump distance a_O away, and ν_O is the jump-attempt frequency of the O^{2-} ions. The extrinsic O^{2-}-ion conductivity is

$$\sigma_O(\text{ext}) = N_O(1 - [V_O''])e\mu_O \tag{3}$$

where N_0 is the density of energetically equivalent oxygen sites on which the ions move. Kingery et al. (20) have demonstrated (*see below*) that O^{2-}-ion conduction in $Zr_{0.85}Ca_{0.15}O_{1.85}$ takes place via the doubly ionized oxygen vacancies V_O'', and we assume this to be true for all O^{2-}-ion conductors considered here. Finally, since a free energy $\Delta G_O = \Delta H_O - T\Delta S_O$ is required to make a jump, the jump frequency is

$$v_O = v_{OO}\exp(-\Delta G_O/kT) = v_{OO}\exp(\Delta S_O/k)\exp(-E_a/kT) \quad (4)$$

where $E_a = \Delta H_O$ is referred to as the activation energy for ionic mobility.

From the factor $(1 - [V_O''])[V_O'']$ appearing in $\sigma_O(\text{ext})$, it is clear that the ionic conductivity vanishes if the N_0 equivalent sites are either completely occupied or completely empty. Such a situation is analogous to a stoichiometric semiconductor or insulator in which intrinsic electronic conductivity requires the creation of mobile electrons and holes by the thermal excitation of electrons from the valence band to the conduction band. In the electronic analogy, the product of the concentrations of intrinsic electrons and holes is

$$[e]_{\text{int}}[h]_{\text{int}} = K_0\exp[-E_g/kT] \quad (5)$$

where K_0 is an equilibrium constant and E_g is the energy gap between the top of the valence band and the bottom of the conduction band. Similarly, intrinsic ionic conductivity requires thermal excitation of O^{2-} ions from normal positions to interstitial positions across an enthalpy barrier ΔH_i:

$$O^{2-} \overset{K_i}{\rightleftharpoons} V_O'' + O_i^{2-} \quad (6)$$

where O_i^{2-} is an interstitial ion and the equilibrium constant is

$$K_i(T) = [V_O''][O_i^{2-}] = K_{i0}\exp(-\Delta H_i/kT) \quad (7)$$

A significant concentration of mobile electronic charge carriers at normal temperatures can be introduced into a semiconductor by a suitable chemical substitution. Similarly, it is customary in ionic conductors to introduce mobile ionic charge carriers by appropriate chemical substitutions. In the stabilized zirconias already mentioned, the following substitutions were made:

$$Zr^{4+} + O^{2-} \text{ by } Ca^{2+} + V_O'' \quad (8a)$$

$$2Zr^{4+} + O^{2-} \text{ by } 2Y^{3+} + V_O'' \quad (8b)$$

An oxygen position adjacent to a substitutional cation is not crystallographically, and hence energetically, equivalent to one that is not. However, with sufficient cation substitution V_O'' is able to migrate to a position neighboring another substitutional cation via a nearest-neighbor jump, and the binding energy between dopant and V_O'' may be neglected at operating temperatures. The situation becomes analogous to extrinsic electronic or hole conduction in a highly doped semiconductor, except that the charge–carrier mobility μ_O is activated (Equations 1–4) as in the case of a small-polaron conductor.

The populations of oxygen vacancies and mobile electrons and holes in such a cation-substituted O^{2-}-ion conductor are also influenced by the chemical equilibrium of the sample with gaseous oxygen $O_2(g)$ (21):

$$\tfrac{1}{2}O_2(g) + V_O'' \overset{K_h}{\rightleftharpoons} O^{2-} + 2h \tag{9a}$$

$$O^{2-} \overset{K_e}{\rightleftharpoons} \tfrac{1}{2}O_2(g) + V_O'' + 2e \tag{9b}$$

From the law of mass action the equilibrium constants are

$$K_h(T) = [h]^2 \cdot [V_O'']^{-1} \cdot p_{O_2}^{-1/2} = K_{ho}\exp(-\Delta H_h/kT) \tag{10a}$$

$$K_e(T) = [e]^2 \cdot [V_O''] \cdot p_{O_2}^{1/2} = K_{eo}\exp(\Delta H_e/kT) \tag{10b}$$

where p_{O_2} is the partial pressure of oxygen and ΔH_h, ΔH_e are the enthalpies for oxidation and reduction, respectively. The corresponding electronic charge-carrier concentrations are

$$[h] = K_h^{1/2}[V_O'']^{1/2}p_{O_2}^{1/4} \tag{11a}$$

$$[e] = K_e^{1/2}[V_O'']^{-1/2}p_{O_2}^{-1/4} \tag{11b}$$

If the valence band of the oxide is an $O^{2-}:2p^6$ band, ΔH_h is large and $[h]$ is negligible. However, if the valence band is a narrow d or f band, as may occur in transition-metal and rare-earth oxides, ΔH_h may be small enough to give a significant concentration of mobile holes $[h]$ at higher oxygen partial pressures. Similarly, if the conduction band associated with the lowest empty orbitals on the cation array is relatively stable, ΔH_e may be small enough to give a significant concentration of mobile electrons $[e]$ at lower oxygen partial pressures.

The total electrical conductivity is the sum of the partial conductivities caused by holes, electrons, and O^{2-} ions:

$$\sigma_h = N_h[h]\, e\mu_h \qquad (12a)$$

$$\sigma_e = N_e[e]\, e\mu_e \qquad (12b)$$

$$\sigma_O = N_O(1 - [V_O''])2e\mu_O + N_i[O_i^{2-}]2e\mu_O^i \qquad (12c)$$

where N_m is the density of crystallographic sites available to charge-carrier species m, e is the magnitude of the electronic charge, and the signs of the mobilities μ_m are so defined (see Equation 1) that the product of the mobility and the charge carried is always positive. The mobility μ_O^i refers to the interstitial O_i^{2-} ions, and N_i is the density of interstitial sites.

The transference number for O^{2-}-ion conduction is defined as

$$t_O = \sigma_O/(\sigma_h + \sigma_e + \sigma_O) \qquad (13)$$

For a solid electrolyte it is desirable to have $t_O = 1$. This ideal may be approached in oxides with large band gaps ($E_g > 100kT$), provided ΔH_h and ΔH_e are also large.

For a given ΔH_h and ΔH_e, the transference number of an extrinsic O^{2-}-ion conductor is

$$t_O = (\alpha p_{O_2}^{1/4}[V_O'']^{-1/2} + \beta p_{O_2}^{-1/4}[V_O'']^{-3/2} + 1)^{-1} \qquad (14)$$

where

$$\alpha = K_h^{1/2}\mu_h/2\mu_O \text{ and } \beta = K_e^{1/2}\mu_e/2\mu_O \qquad (15)$$

Figure 1 shows schematically the dependence of t_O on the oxygen partial pressure p_{O_2} (22). Significantly, the range of oxygen pressures at which

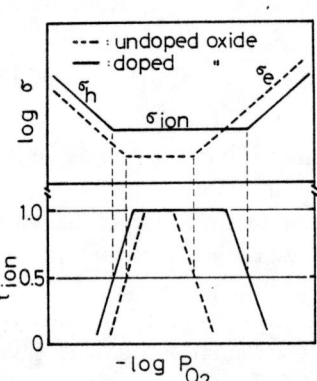

Figure 1. Schematic showing the effect of doping on conductivity and ionic transference number

conduction is predominantly ionic can, for a given ΔH_h and ΔH_e, be widened by increasing $[V_O'']$ with chemical doping.

Schmalzried (23) has defined p_h^* and p_e^* as the oxygen partial pressures at which $\sigma_h = \sigma_0$ and $\sigma_e = \sigma_0$ (i.e. $t_0 = 0.5$). Then

$$t_0 = \{1 + (p_{O_2}/p_h^*)^{1/4} + (p_{O_2}/p_e^*)^{-1/4}\}^{-1} \quad (16)$$

and it is sufficient to know p_h^* and p_e^* to calculate t_0 at any given oxygen partial pressure. Conversely, p_h^* and p_e^* can be obtained from conductivity measurements that give t_0 under a few different oxygen partial pressures (24, 25). Schmalzried's (23, 26) original equation to estimate these parameters is based on the emf of an oxygen concentration cell:

$$\epsilon = \frac{RT}{F}\left\{\ln\frac{p_h^{*n} + p_{O_2}'^n}{p_h^{*n} + p_{O_2}''^n} + \ln\frac{p_e^{*n} + p_{O_2}''^n}{p_e^{*n} + p_{O_2}'^n}\right\} \quad (17)$$

where F is Faraday's constant and p_{O_2}', p_{O_2}'' are, respectively, the oxygen partial pressures at the anode and cathode ($p_{O_2}'' > p_{O_2}'$). If the ionic defects are fully ionized, as is generally the case at elevated temperatures (20), $n = \frac{1}{4}$.

The "electrolytic domain" is defined as the domain of oxygen partial pressures in which t_0 is greater than some specified value such as 0.99 (27, 28). Most O^{2-}-ion conductors have an $O^{2-}:2p^6$ valence band and hence a large ΔH_h. In these cases it is practically more important to determine the value of p_e^* in order to check the suitability of a material for a fuel/electrolysis cell or an oxygen sensor (29, 30, 31).

In addition to chemical stability against oxidizing and reducing atmopsheres (large ΔH_h and ΔH_e) and a large intrinsic electronic energy gap E_g, all to minimize any electronic contribution to the conductivity, a good O^{2-}-ion electrolyte must have a high μ_O, and hence a low activation energy E_a for ionic mobility. Extensive studies on O^{2-}-ion electrolytes have shown that the O^{2-}-ion mobilities may vary greatly with crystal structure. Although little attention has been given to learning why certain crystal structures are more favorable than others for high ionic mobility, it has been found empirically that most of the good O^{2-}-ion conductors known crystallize in the fluorite, perovskite, or pyrochlore structures. The Bi_2O_3 and C-type lanthanide structures, which are related to fluorite with V_O'' in an ordered array, are also good host structures. In what follows, representative solid electrolytes from each structural group are discussed in some detail.

Fluorites. STRUCTURE. The cubic fluorite structure corresponds to chemical formula MX_2. It consists of a face-centered-cubic array of cations M with anions X occupying all of the tetrahedral interstices. The octahedral sites are empty; they represent interstitial positions for the anions. Introduction of vacancies into the tetrahedral sites permits anion mobility, and the most probable jump path is via an octahedral site that shares common faces with the two neighboring tetrahedral sites. The activation energy E_a increases as the energy required to place an anion into an adjacent interstitial site increases.

The C-type lanthanide structure (M_2X_3) contains 25% anion vacancies that are ordered among the tetrahedral sites. In such an ordered structure the tetrahedral-site vacancies are not crystallographically equivalent and must be considered interstitial positions.

DOPED ZrO_2. Pure ZrO_2 has two crystallographic modifications: it is monoclinic at low temperatures and tetragonal at high temperatures (32, 33, 34). Intrinsic electronic conductivity competes with intrinsic ionic conductivity, and the electronic mobilities are much greater than the ionic mobilities. Electronic conductivity dominates the monoclinic phase (35, 36, 37); the tetragonal modification shows a mixed ionic/electronic conduction of comparable magnitude (38). Figure 2 shows the ionic and electronic conductivities of pure ZrO_2 as a function of the partial pressure of oxygen p_{O_2} for two temperatures, 600°C and 900°C. At higher p_{O_2}, interstitial oxygen ions O_i^{2-} are introduced; these are the mobile ionic species. At lower p_{O_2}, oxygen vacancies V_O'' are introduced, and the normal-site O^{2-} ions are mobile. At intermediate p_{O_2} both the ionic and electronic conductivities are intrinsic.

On the other hand, ionic conductivity dominates in cation-substituted ZrO_2, even in the monoclinic phase. For example, $0.01Y_2O_3 0.99ZrO_2$ has a transference number $t_O > 0.9$ at 600°C and a total conductivity nearly three orders of magnitude larger than that of pure ZrO_2 (37).

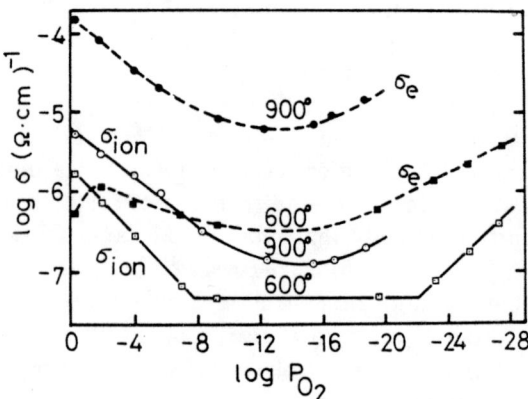

Figure 2. Electronic and ionic partial conductivity isotherms for pure ZrO_2. Adapted from Ref. 37.

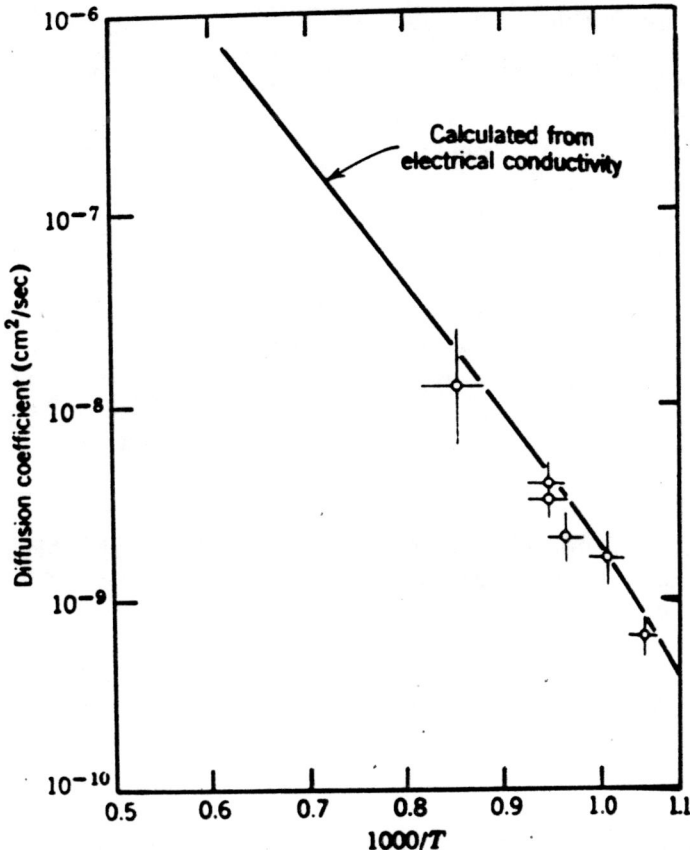

Figure 3. Comparison of directly measured diffusion constants with a line calculated from electrical conductivity and the Nernst–Einstein relation. Adapted from Ref. 20.

Kingery et al. (20) have demonstrated that the high O^{2-}-ion conductivity in $Zr_{0.85}Ca_{0.15}O_{1.85}$ is caused by normal-site O^{2-} ions hopping to doubly ionized oxygen vacancies $[V_O'']$. They measured the oxygen mobility at 700°–1100°C by introducing O^{18} via ion exchange and employing mass-spectrometer analysis. They also measured the total electrical conductivity and compared it with the ionic conductivity calculated from the measured ionic mobility and Equation 3. The comparison is shown in Figure 3. Within experimental error, the entire electrical conductivity can be attributed to an extrinsic ionic conductivity in which the ionic mobility is proportional to the concentration of doubly ionized vacancies, $[V_O'']$. Simpson and Carter (18) obtained similar results with single-crystal and polycrystalline $Zr_{0.858}Ca_{0.142}O_{1.858}$.

With larger cation substitutions a cubic fluorite structure is stabilized relative to the monoclinic and tetragonal modifications of pure ZrO_2. The anion-deficient fluorite structure has a significantly larger O^{2-}-ion conductivity, and it is this phase that is of interest for solid electrolytes. In the system $Ca_xZr_{1-x}O_{2-x}$, the cubic phase is stabilized above 1700°C for $x >$ 0.12, above 1300°C for $x > 0.16$ (39). Maximum conductivities are observed in the range $0.13 > x > 0.15$ (40). The systems $M_{2x}Zr_{1-2x}O_{2-x}$, where M is a trivalent cation of comparable size (M = Sc, Y, Nd, Sm, Gd...) have also been extensively investigated (41). Figure 4 shows conductivity vs. dopant concentration at 800°C for several of these systems (42). In most cases the conductivity maximum occurs at the minimum dopant level needed to stabilize the cubic phase. This observation seems to indicate that the magnitude of the mobility is not obtained by optimizing the product $[V_O''](1-[V_O''])$ with the simplified assumption of random oxygen vacancies in the fluorite structure (43, 44). In $Ca_{0.15}Zr_{0.85}O_{1.85}$, for example, the oxygen-vacancy concentration is $[V_O''] = 0.15/2 = 0.075$. At the large vacancy concentrations required for stabilization of the cubic phase, the dilute-solution approximation, which neglects interactions between the vacancies, may not hold.

Indeed it has been suggested that a cubic phase with disordered oxygen vacancies is not stable below about 1000°C (45, 46, 47, 48, 49). Baukal (50) found that a decrease with time in the conductivity of $(ZrO_2)_{0.91}(Y_2O_3)_{0.09}$ on lowering the temperature to 800°C could be described by first-order kinetics and that the activation energy E_a for ionic conduction remained unchanged during the aging process. This result provides clear evidence for some kind of ordering of the oxygen vacancies, and this inference has been partially confirmed by direct observation from x-ray and neutron-diffraction studies for the aged specimens (40, 48, 49). This aging effect, observed for many systems, appears below a critical temperature but has a cutoff at low temperatures where the ionic mobility becomes too low for aging to take place. In the system $Ca_xZr_{1-x}O_{2-x}$, ordering occurs more readily at higher values of x (47, 48, 51, 52); it has a maximum rate of 1000°C (53), it does not occur for $T > 1250$°C, and it is too slow to be observable for $T < 650$°C (54).

For a given dopant concentration the ionic mobility is dependant upon the lattice parameter and hence on the ionic radius. Table I summarizes the conductivities of several $M_{2x}Zr_{1-2x}O_{2-x}$ systems at 1000°C for several dopant concentrations (12). The ionic radii of the dopants decrease vertically in the table, and it is apparent that the ionic conductivity increases with decreasing radius of the dopant ion. Presumably the difference in tetrahedral vs. octahedral site-preference energy decreases with decreasing lattice parameter. Maximum conductivities occur

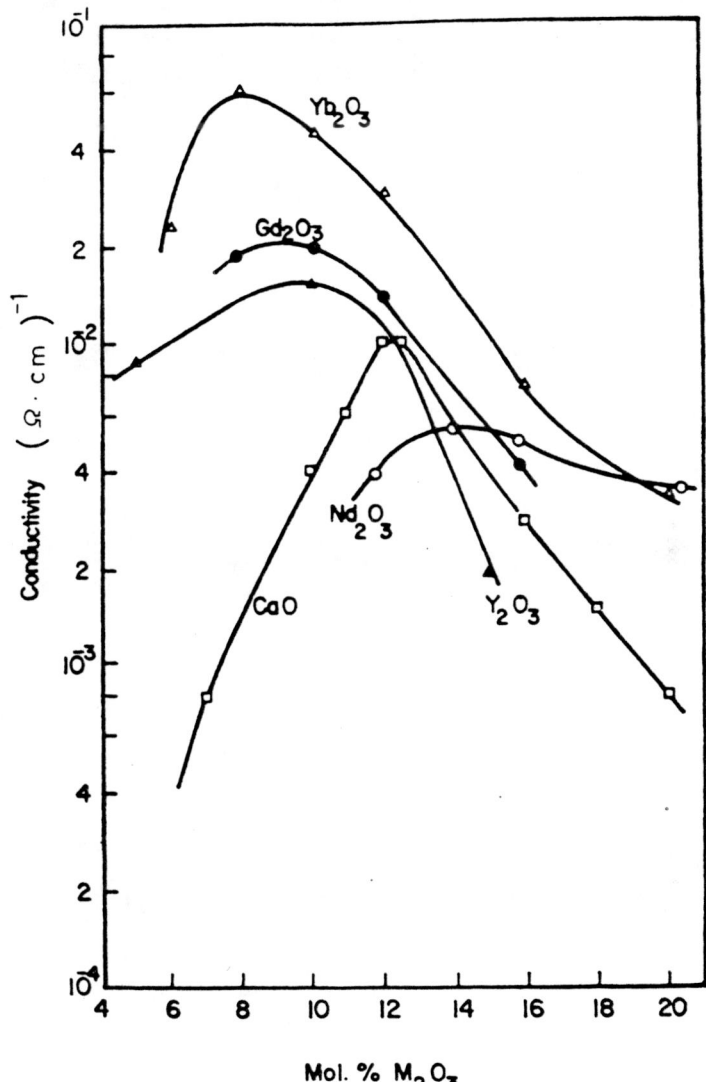

Figure 4. Conductivities of the ZrO_2–M_2O_3 system at 800°C. Adapted from Ref. 42.

in Sc_2O_3-doped ZrO_2, where activation energies as low as $E_a = 0.8$ eV have been obtained.

Figure 5 summarizes measurements of p_e^* for the system $Ca_xZr_{1-x}O_{2-x}$. For a given p_{O_2}, a plot of log p_e^* vs. $1/T$ is theoretically a straight line. Although the values measured by different workers (55–60) are scattered, they fall in the range $-30 < \log p_e^* < -23$ atm at 1000°C, which

Table I. Ionic Conductivities $((\Omega \text{ cm})^{-1} \times 10^2)$ for ZrO_2–M_2O_3 Electrolytes at $1000°C$[a]

	Composition (mol % Mn_2O_3)[b]				
	8	9	10	15	Ref.
Nd_2O_3			0.60		54a
				1.7	54b
				1.4 (24.9)	54c
Sm_2O_3			5.8 (22.0)	2.3 (26.1)	54d
Gd_2O_3	11		11	3.1	54e
Yb_2O_3		15		3.2	54a
	8.8 (17.3)		11 (19.6)	3.9 (26.1)	54d
				4.9	54e
Lu_2O_3	1.5			1.2	54f
Sc_2O_3			24	13	54a
	2.5		25 (14.9)	15 (15.8)	54d
	1.1			0.84	54f

[a] Adapted from Ref. *12*.
[b] Activation energies (kcal/mol) are in parentheses.

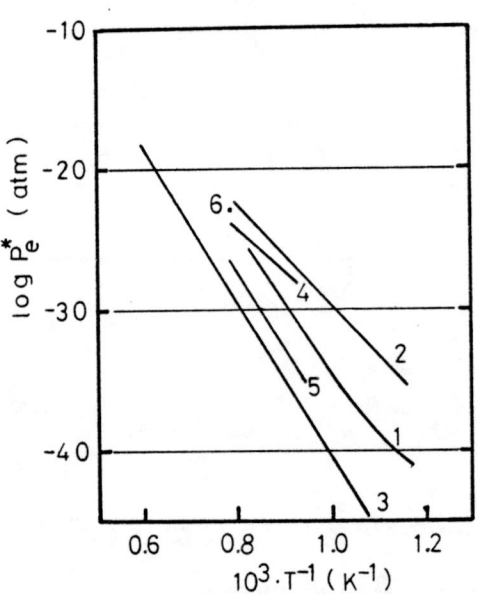

Figure 5. The P_e^* values for CaO-stabilized ZrO_2's. Curves 1–5 are adapted from Refs. 55, 51, 57, 58, and 59, respectively.

guarantees an acceptably high t_0 for high-temperature applications as a solid electrolyte (61, 31).

DOPED CeO_2. Pure CeO_2 already crystallizes in the fluorite structure. In this compound, a localized $4f^1$ level lies in the energy gap between the $O^{2-}:2p^6$ valence band and the broad $Ce^{4+}:5d, 6s$ conduction band. Consequently, it is relatively easy to reduce CeO_2 with the introduction of localized $4f^1$ electrons to form Ce^{3+} ions. These electrons are mobile, but the transition time for an electron transfer $Ce^{3+} + Ce^{4+} \rightarrow Ce^{4+} + Ce^{3+}$ is long compared with the period of an optical-mode lattice vibration; therefore the electrons become trapped at individual cations by a local lattice relaxation. The electron and its local lattice deformation is called a small polaron. Since the lattice deformation introduces an activation energy E_p for electron transfer, the mobility of the small polaron is described by diffusion theory.

$$\sigma_p = (Ne^2/kT)[p](1 - [p])a^2\nu_{op}\exp(\Delta S_p/k)\exp(-E_p/kT) \quad (18)$$

where [p] is the concentration of small polarons. This expression is directly comparable with that for $\sigma_0(\text{ext})$ obtained from Equations 1–4. Because CeO_2 is readily reduced, "pure" CeO_2 exhibits n-type electronic (small-polaron) conduction.

Investigations (62, 63, 64) of the conduction mechanisms in CeO_2 have given conflicting results, which is not surprising in view of the sensitivity of the oxygen content to p_{O_2}. Kevane et al. (65) measured the electrical conductivity as a function of temperature (250°–1500°C) and oxygen pressure ($p_{O_2} = 1\text{–}10^{-4}$ atm); they reported mixed electronic/ionic conduction. Noddak and Walch (66), Rudolph (67), Greener et al. (68), and Blumenthal and Pinz (69) found no evidence of ionic conductivity in their measurements. The lack of any measurable ionic conductivity implies no V_O'' in these samples, and indeed interstitial Ce ions have been identified as native defects in pure and nonstoichiometric CeO_2 (70, 71). Blumenthal and Hofmaier (72) showed that their values of $\log \sigma$ vs. $1/T$ for several values of x in CeO_{2-x} (see Figure 6) could be well interpreted by a $\sigma \propto x \exp(-E_p/kT)$, which agrees with Equation 18 for small $x \propto [p]$.

On the other hand, divalent or trivalent dopant cations in cation-substituted CeO_2 introduce oxygen vacancies V_O'' just as in doped ZrO_2. With the introduction of a finite $[V_O'']$, the O^{2-}-ion conduction becomes significant, and transference numbers $t_0 > 0.8$ are obtained.

The system $La_{2x}Ce_{1-2x}O_{2-x}$ has been studied by Croatto and Mayer (73), Takahashi et al. (8, 74, 75), and Singman (76). Figure 7 shows typical conductivity isotherms vs. x. All the specimens were sintered in air at 1600°C, and the oxygen transference numbers were $t_0 > 0.8$ over

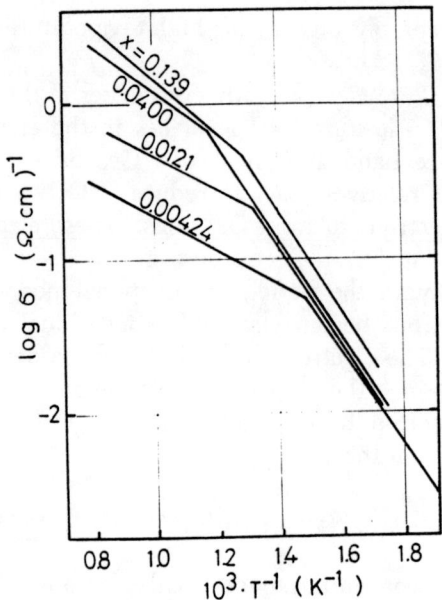

Figure 6. Conductivities of nonstoichiometric CeO_{2-x} for several values of x. Adapted from Ref. 72.

the temperature range 600°–800°C for $x > 0.1$. In the composition range $x < 0.1$, t_0 increases sensitively with x.

Figures 8(a) and 8(b) show curves of resistivity vs. reciprocal temperature for the systems $Ln_{2x}Ce_{1-2x}O_{2-x}$, $x = 0.15$, where Ln is a trivalent rare-earth ion (77). All specimens were, in this case, sintered at 1800°C for three hours, and the transference numbers were $t_0 > 0.95$ at $T > 700°C$ for Ln = Nd, Sm, Gd, Dy, and Er, as can be seen from Figure 9. In fact, for Ln = Er and Gd a $t_0 \simeq 1$ was achieved at temperatures $t > 600°C$. The change in resistivity with x is illustrated in Figure 10 for Ln = Gd. The smallest resistivity occurred for $x = 0.115$; at 750°C the value is comparable with the best stabilized zirconia at 1000°C.

Studies of other dopants include CaO (78, 79), Y_2O_3 (80), BeO, MgO, SrO, and BaO (81). Figure 11 compares the conductivities vs. reciprocal temperature of CaO-doped ZrO_2 and CeO_2 doped with CaO and Y_2O_3. Tuller and Nowick (80) measured the dependence of total conductivity on p_{O_2} for $(CeO_2)_{0.95}(Y_2O_3)_{0.05}$ over the temperature range 635°–1150°C. They also measured the ionic transference numbers, from which they obtained the plots of constant t_0 as a function of p_{O_2} and $1/T$ shown in Figure 12. The curve for $t_0 = ½$ gives p_e^*. Kudo and Obayashi (82) used emf measurements to obtain p_e^* from Equation 17 for the

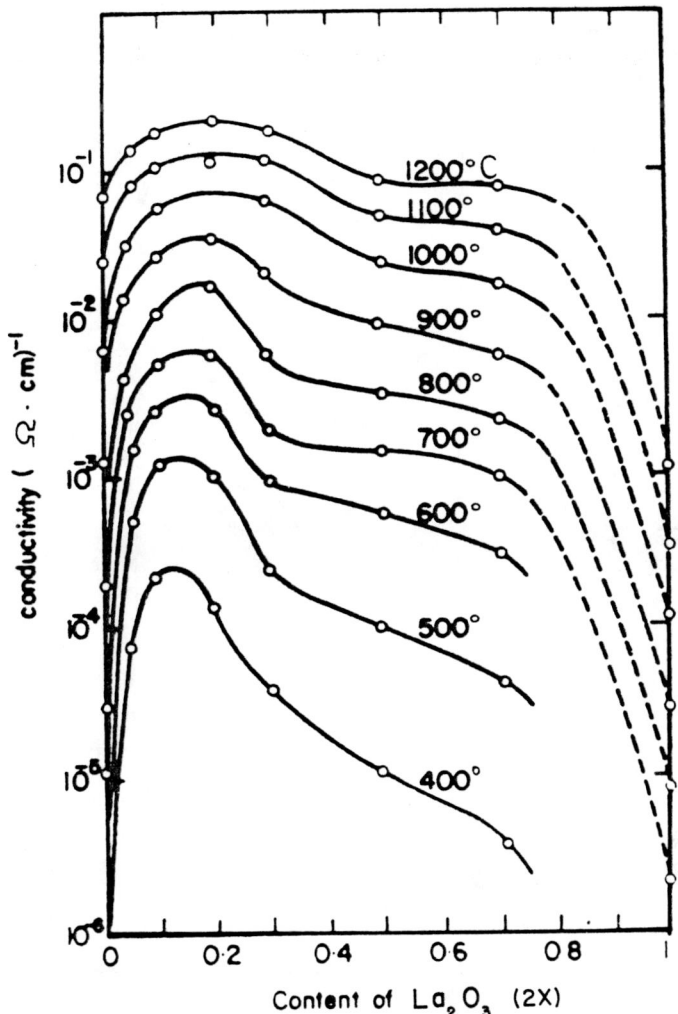

Figure 7. Conductivity isotherms for $(CeO_2)_{1-2x}(La_2O_3)_x$ as a function of x. Adapted from Ref. 75.

system $Ce_{1-2x}Gd_{2x}O_{2-x}$ (0.05 < x < 0.25), also shown in Figure 12. Comparison of Figures 5 and 12 shows that the doped CeO_2 systems are less resistive to reducing atmospheres than the ZrO_2 systems, as must be expected with the availability of a $4f^1$ acceptor level at the cerium. However, as pointed out by Tuller and Nowick (80) and Kudo and Obayashi (77, 82), the compounds could be used as electrolytes for practical applications provided sufficient oxidizing conditions are maintained. This shortcoming of the ceria-based compounds may be well compensated by the much higher ionic conductivities.

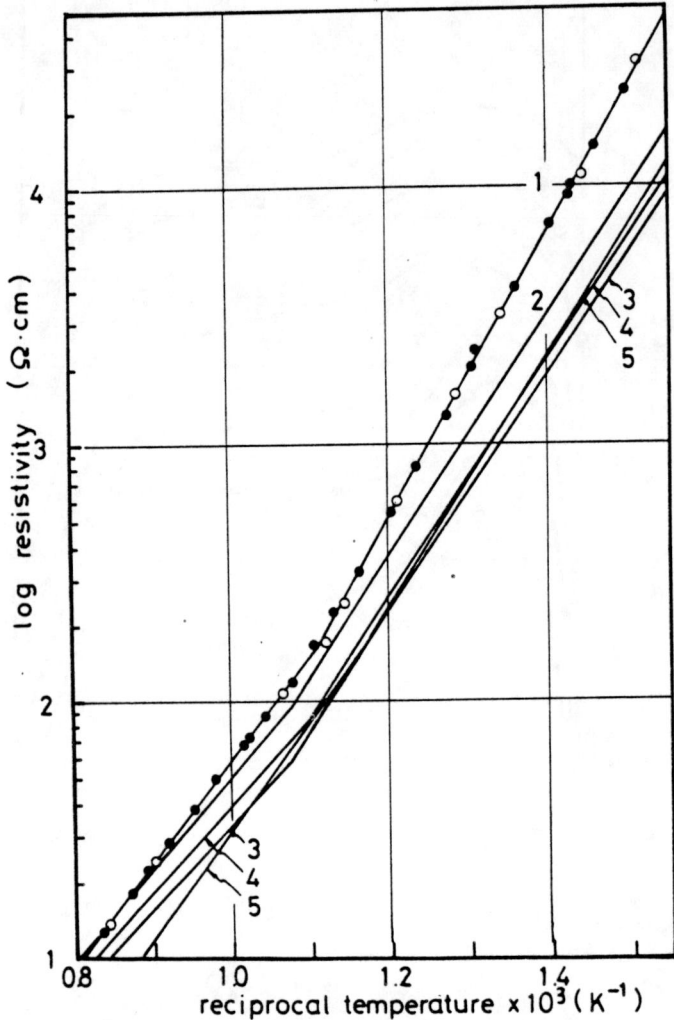

Figure 8a. Resistivities for $Ce_{1-2x} Ln_{2x} O_{2-x}$ (x = 0.15). Lines 1–5 represent Ln = Y, La, Nd, Sm, and Eu. Filled circles represent rising temperature; open circles represent falling temperature.

Takahashi et al. (8, 83) have analyzed the relation between conductivity, t_0, and the performance of a high-temperature fuel cell. The dependence of the fuel-cell energy efficiency η on output current i is given by

$$\eta = \frac{\sigma E_0 \cdot t_0 \cdot i - i^2}{\sigma E_0 [t_0 (1 - t_0) \sigma E_0 + t_0 i]} \tag{19}$$

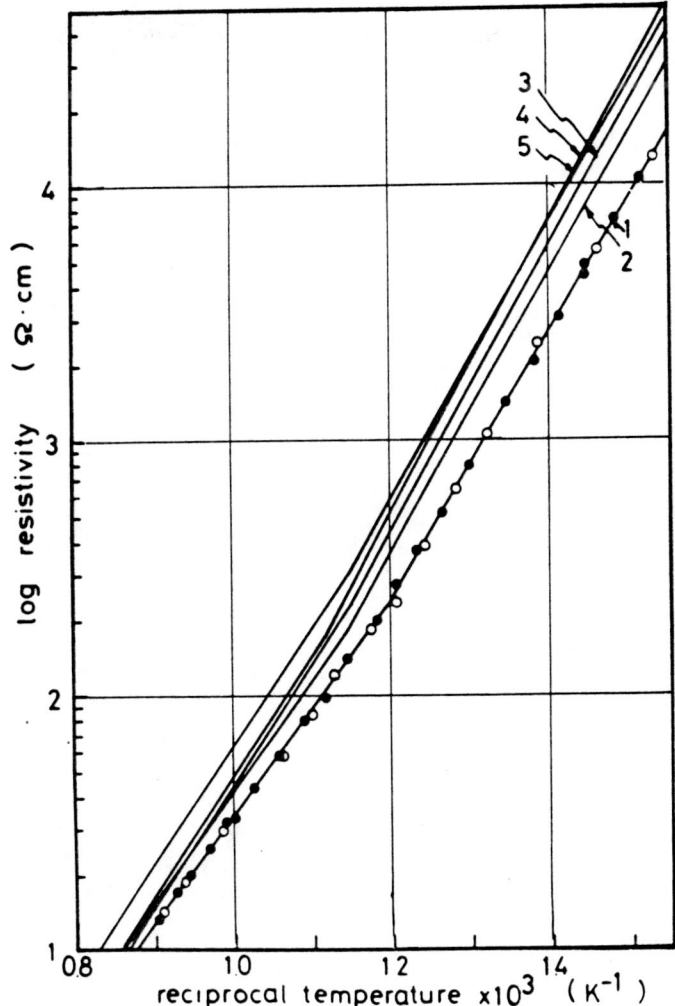

Figure 8b. Resistivities for $Ce_{1-2x}Ln_{2x}O_{2-x}$ ($x = 0.15$). Lines 1–5 represent $Ln = Gd$, Dy, Ho, Er, and Yb.

where σ is the total conductivity and E_0 is a thermodynamically calculated emf. With t_0 as a parameter and a fixed σE_0, optimization of η with respect to i gives

$$\eta_{max} = [1-(1-t_0)^{1/2}]/[1+(1-t_0)^{1/2}] \quad (20)$$

at the optimum current

$$i(\eta_{max}) = \sigma E_0 [(1-t_0)^{1/2} - (1-t_0)] \quad (21)$$

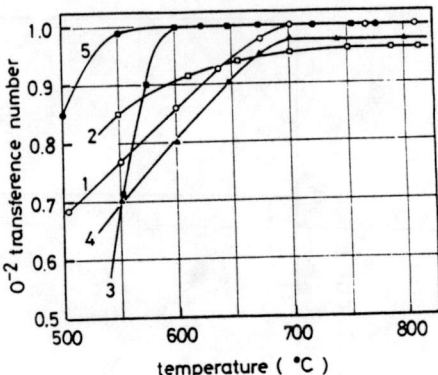

Figure 9. Oxygen-ion transference numbers for $Ce_{1-2x}Ln_{2x}O_{2-x}$ (x = 0.15). Curves 1-5 represent Ln = Nd, Sm, Gd, Dy, and Er.

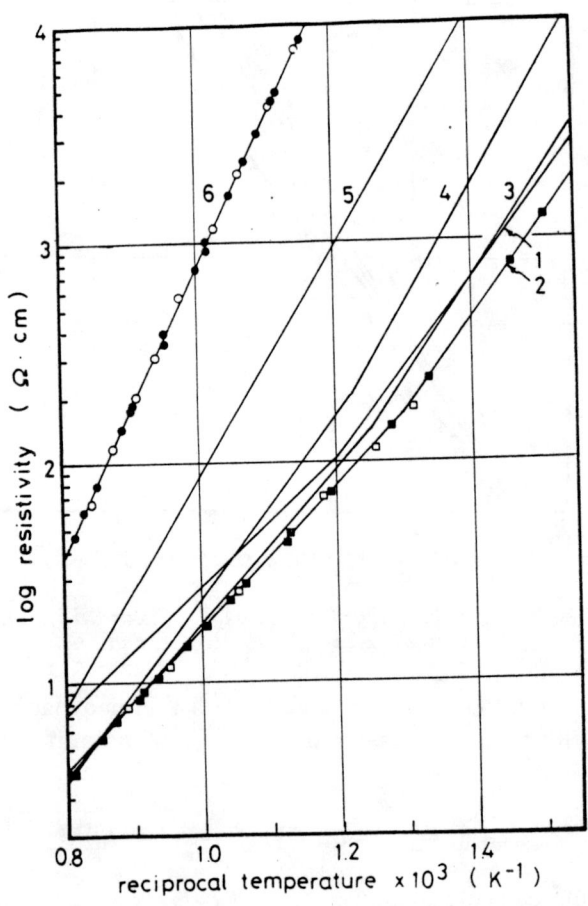

Figure 10. Resistivities for $Ce_{1-2x}Gd_{2x}O_{2-x}$. Lines 1-6 represent 2x = 0.1, 0.2, 0.25, 0.3, 0.4, and 0.5.

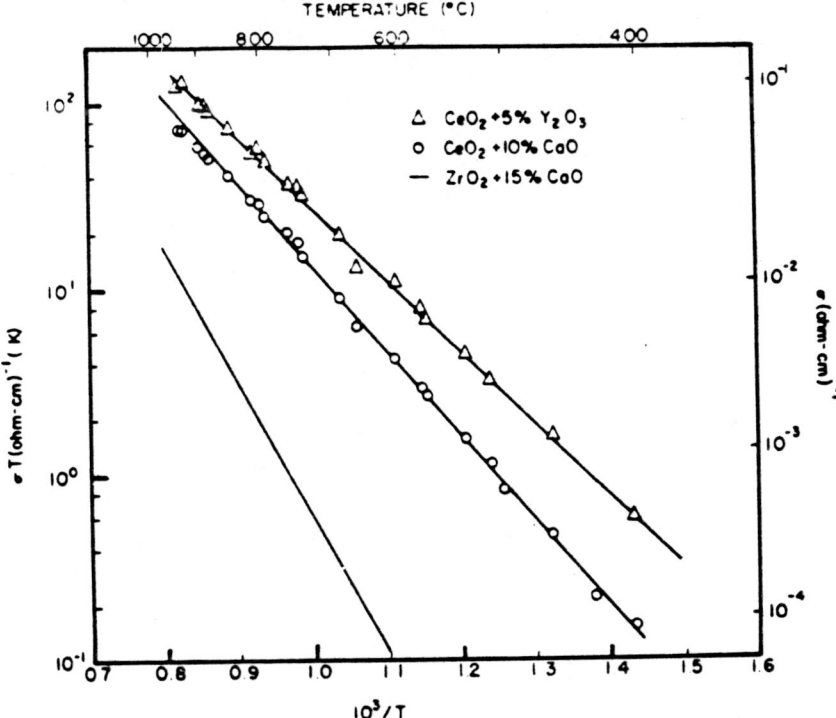

Figure 11. Conductivity for CeO_2 doped with 5 mol % Y_2O_3 and with 10 mol % CaO at $p_{O_2} = 1$ atm. Adapted from Ref 80 for $(CeO_2)_{0.95}(Y_2O_3)_{0.05}$.

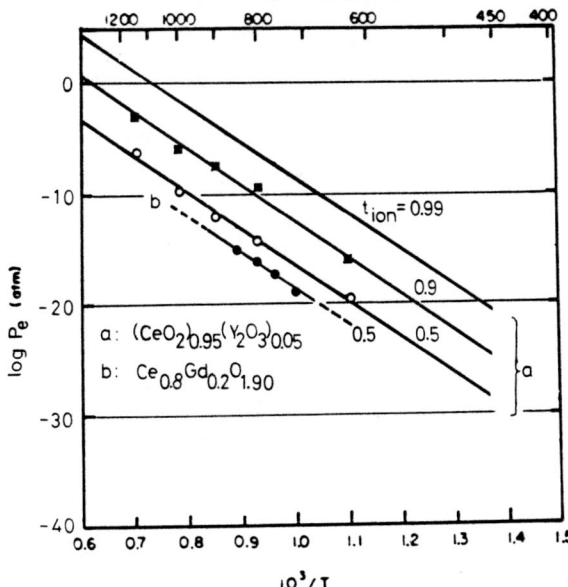

Figure 12. Parameters p_e as a function of temperature for $(CeO_2)_{0.95}(Y_2O_3)_{0.05}$ and $Ce_{0.8}Gd_{0.2}O_{1.90}$. Adapted from Ref. 80.

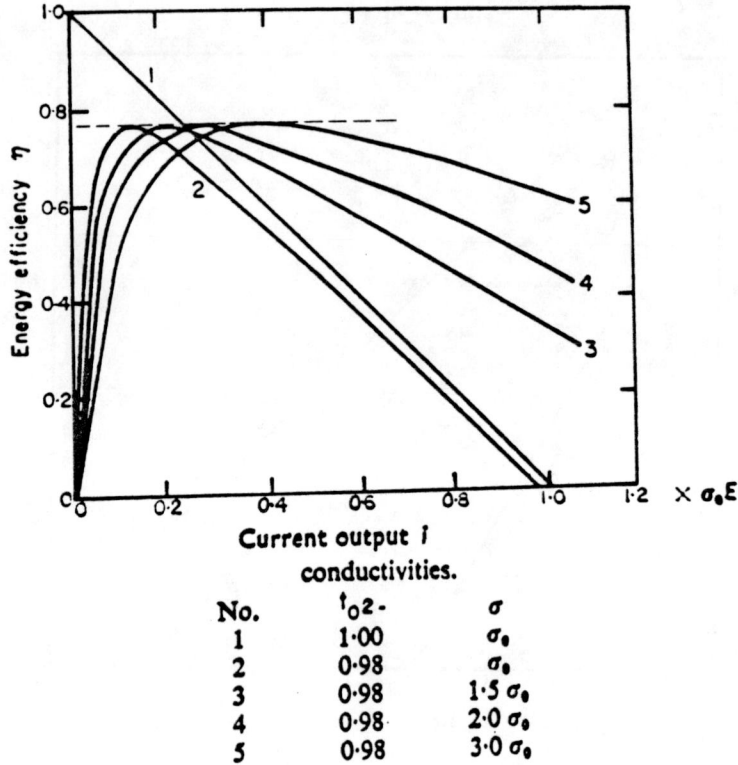

Figure 13. The effect of increasing electrolyte conductivity on the optimum current output of a high-temperature cell. Adapted from Ref. 83.

and η_{max} depends only on t_O, becoming unity for $t_O = 1$. However the dependence of $i(\eta_{max})$ on σ shows that the maximum power efficiency may occur for a smaller t_O and a larger σ. Therefore, the ceria-based electrolytes may prove competitive in high-temperature fuel/electrolysis cells.

DOPED ThO_2. Pure thoria has the cubic fluorite structure and is a mixed O^{2-}-ion/electronic conductor at higher temperatures (60, 84, 85). Figure 14 shows the ac (1592 Hz) conductivity as measured by Choudhury and Patterson (86). At high p_{O_2} ($> 10^{-5}$ atm), interstitial O^{2-} ions are introduced, and the electronic conduction is p-type; at lower p_{O_2} ($< 10^{-28}$ atm at 900°C) oxygen vacancies are introduced and the electronic conduction is n-type. At intermediate partial pressures of oxygen the conductivity is intrinsic. The conductivity of pure ThO_2 is more than one order of magnitude smaller than that of pure ZrO_2 (see Figure 2).

Substitution of Ca, Y, or Ln for Th introduces a $[V_O'']$, and the O^{2-}-ion conductivity increases sharply (87, 88). Figure 15 shows the partial conductivities for $Th_{0.86}Y_{0.14}O_{1.93}$ as a function of p_{O_2} as measured by Etsell (89). Data for pure ThO_2 (dashed line in Figure 15) is taken from Figure 14. As can be seen, the ionic conductivity at 1000°C is increased by nearly four orders of magnitude, and the electrolytic region (large t_O) is widened. The shift of the p-n transition and of p_e^* to higher p_{O_2} with increasing temperature is caused by the stronger temperature dependence of the n-type conductivity. Figure 16 compares the p_e^* and p_h^* values of some Y_2O_3-doped ThO_2 samples with those for pure ThO_2 (86, 90–94).

Other systems studied include $M_xTh_{1-x}O_2$, where M = Ba, Sr, Mg, Be (90, 91, 92). The relatively poor ionic conductivity of the thoria-based systems will restrict their use as solid electrolytes to oxygen sensors in atmospheres that are too reducing for ZrO_2-based electrolytes to have a high t_O.

DOPED Bi_2O_3. The compound Bi_2O_3 has three crystallographic modifications: monoclinic (α), tetragonal (β), and cubic (γ) (95). The $\alpha \rightarrow \beta$ transition at 710°C is not disputed, but the existence of a b.c.c. γ phase is open to question. If Bi_2O_3 is fused for a long time in a porcelain crucible, the simple-cubic structure of Figure 17a is stabilized by a small amount of impurity (95). The volume of the tetragonal β phase ($a = 10.93$ Å and $c = 5.67$ Å) is four times that of the simple-cubic structure

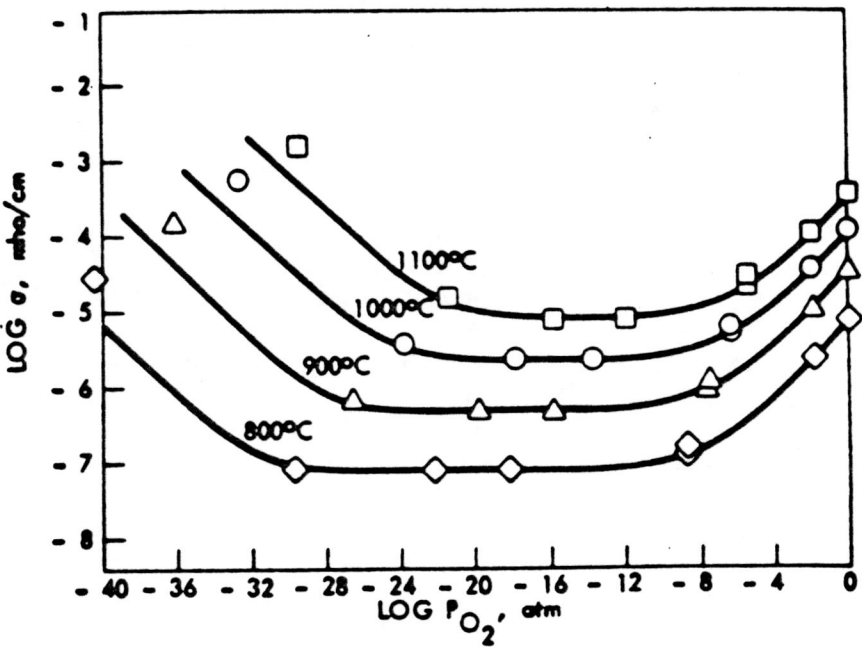

Figure 14. The ac conductivity isotherms for pure ThO_2. Adapted from Ref. 86.

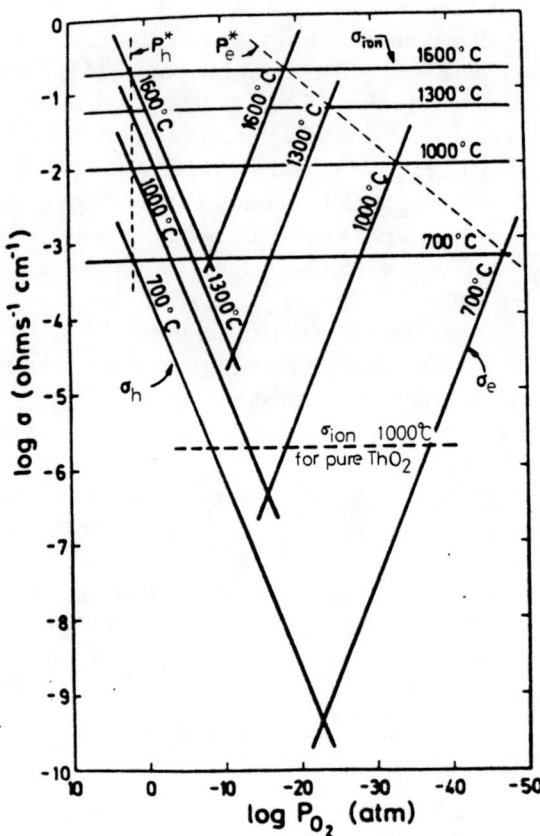

Figure 15. Partial conductivity isotherms for $Th_{0.86}Y_{0.14}O_{1.93}$. For comparison, σ_{ion} for pure ThO_2 is inserted. Adapted from Ref. 89.

($a = 5.5$ Å), and the atoms are only slightly displaced from their positions in the idealized cubic structure. As can be seen from Figure 17b, the idealized cubic structure of Bi_2O_3 is formed from the fluorite structure by removing one-quarter of the O^{2-} ions in an ordered manner. Thus the tetragonal β phase also corresponds to an ordered-vacancy fluorite structure, with cooperative displacements of the atoms about the vacancies lowering the symmetry from cubic to tetragonal. Such a structure is a promising candidate for O^{2-}-ion conduction.

Takahashi et al. (96, 97, 98) have investigated the conductivities of Bi_2O_3 and doped Bi_2O_3. The tetragonal phase has a relatively high conductivity in the temperature interval $400 < T < 800°C$, as illustrated in Figure 18 for SrO-doped Bi_2O_3 (96). The high-temperature region, which has a high conductivity and low activation energy, corresponds to the tetragonal β phase. Specimens with a SrO content between 20 and 40 mol % showed the highest conductivities over the entire temperature range studied; at these concentrations the structure is rhombohedral at lower temperatures. A concentration cell utilizing SrO-doped Bi_2O_3 as the electrolyte gave a nearly theoretical emf and a stable current flow, indicating that the conductivity is caused primarily by mobile O^{2-} ions.

Takahashi and Iwahara (97) attempted to stabilize the tetragonal β phase to lower temperatures by doping with Y_2O_3, Nb_2O_5, or Ta_2O_5. The conductivities of representative samples are shown in Figure 19; the activation energy increases with dopant concentration, but the α-β phase transition is suppressed. The lower activation energies associated with Bi^{3+} ions appear to be caused by the polarizability of the $6s^2$ core. Unfortunately, the Bi^{3+} ions are not sufficiently basic (large electron affinity) to be resistive to reduction at lower oxygen pressures. Nevertheless, the Bi_2O_3-based systems doped with CdO or a transition-metal monoxide seem to be potential cathode materials (98).

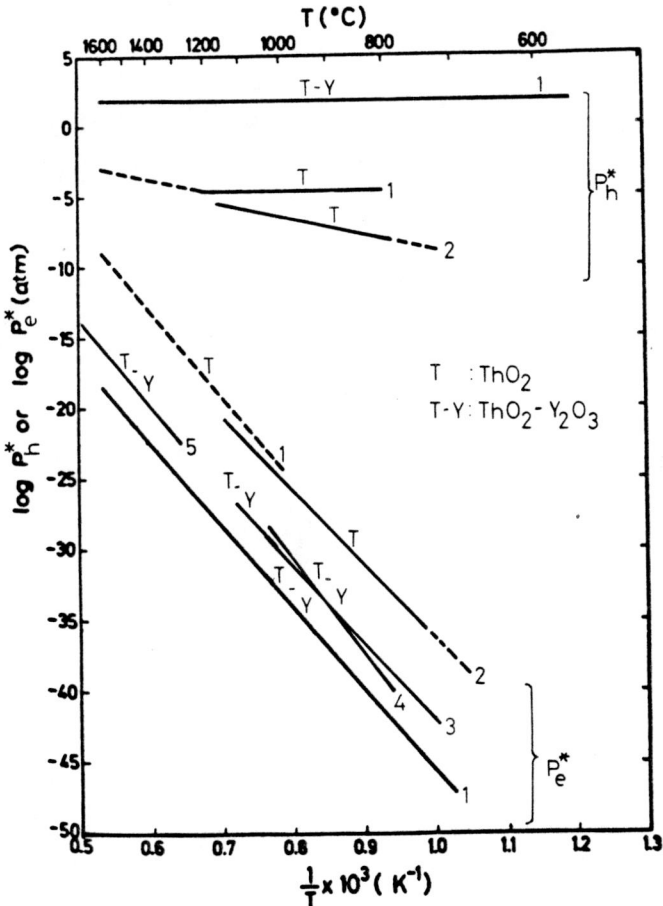

Figure 16. The parameters p_h^* and p_e^* for ThO_2 and ThO_2–Y_2O_3. Lines 1–5 are adapted from Refs. 89, 86, 94, 93, and 92, respectively.

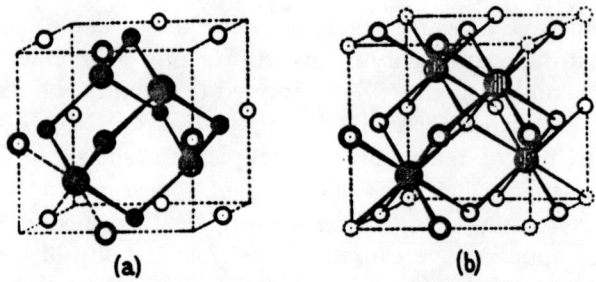

Figure 17. Relation between (a) the Bi_2O_3 and (b) the fluorite structures. Note that dotted circles (oxygen ions) in (b) are missing in Bi_2O_3. Adapted from Ref. 95.

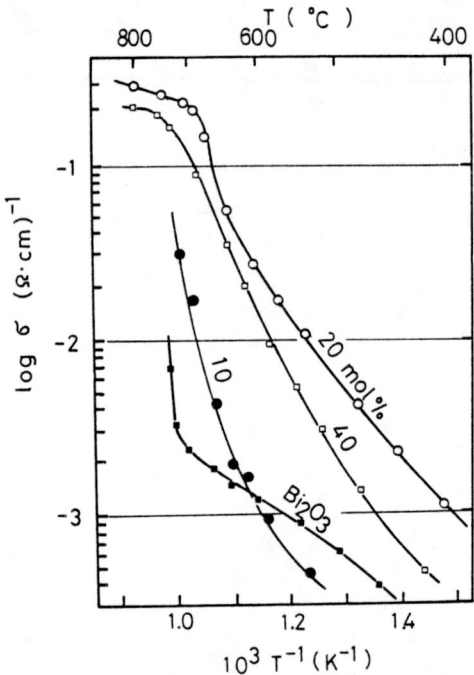

Figure 18. The conductivities of SrO-doped Bi_2O_3. Adapted from Ref. 96.

Perovskites. Oxides with the perovskite structure correspond to the chemical formula ABO_3, where B is an octahedral-site cation and A is a larger cation. Excellent reviews by Goodenough and Longo (99) and Galasso (100) summarize the electrical, magnetic, structural, and related data on the large number of oxides with perovskite structure that had been investigated up to 1972.

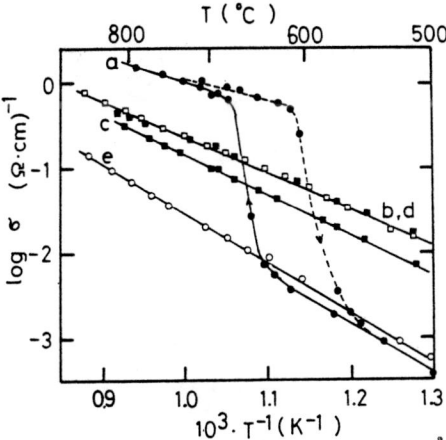

Figure 19. The conductivities of doped Bi_2O_3. (a) $(Bi_2O_3)_{0.95}$ $(Y_2O_3)_{0.05}$; (b) $(Bi_2O_3)_{0.75}$ $(Y_2O_3)_{0.25}$; (c) $(Bi_2O_3)_{0.70}$ $(Y_2O_3)_{0.30}$; (d) $(Bi_2O_3)_{0.85}$ $(Nb_2O_5)_{0.15}$; (e) $(Bi_2O_3)_{0.80}$ $(Ta_2O_5)_{0.20}$. Adapted from Ref. 97.

In general, perovskites consist of close-packed AX_3 layers, where X represents the anion, with B cations occupying octahedral interstices of X ions. "Cubic" perovskites have cubic stacking of the AX_3 layers; hexagonal perovskites ($CsNiCl_3$ structure) have hexagonal stacking of these layers. Perovskite polytypes having different sequences of cubic and hexagonal stacking are also known. Of particular interest for O^{2-}-ion conductors are oxygen-deficient "cubic" perovskites with chemical formula $ABO_{3-\delta}$. The "cubic" perovskites contain a rigid AB network, which has a CsCl-type configuration, with O^{2-} ions at the sites bounded by two B cations and four A cations. The tetragonal octahedra occupied by O^{2-} ions form a face-shared network, so that an O^{2-} ion can jump directly to a V_O'' via a common octahedral-site face. Like the fluorites, the oxygen-deficient perovskites contain a rigid cation network, and O^{2-}-ion conduction occurs via the $[V_O'']$.

Takahashi et al. (101, 102, 103) have shown that O^{2-}-ion conduction in "cubic" perovskites is strongly enhanced by substituting cations of lower formal valence for the A or B cations so as to introduce a finite $[V_O'']$. Figure 20 illustrates typical curves of conductivity vs. reciprocal temperature for some "cubic" perovskites. Takahashi et al. (152) measured the transference numbers t_O in two types of cells:

[I] O_2(air), Pt/specimen electrolyte/Pt, O_2(1 atm)

[II] $H_2 + H_2O$, Pt/specimen electrolyte/Pt, O_2 (1 atm)

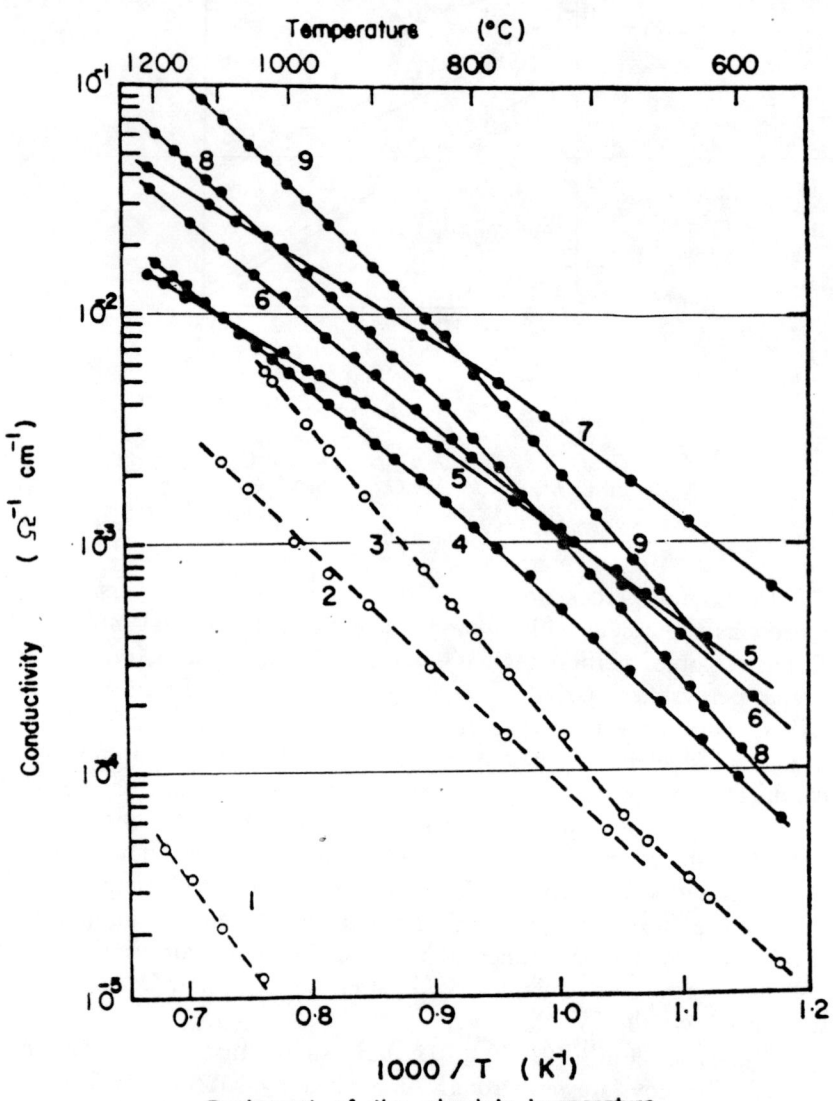

Figure 20. Conductivities of some perovskite oxides with nominal compositions. (1) $LaAlO_3$; (2) $CaTiO_3$; (3) $SrTiO_3$; (4) $La_{0.7}Ca_{0.3}AlO_3$; (5) $La_{0.9}Ba_{0.1}AlO_3$; (6) $SrTi_{0.9}Al_{0.1}O_3$; (7) $CaTi_{0.95}Mg_{0.05}O_3$; (8) $CaTi_{0.5}O_3$; (9) $CaTi_{0.7}Al_{0.3}O_3$. Adapted from Ref. 152.

The larger the p_{O_2}, the smaller t_O, indicating that the electronic component of the conductivity is caused by mobile holes. Presumably the concentration of oxygen vacancies $[V_O'']$ tends to be reduced by a p_{O_2}. The type-II cells showed transference numbers $t_O > 0.9$. The conductivities of nominal $CaTi_{0.7}Al_{0.3}O_{2.85}$ and $CaTi_{0.95}Mg_{0.05}O_{2.95}$ are comparable with those of ZrO_2 stabilized in the fluorite structure, but in each case t_O is relatively low. $La_{0.7}Ca_{0.3}AlO_{2.85}$, on the other hand, has a high t_O and a relatively low conductivity. The nominal compounds $BaZr_{0.9}Bi_{0.1}O_{2.95}$, $BaCe_{0.9}Bi_{0.1}O_{2.95}$, and $BaTh_{0.95}La_{0.05}O_{2.975}$ show high ionic conductivity, but are reactive with the electrode materials.

Matsuo and Sasaki (104) measured $PbTiO_3$–$La_{2/3}TiO_3$ ceramics and suggested that O^{2-}-ion conduction takes place in this system via the A-cation vacancies (since there is no $[V_O'']$ where the electronic conductivity is small), but details of these experiments are not available.

Schwarz and Anderson (105) have measured on single-crystal $SrTiO_3$ an oxygen diffusion constant $D_O = 10^{-5} - 10^{-4} cm^2/sec$ over the temperature range $700° < T < 975°C$ at $p_{O_2} = 10^{-3}$ atm. Unfortunately the ionic conductivity of this specimen was not measured.

Stephenson and Flanagan (106) have reported evidence for some O^{2-}-ion conduction from emf measurements of the cell Pt, Pb, PbO_2/$PbZr_{0.53}Ti_{0.47}O_3$/Cu, CuO_2, Pt. This result could reflect the existence of some Pb^{4+} ions on the B sublattice, which would create oxygen and A-site vacancies.

Pyrochlores. The cubic pyrochlore structure, corresponding to chemical formula $A_2B_2X_6X'$, may be derived from the fluorite structure by an ordering of A and B cations within the cation array and of anion vacancies within the anion array. Cooperative atomic displacements accompany the ordering (107, 108).

This structure has not been studied extensively for ionic conduction. Mazelsky and Kramer (109) report an ionic transference number $t_O = 0.57$ for $Cd_2Nb_2O_7$ in air at 850°C, typical of a mixed electronic/ionic intrinsic conductor. On the other hand, they found the conductivity of $Pb_{1.5}Nb_2O_{6.5}$ to be almost exclusively electronic.

C-type Ln_2O_3. Like tetratgonal Bi_2O_3, the cubic structure of C-type rare-earth sesquioxides can be derived from the fluorite structure by a cooperative ordering of one-quarter oxygen vacancies, but in the manner illustrated in Figure 21 (110). This has suggested to several workers the possibility of good O^{2-}-ion conduction.

Tare and Schamlzried (26) have studied ionic conduction in Gd_2O_3, Dy_3O_3, Sm_2O_3, and Y_2O_3 over the temperature range 650°–900°C and pressure range $1 < p_{O_2} < 10^{-2}$ atm. By measuring the emf of an oxygen concentration cell, they used Equation 17 to evaluate p_h^* and p_e^*. Their results are given in Table II.

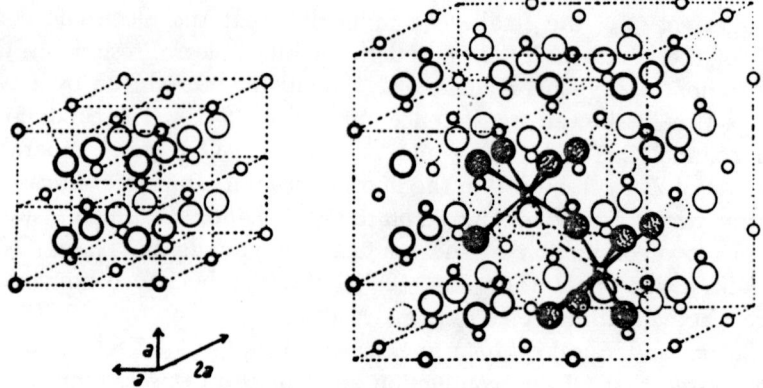

Figure 21. Relation between (a) the fluorite and (b) the C-type Ln_2O_3 structures. Adapted from Ref. 95.

Etsell and Flengas (111) measured the conductivities of compositions in the La_2O_3–CaO system as a function of temperature and oxygen partial pressure. At a $p_{O_2} < 10^{-7}$ atm, O^{2-}-ion conduction was found; at higher p_{O_2} a p-type conductivity varying as $p_{O_2}^{1/4}$ was observed. The O^{2-}-ion conduction increases by over two orders of magnitude with the first several mol % CaO but changes only marginally thereafter up to 50 mol % CaO. Although its O^{2-}-ion conduction is relatively poor, this system is extremely resistant to reduction and may therefore be useful for measuring the thermodynamic properties of systems exerting very low oxygen potentials.

Electrodes and Interconnectors

Cathodes. Besides the solid electrolyte, the cathode is the most significant materials limitation to the development of high-temperature fuel/electrolysis cells. For laboratory-scale evaluation of high-temperature cells, the noble metals Pt, Pd, or Ag have been used. However, these are too expensive for commercial purposes.

According to Tedmon et al. (112) four classes of criteria must be satisfied before a potential cathode material can be considered for practical fuel cells: chemical, electrochemical, mechanical, and economic.

(1) The chemical requirements, which must allow for the severe oxidizing environment in which the cathode operates, include:
    (a) chemical stability (against oxidation or nitrification) in air at and above 700°C
    (b) phase stability over the temperature range of operation
    (c) low vapor pressure, to minimize volatilization loss that would increase the electrical resistance
    (d) chemical inertness with respect to the electrolyte and lead

wires while maintaining adherence to the electrolyte over the temperature range of operation.

(2) The electrochemical requirements include:
    (a) small back diffusion and overpotentials so that the reduction/oxidation of gaseous oxygen to/from ionic oxygen occurs smoothly and rapidly; and
    (b) good electrical conductivity to minimize ohmic loss at the electrode.

(3) The mechanical requirements are:
    (a) maintenance of any porosity required for oxygen transport to the electrolyte; and
    (b) maintenance of good electrical and physical contact between the electrolyte and the lead wire, which means good matching of the thermal expansions of electrode and electrolyte. (Mismatch of the thermal expansions may cause cracking of the ceramic electrolyte as well as pealing of the electrode.

(4) The economic requirement is straightforward. Although the use of noble metals such as Ag or Pt in electrodes or lead wires may be justified on the basis of higher cell performance (*113*) for some specialty applications, it may not be possible to bear the resulting increase in the cost of the system.

The wide variety of materials that have been proposed and tested for cathodes may be grouped into three classes: metals, composites, and metallic oxides.

Only the noble metals can be used in the highly oxidizing cathodic environment (*114, 115, 116*), and economic constraints limit their use. Moreover, Au and Ag are further limited by their high volatility at elevated temperatures (*117, 118, 119*). Since composite systems, such as porous stabilized zirconia over a Pt grid, must also use a noble-metal current collector, they too are considered impractical (*112*). A similar judgment would undoubtedly be made on a cermet (ceramic/metal mixed system) electrode. Therefore, the most promising class of cathode

Table II. Parameters P_h^* and P_e^* for C-type Ln_2O_3 at Several Temperatures[a]

Compound	T (K)	log P_h^* (atm)	log P_e^* (atm)
	1121	−4.1	−19.2
Sm_2O_3	1059	−3.8	−21
	945	−2.6	−23
	1203	−3.7	<−20
Gd_2O_3	1078	−2.0	<−20
	1003	−1.5	<−20
	1133	−8.3	<−20
Dy_2O_3	1041	−9.1	<−20
	940	−9.1	<−20
Y_2O_3	1098	−4.2	−21.5

[a] Adapted from Ref. *26*.

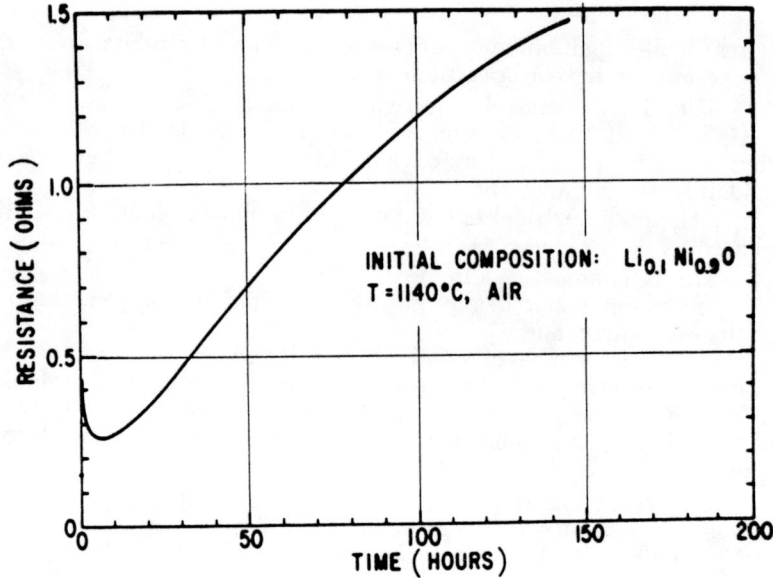

Figure 22. Resistance change of a lithiated nickel oxide film showing the effect of Li evaporation at 1140°C. Adapted from Ref. 112.

materials is the metallic oxides. Goodenough (120) has given an important survey of the metallic oxides, although not with the cathodic application in view.

The large number of presently known metallic oxides have yet to be adequately explored for use as cathode materials. The electronically conducting oxides that have been studied include Li-doped NiO (112, 121, 122), doped ZnO (123, 124), SnO_2 doped with Sb or Te (125), nonstoichiometric perovskites such as $PrCoO_3$ (112, 126, 127), and In_2O_3 doped with Sn, Sb, or Te (13, 128).

The system $Li_xNi_{1-x}O$ contains x holes per molecule in the $Ni^{2+}:3d^8$ valence band of the rocksalt structure. At high temperatures the holes are ionized from the Li^+ acceptor centers. They are low-mobility, itinerant holes (120, 129), not mobile small polarons associated with distinguishable Ni^{3+} ions. Unfortunately this system suffers from the volatilization of lithium at high temperatures. Tedmon et al. (112) have observed a tripling of the resistance of a porous $Li_{0.1}Ni_{0.9}O$ film after 150 hr at 1140°C in air (Figure 22).

Takahashi et al. (124) have demonstrated excellent cathodic properties of the ZnO-based materials $(ZnO)_{0.97}(Al_2O_3)_{0.03}$ and $(ZnO)_{0.95}(ZrO_2)_{0.05}$. Electrical conductivities of 1–10 $(\Omega cm)^{-1}$ were found in the temperature range 700°–1100°C, while the adhesivity to CaO-stabilized ZrO_2, unlike that of pure ZnO, is excellent; it is also nonreactive to the electrolyte. Figure 23 shows the polarization characteristics of the cath-

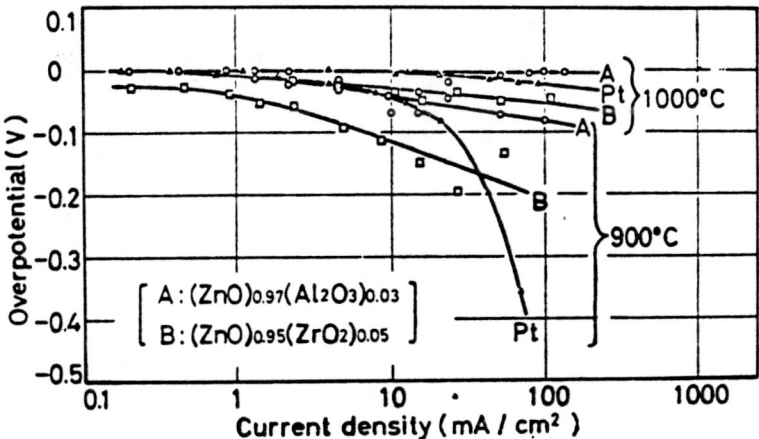

Figure 23. The IR-free cathodic polarization of doped ZnO cathode materials: $(ZrO_2)_{0.85}(CaO)_{0.15}$ electrolyte and pure oxygen as cathode gas. Adapted from Ref. 124.

ode, with the resistance overpotentials subtracted out. These data indicate that $(ZnO)_{0.97}(Al_2O_3)_{0.03}$ and $(ZnO)_{1-x}(ZrO_2)_x$ can replace Pt as stable cathode materials in the temperature range 800°–1000°C.

Böhm and Kleinschmager (125) and Sverdrup et al. (13, 128) have investigated porous films of SnO_2 and Sb-doped SnO_2 vapor-desposited on stabilized ZrO_2. Such films have been used as transparent conductors. The conductivity of SnO_2 at 1000°C in air is $10^3(\Omega cm)^{-1}$. Figure 24 shows the voltage–current relationship of a cell (125).

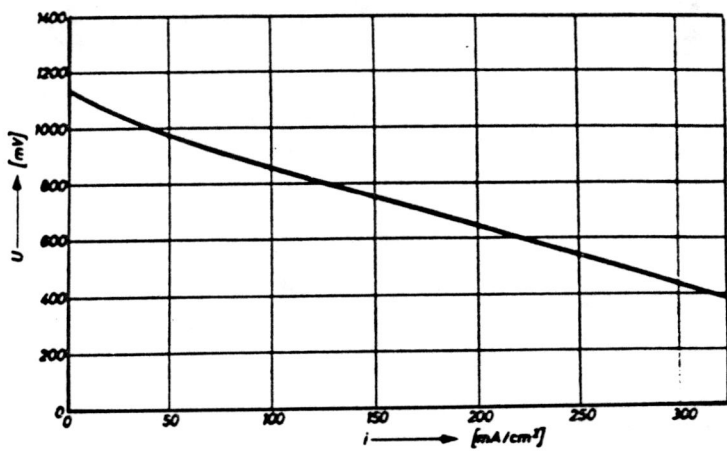

Figure 24. The voltage–current relation of a cell: $H_2 + 3\% H_2O$ (Ni) / doped ZrO_2 / (SnO_2 – 8.9 at % Sb), O_2. Adapted from Ref. 125.

$$H_2 + 3\% H_2O\,(Ni)/ZrO_2\,(doped)/(SnO_2 - 8.9\;at.\%\,Sb)\,O_2$$

where the electrode materials appear in parentheses. Unfortunately a mismatch of thermal-expansion coefficients ($\alpha_{ZrO_2} \simeq 11 \times 10^{-6}$ and $\alpha_{SnO_2} = 4.5 \times 10^{-6}\,°C^{-1}$), although slightly relieved by Sb doping (*13*), places this cathode material at a serious disadvantage.

Tedmon et al. (*112*) have evaluated the perovskite electrodes $LaCoO_3$ and $PrCoO_3$ obtained from a slurry and sintered in situ as the cell was heated to operating temperatures. With H_2 as the fuel and air as the oxidant, power densities of 300 mW/cm² at 1000°C and 600 mW/cm² at 1100°C were obtained for periods in excess of 5000 hr.

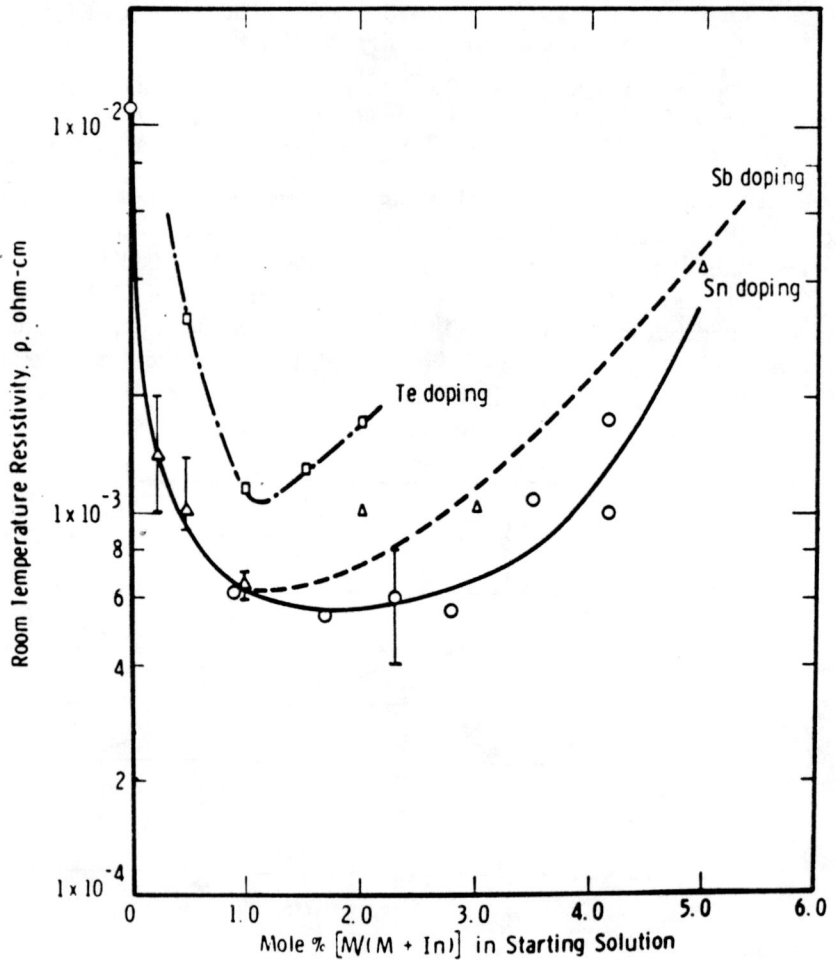

Figure 25. Effect of doping on the resistivity of In_2O_3. Adapted from Ref. 13.

Unfortunately, $PrCoO_3$ reacts with ZrO_2 in a complex manner, and a thermal-expansion mismatch ($\alpha_{PrCoO_3}/\alpha_{ZrO_2} > 2.5$) causes spalling of the cathode during cooling.

Shirai et al. (130) investigated CeO_2 doped with Y_2O_3, Fe_2O_3, La_2O_3, Cr, Co, and Ni as well as La_2O_3 doped with Cr, Co, and Ni. A 60–40 wt % La_2O_3–Co sample had the lowest resistivity, and adhesiveness to ZrO_2 was much improved by substituting Y for La in the system $La_{2-x}Y_xO_3$–Co. They claim that a fuel cell employing this cathode could be operated above 1000°C, but details were not given. It would appear that their composition contains $LaCoO_3$, in which case reactivity with the solid would be a problem.

Sverdrup et al. (13, 128) found a resistivity minimum near 1–2 mol % Sn, Sb, or Te as a dopant in In_2O_3 (see Figure 25). In_2O_3 is stable at 1000°C, and Sn-doped In_2O_3 is used as a transparent electronic conductor. It has the C-type Ln_2O_3 structure of Figure 21, which is closely related to fluorite. At 25°C, its pseudofluorite lattice parameter, a_o = 5.06 Å, matches well that of ZrO_2 (a_o = 5.10 Å), and the thermal expansions of the two compounds are similar (131), as seen in Figure 26. Moreover, In_2O_3 is essentially unreactive with ZrO_2 at 1000°C. Polarization losses at the electrode can be significantly reduced by a simple reverse-current treatment, as can be seen in the voltage–current curves of Figure 27 for a film of In_2O_3 deposited on stabilized ZrO_2 by chemical vapor transport. Polarization losses were also reduced with a composite electrode consisting of a doped-In_2O_3 current collector covered by a

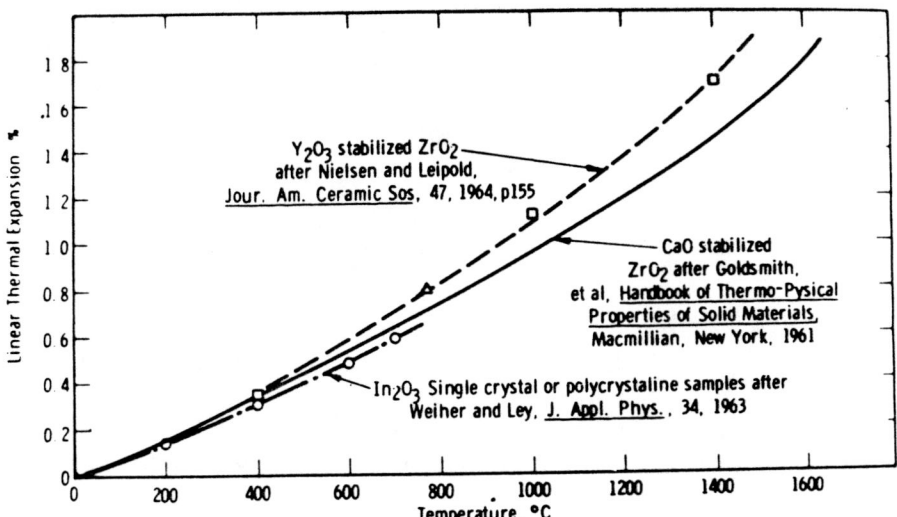

Figure 26. A comparison of the linear thermal expansion characteristics of In_2O_3 and stabilized zirconia. Adapted from Ref. 13.

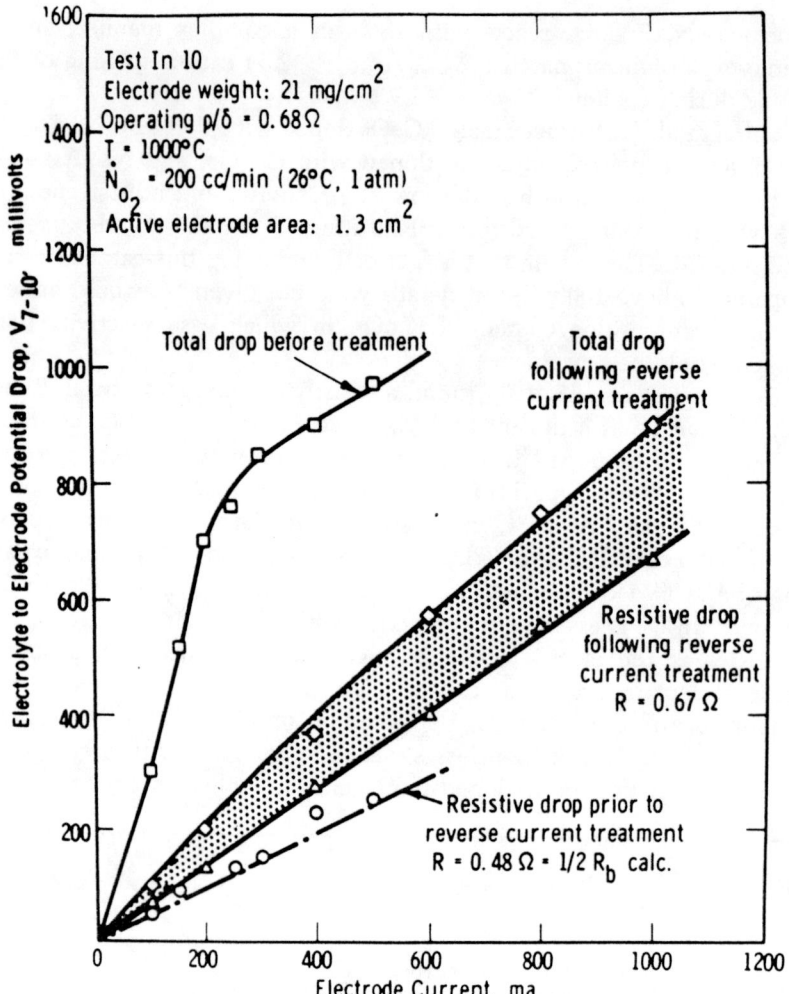

Figure 27. Voltage–current characteristic of In_2O_3 air electrodes. "Reverse-current" treatment reduces the potential drop. Adapted from Ref. 13.

porous layer of ZrO_2 impregnated with one of the "cubic" perovskites $PrCoO_3$, $NdCoO_3$, $PrNiO_3$, or $NdNiO_3$ (13). From these observations, doped In_2O_3 is a promising air electrode for high-temperature fuel/electrolysis cells operating at about 1000°C.

An alternative to a porous electrode is a mixed electronic/ionic conductor. As illustrated in Figure 28 for an anode (a similar argument applies to cathodes), Takahashi et al. (124) have pointed out that with such a nonporous electrode the electrode reaction takes place over the entire electrode surface; it is not confined to the portion adjacent to the electrolyte. This situation has been demonstrated with a molten Ag cathode having a diffusion constant $D_O = 10^{-5} - 10^{-4}$ cm^2/sec (*117, 118, 119*). If the D_O of the electrode is comparable with that of the electrolyte, O^{2-}-ion diffusion through the electrode is not rate-determining, and increasing the effective electrode area reduces the polarization losses. Since doped and nonstoichiometric perovskites have been shown to have high O^{2-}-ion conductivities (*101, 102, 103*) and PrCoO$_3$ to be good cathode materials, it is reasonable to suspect that "PrCoO$_3$" is a mixed electronic/ionic conductor.

Kudo et al. (*132*) have shown that electronically conducting Nd$_{1-x}$Sr$_x$CoO$_{3-\delta}$ and similar perovskite systems are good cathode materials

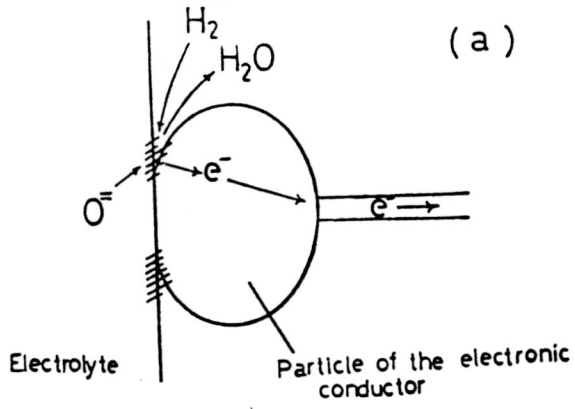

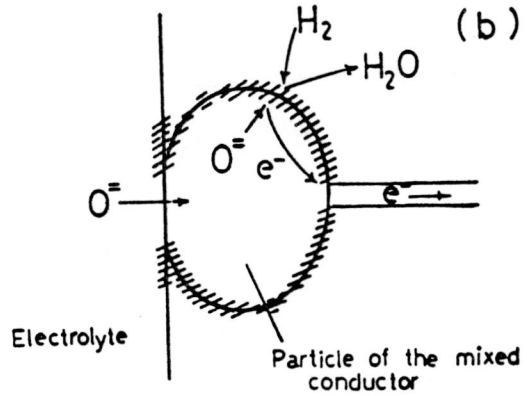

Figure 28. Models of the anodic reaction zone showing the superiority of a mixed conductor: (a) electronic conductor, (b) mixed conductor. (The hatched parts indicate the active reaction zone.) A similar model is applicable to the cathodic reaction. Adapted from Ref. 124.

for room-temperature, alkaline-solution air batteries; and Obayashi et al. (133) have measured O^{2-}-ion diffusivities at 25°C in the range 10^{-13}–10^{-11} cm²/sec, which are very much larger than those of ordinary oxides. Similar results have been obtained for "LaCoO$_3$" (134). It is reasonable to assume that the O^{2-}-ion conduction is high enough at elevated temperatures to characterize these systems as mixed electronic/ionic conductors. Schwarz and Anderson (105) have reported a $D_0 = 10^{-5}$–10^{-4} cm²/sec in reduced rutile in the interval $700° < T < 950°C$ (see above).

Obayashi and Kudo (135) have shown that LaNiO$_3$ loses oxygen stepwise on raising the temperature in a reversible, very fast reaction, suggesting that this metallic compound is also a mixed conductor. It has been successfully used at 800°C as the cathode of a flow-through oxygen meter using stabilized ZrO$_2$ as the solid electrolyte. Figure 29 compares the aging of LaNiO$_3$ and porous-Pt electrodes (136). The deviation with time of the measured from the theoretical emf at the porous-Pt electrodes is caused by aggregation of the platinum. Nonporous LaNiO$_3$ cathodes can be used successfully for more than 10^4 hours.

Anodes. The conditions under which anode materials operate are much less severe, and good performance has been reported for various metals such as Ti, Mn, Fe, Ni, Cu, and Pt (137) as well as several oxides such as NiO, CrO$_3$, CoO, Fe$_2$O$_3$, reduced TiO$_2$ (138), V$_2$O$_3$ (138), and UO$_2$ (139). Takahashi et al. (140) and Tedmon et al. (138) have reported excellent depolarization in the mixed electronic/ionic conductors CeO$_2$–La$_2$O$_3$ and CeO$_2$–Y$_2$O$_3$, as shown in Figure 30.

Interconnectors. In practice, individual electrolysis/fuel cells must be connected in series to make a cell stack (8, 13). Figure 31 illustrates one possible configuration (141). The material used to interconnect the

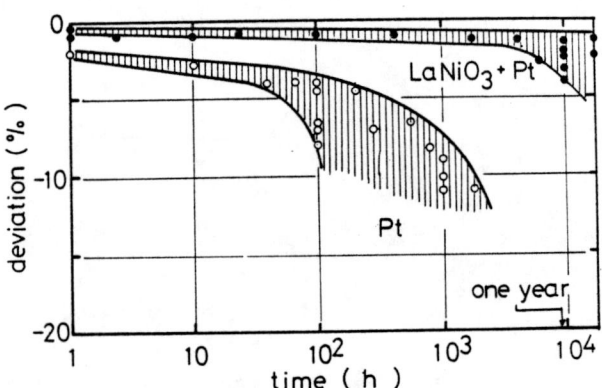

Figure 29. Comparison of a LaNiO$_3$ and a porous-platinum cathode: deviation of observed emf's from the theoretical value of the cell $p_{O_2} = 10^{-4}$ atm, (Pt)/Zr$_{0.85}$ Ca$_{0.15}$O$_{1.85}$/cathode, air.

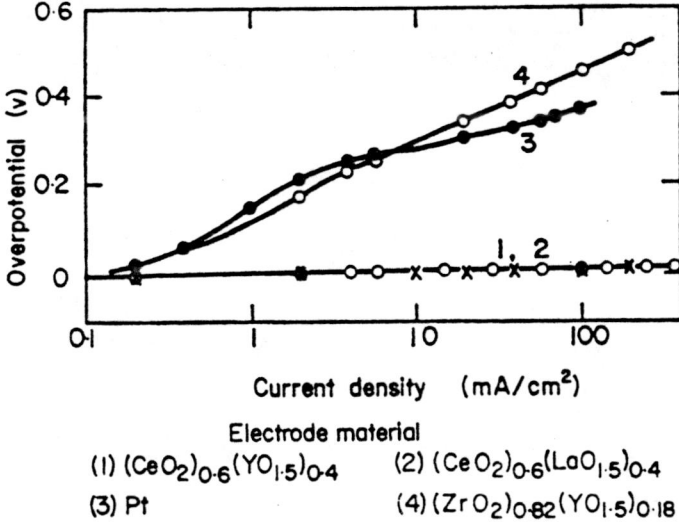

Electrode material
(1) $(CeO_2)_{0.6}(YO_{1.5})_{0.4}$ (2) $(CeO_2)_{0.6}(LaO_{1.5})_{0.4}$
(3) Pt (4) $(ZrO_2)_{0.82}(YO_{1.5})_{0.18}$

Figure 30. Anodic polarization characteristics of various electrode materials at 1000°C. Adapted from Ref. 124.

air and fuel electrodes of adjacent cells must be a good electronic conductor stable at operating temperatures in both the reducing atmosphere of the anode and the oxidizing atmosphere of the cathode. It must also be readily made non-porous, since neither fuel nor air should penetrate through it.

Sverdrup et al. (*13*) have studied the normal spinel $CoCr_2O_4$, which is stable at oxygen partial pressures $p_{O_2} > 10^{-15}$ atm. At high p_{O_2} it becomes a p-type semiconductor; at low p_{O_2} it is an n-type semiconductor. Figure 32 shows resistivity isotherms at 1000°C as a function of p_{O_2} (*141, 142*). Doping with V or Mn lowers the resistivity one to two orders of magnitude, especially at lower p_{O_2}. An irreversible loss of V occurs for $p_{O_2} > 10^{-10}$ atm via oxidation and vaporization, which raises

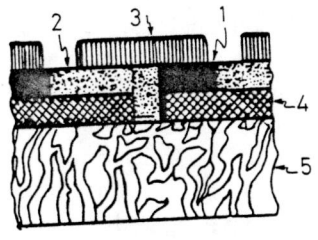

1 interconnector
2 electrolyte
3 cathode
4 anode
5 support

Figure 31. Possible series connection of a high-temperature cell. Adapted from Ref. 141.

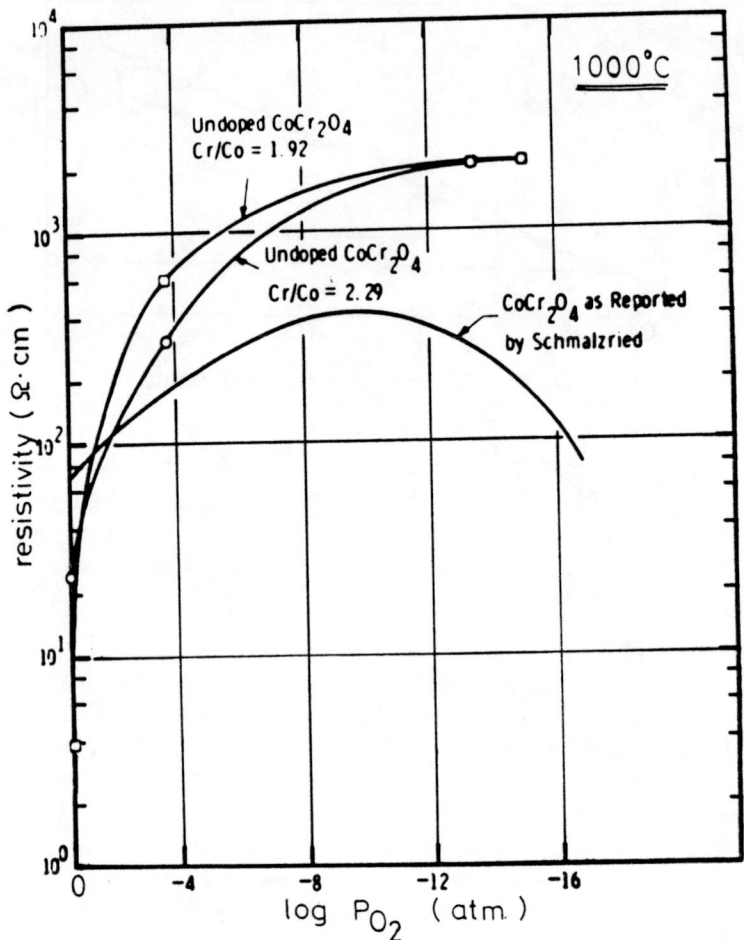

Figure 32. Comparison of resistivities of two types of cobalt chromite. Adapted from Ref. 141.

the resistivity. On the other hand, Mn-doped $CoCr_2O_4$ (2–4 mol % Mn) was stable for over 26 days in the reducing fuel atmosphere, and its resistivity was independent of cycling to an oxidizing atmosphere (*see* Figure 33) (*141*). It appears that Mn-doped $CoCr_2O_4$ has the chemical stability and a nearly tolerable resistivity at 1000°C for use as an interconnector material in a high-temperature electrolysis/fuel cell.

Kleinschmager and Reich (*143*) have investigated the electronically conducting perovskites doped and/or reduced $LaNiO_3$ and $LaCrO_3$ as well as the interlayer compound La_2NiO_4. At 1000°C the resistivities of the nominal compositions $La_{0.8}Sr_{0.2}Cr_{0.8}Ni_{0.2}O_3$ and $La_{1-\delta}Ni_{0.9}Co_{0.1}O_{3-\delta}$ changed very little over the range of oxygen partial pressures 10^{-17} <

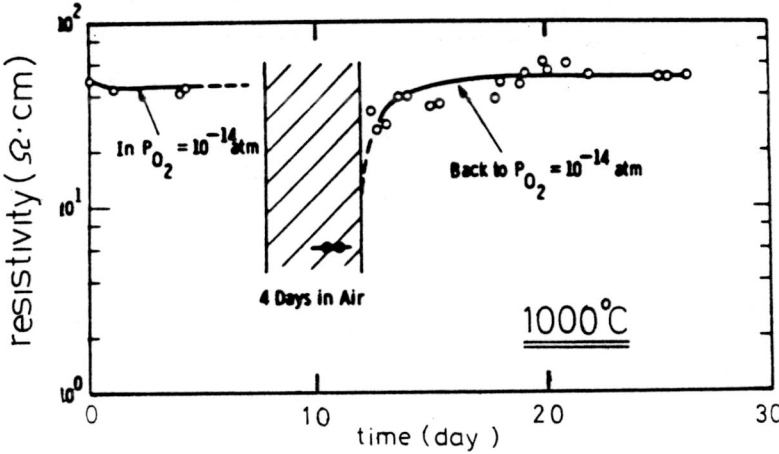

Figure 33. Resistivity of 2 mol % Mn-doped $CoCr_2O_4$ as a function of testing time. Adapted from Ref. 141.

$p_{O_2} < 1$ atm. A stability test for $La_{0.8}Sr_{0.2}Cr_{0.8}Ni_{0.2}O_3$ is recorded in Figure 34.

There are many metallic oxides that have yet to be investigated.

Cell Technology

Fuel Cells. Several types of fuel-cell designs have been considered. Among the paper studies, Ball and Bhada (*144*) have evaluated a planar-cell design, Leibhafsky and Cairns (*1*) have discussed the influence of fuel-composition change on cell voltage, Kudo and Obayashi (*145*) have calculated cell performance as a function of both cell geometry and fuel composition. The technical feasibility of coal-reacting fuel-cell power

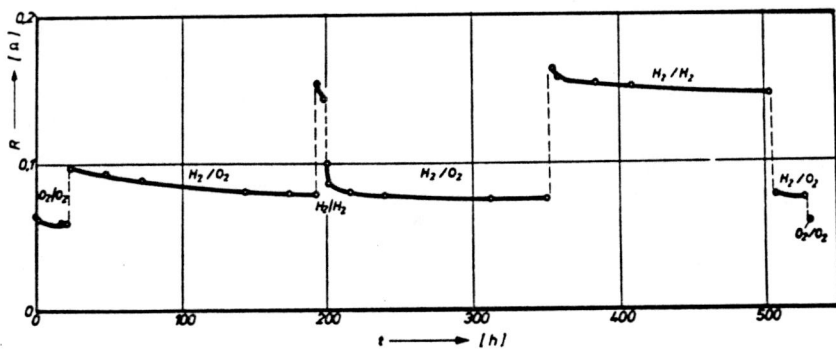

Figure 34. Resistivity of doped $LaCrO_3$ at 1000°C as a function of testing time. Adapted from Ref. 143.

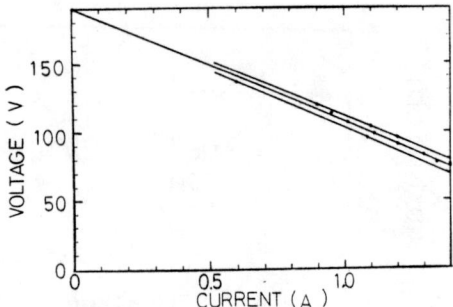

Figure 35. Performance of a 100-W solid-electrolyte fuel-cell battery operated at 1020°C. Adapted from Ref. 13.

systems have been evaluated by Westinghouse Electric Co. on a 100-W scale (*13, 141*).

The institutions where cell-performance studies have been carried out include General Electric Co., Westinghouse Electric Co., Brown–Boveri Co., Battelle Laboratories at Frankfurt and Geneva, and A.E.R.E. Harwell. We select the fuel cells fabricated by Westinghouse Electric Co. and Brown–Boveri Co. to illustrate cell-performance data.

The Westinghouse unit consisted of two cell stacks connected in parallel to supply power to a load. Each stack contained 200 series-connected cells. From the performance data of Figure 35, the open-circuit voltage at 1020°C is 190 V, and the maximum power is 110 W.

The Brown–Boveri Co. has developed both planar and cylindrical fuel cells, especially the latter. The cells use a stabilized zirconia electrolyte ($ZrO_2 - Y_2O_3 - Yb_2O_3$) with a Ni anode and a complex-oxide cathode. Figure 36 illustrates a cylindrical structure of 100 series-connected cells (*146, 147*). Air is introduced to the inner side of the electrolyte tubes and purged at the top of the cell stack. Fuel gas, which is oxidized non-electrochemically, is introduced to the outer compartment. Both the inlet and the fuel gas are preheated in a counter-flow heat exchanger by the exhaust. Figure 37 gives the cell characteristic; power densities of 105 mW/cm² decreased to 35 mW/cm² after 10^4 hours of operation.

Electrolysis Cells. Operation of a high-temperature electrolysis cell to obtain hydrogen from water is essentially the same as that of a fuel cell, but in reverse. Water vapor is introduced into the "fuel" compartment, and an external voltage is applied across the electrodes to give the following reactions.

$$\text{cathode: } H_2O + 2e \rightarrow H_2 + O^{2-}$$

$$\text{anode: } O^{2-} \rightarrow \tfrac{1}{2}O_2 + 2e$$

Such cells have been demonstrated by many workers (*148, 149, 150*).

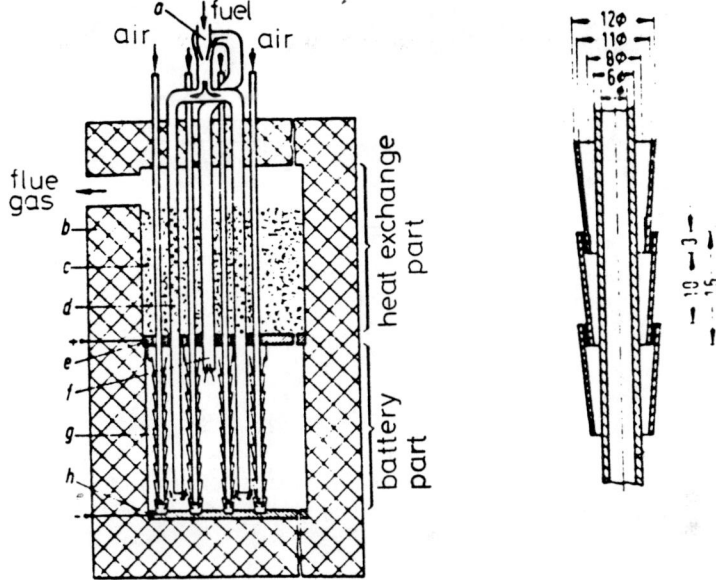

Figure 36. Structure of a stack containing 100 series-connected cells developed by Brown–Boveri Co. (a) Mixing nozzle, (b) thermal isolation, (c) afterburning catalyst layer, (d) reforming catalyst layer, (e,h) current collector, (f) flue gas recycling, (g) fuel-cell module.

Browall and Hanneman (*151*) have used a mixed electron/O^{2-}-ion conductor and a reducing gas to produce hydrogen from water without the need for electrodes. They have named the process GEZRO. The function of the reducing gas is to lower the temperature at which water dissociates. With CO, for example, the reactions

$$\text{Inside: } H_2O + 2e \rightarrow H_2 + O^{2-}$$

$$\text{Outside: } CO + O^{2-} \rightarrow CO_2 + 2e$$

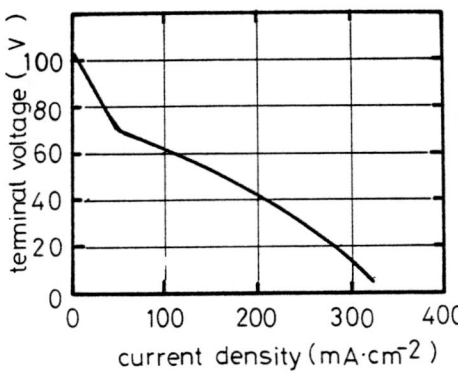

Figure 37. Performance of H_2 + 3% H_2O/air high-temperature fuel cell using a ZrO_2–Y_2O_3–Yb_2O_3 solid electrolyte 0.8 mm thick operated at 830°C. Adapted from Ref. 147.

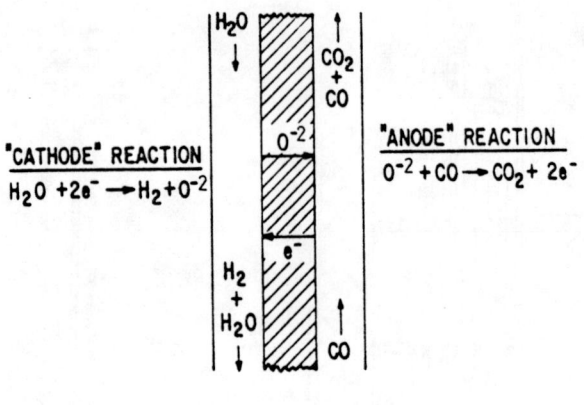

Figure 38. Schematic of self-driven mode of hydrogen production using an oxygen-ion/electron mixed conductor. Adapted from Ref. 151.

proceed at 1000°C without any applied voltage. A schematic of such a "self-driven" cell is shown in Figure 38. Overall thermal efficiencies of 60–70% are anticipated. The optimum transference numbers for the mixed electronic/O^{2-}-ion conductor are $t_O = t_e = 0.5$, but little work has been done to obtain an optimum material.

Summary

Realization of commercial high-temperature cells awaits the solution of several materials problems. Ideally, high-temperature cells would be used for electrolysis, medium-temperature cells for power generation. Development of a medium-temperature cell utilizing a solid electrolyte probably awaits the discovery and development of a suitable proton conductor, a problem not discussed in this paper. However, with the rising cost of fossil fuels, the motivation for developing a high-temperature electrolysis cell is strong, and such cells can provide a test-bed for optimizing cell designs.

The O^{2-}-ion conductors used as solid electrolytes are not satisfactory at present. The most widely investigated, the stabilized zirconias, have too high a resistivity (10–20 Ω-cm at 1000°C), and efforts to find alternate electrolytes of lower resistivity have been only partially successful. The doped cerias such as $Ce_{1-2x}Gd_{2x}O_{2-x}$ ($x = 0.115$) have better O^{2-}-ion conduction but are not sufficiently resistive to reducing atmospheres. A satisfactory electrolyte must have the following properties:

(a) an O^{2-}-ion conduction at 700°C greater than the best $Ce_{1-2x}Gd_{2x}O_{2-x}$ and with an oxygen transference number $t_O \rightarrow 1.0$;

(b) chemical stability over a wide range of oxygen partial pressures, $10^{-30} \lesssim p_{O_2} \lesssim 1$ atm;

(c) mechanical stability, especially no crystallographic modification over a wide range of temperatures;

(d) chemical inertness to fuel and electrodes as well as a thermal expansion matching that of the electrodes; and

(e) invariance during current transfer.

The choice of electrode materials is influenced by the electrolyte used. In addition to matched thermal expansions the electrodes must be chemically stable and catalytically reactive at the operating temperature of the cell. Known electrode materials are acceptable for operating temperatures of 700°C, but at 1000°C their chemical stabilities are inadequate. The most promising cathode materials are electronically conducting oxides, and the mixed electron/O^{2-}-ion conductors give excellent depolarization. A search for more suitable mixed conductors should be encouraged.

Interconnectors must be chemically stable over a wide range of oxygen partial pressures, and a search for acceptable materials has barely begun. The conductivity of doped $CoCr_2O_4$ is much too low. Although several institutions have studied fuel-cell performance, cell design has not been optimized for heat transfer and mechanical strength.

High-temperature and medium-temperature cells utilizing solid electrolytes offer great potential for the economic manufacture of synthetic fuels and for more efficient power generation with built-in storage capacity for load averaging. The required materials research should be supported.

Acknowledgment

H. Obayashi thanks the Petroleum Research Fund of the American Chemical Society for financial assistance in attending the New York Meeting of the Society.

Literature Cited

1. Liebhafsky, H. A., Cairns, E. J., "Fuel Cells and Fuel Batteries—A Guide to Their Research and Development," Wiley, New York, 1968.
2. Young, G. J., Linden, H. R., Eds., "Fuel Cell Systems," *Advan. Chem. Ser.* (1965) **47**.
3. Young, G. J., Ed., "Fuel Cells," Vol. 1, Reinhold, New York, 1960.
4. Young, G. J., Ed., "Fuel Cells," Vol. 2, Reinhold, New York, 1963.
5. Baker, B. S., Ed., "Fuel Cell Systems—II," *Advan. Chem. Ser.* (1969) **90**.
6. Rogers, L. J., "Status of the 4.8 Megawatt Demonstration Program," National Fuel Cell Seminar, Boston, June 21–23, 1977.
7. Markin, T. L., in "Power Sources 4, Research and Development in Non-Mechanical Electrical Power Sources," D. H. Collins, Ed., p. 583, Oriel Press, New Castle upon Tyne, 1973.

8. Takahashi, T., in "Physics of Electrolytes," J. Hladik, Ed., p. 989, Academic, London, 1972.
9. Takahashi, T., Denki Kagaku (1976) 44, 78.
10. Nernst, W., Z. Elektrochem. (1899) 6, 41.
11. For a detailed list of oxide electrolytes and historical aspect see Ref. 1.
12. Etsell, T. H., Flengas, S. N., Chem. Rev. (1970) 70, 339.
13. Sverdrup, E. F., et al., Westinghouse Electric Corp., "1970 Final Report Project Fuel Cell R & D Report No. 57," U.S. Government Printing Office, Washington, D.C., 1970.
14. Antonsen, O., Baukal, W., Fischer, W., Brown Boveri Rev. (1966) 53, 21.
15. Fischer, W., Kleinschmager, H., Rohr, F. J., Steiner, R., Eysel, H. H., Chem. Ing. Tech. (1972) 44, 726.
16. Markin, T. L., Bones, R. J., Dell, R. M., in "Superionic Conductors," G. D. Mahan and W. L. Roth, Eds., p. 15, Plenum, New York and London, 1976.
17. Rhodes, W. H., Carter, R. E., J. Am. Ceram. Soc. (1966) 49, 244.
18. Simpson, L. A., Carter, R. E., J. Am. Ceram. Soc. (1966) 49, 139.
19. Oishi, Y., Ando, K., in "Reactivity of Solids," Chem. Rev. Ser. No. 9 Tokyo University Press, Tokyo, 1975, p. 31.
20. Kingery, W. D., Pappis, J., Doty, M. E., Hill, D. C., J. Am. Ceram. Soc. (1959) 42, 393.
21. Kröger, F. A., Vink, H. J., in "Solid State Physics," Vol. 3, F. Seitz and D. Turnbull, Eds., p. 307, Academic Press, New York, 1956.
22. Lasker, M. F., Rapp, R. A., Z. Phys. Chem. (1966) 49, 198.
23. Schmalzried, H., Z. Phys. Chem. (1963) 38, 87. We replace the notation p_+ and p_- by p_h° and p_e°.
24. Tuller, H. L., Nowick, A. S., J. Electrochem. Soc. (1975) 122, 255.
25. Kofstad, P., "Nonstoichiometry, Diffusion, and Electrical Conductivity in Binary Metal Oxides," Wiley–Interscience, New York, 1972.
26. Tare, V. B., Schmalzried, H., Z. Phys. Chem. (1964) 43, 30.
27. Patterson, J., in "Physics of Electronic Ceramics," Vol. 1., Chap. 5, L. L. Hench and D. B. Dove, Eds., Marcel Dekker, New York, 1971.
28. Patterson, J. W., J. Electrochem. Soc. (1971) 118, 1033.
29. Choudhury, N. S., Patterson, J. W., J. Electrochem. Soc. (1971) 118, 1107.
30. Choudhury, N. S., Patterson, J. W., J. Electrochem. Soc. (1970) 117, 1384.
31. Saito, Y., in "Nonstoichiometric Metal Oxides," S. Takeuchi, Ed., p. 423, Metallurgical Society of Japan, Maruzen, Tokyo, 1975 (in Japanese).
32. McCullough, J. D., Trueblood, K. N., Acta Cryst. (1959) 12, 507.
33. Domagala, R. F., McPherson, D. J., J. Met. (1954) 6, Trans. Am. Inst. Min., Metall. Pet. Eng. (1954) 200, 238.
34. Ruh, R., Garrett, H. J., J. Am. Ceram. Soc. (1967) 50, 257.
35. Vest, R. W., Tallan, N. M., Tripp, W. C., J. Am. Ceram. Soc. (1964) 47, 635.
36. Kumar, A., Rajdev, D., Douglass, D. L., J. Am. Ceram. Soc. (1972) 55, 439.
37. Nasrallah, M. M., Douglass, D. L., J. Electrochem. Soc. (1974) 121, 255.
38. Vest, R. W., Tallan, N. M., J. Am. Ceram. Soc. (1965) 48, 472.
39. Garvie, R. C., J. Am. Ceram. Soc. (1968) 51, 553.
40. Carter, R. E., Roth, W. L., G. E. Res. Rep. No. 63-RL-3479M, 1963.
41. For work done before 1970 see Ref. 14 for details.
42. Tannenberger, H., Schachner, H., Kovas, P., Proc. J. Int. Etude Piles Combust., Brussels, June, 1965, p. 19.
43. Heyne, L., Electrochem. Acta (1970) 15, 1251.
44. Heyne, L., in "Mass Transport in Oxides," J. B. Wachtman and A. D. Franklin, Eds., p. 149, NBS Special Publ. 296, 1968.

45. Michel, D., Jorba, M. P., Collonges, R., *C. R. Acad. Sci.* (1968) **266**, 1602.
46. Rhodes, W. H., Carter, R. E., *J. Am. Ceram. Soc.* (1966) **49**, 244.
47. Tien, T. Y., *J. Am. Ceram. Soc.* (1964) **47**, 430.
48. Tien, T. Y., Subbarao, E. C., *J. Chem. Phys.* (1963) **39**, 1041.
49. Subbarao, E. C., Sutter, P. H., *J. Phys. Chem. Solids* (1963) **24**, 128.
50. Baukal, W., *Electrochim. Acta* (1969) **14**, 1071.
51. Cocco, A., Danelon, M., *Ann. Chim. (Rome)* (1965) **55**, 1313.
52. Wachtman, J. B., Corwin, W. C., *J. Res. Nat. Bur. Stand.* (1965) **694**, 457.
53. White, D. W., *Rev. Energ. Primaire* (1966) **2**, 10.
54. Carter, R.E., Roth, W. L., in "Electromotive Force Measurements in High-Temperature Systems," C. B. Alcock, Ed., p. 125, Institution of Mining and Metallurgy, London, 1968.
54a. Dixon, J. M., LaGarange, L. D., Merten, U., Miller, C. F., Porter, J. T., *J. Electrochem. Soc.* (1963) **110**, 276.
54b. Nuimin, A. D., Pal'guev, S. F., *Tr. Inst. Elektrokhim., Akad. Nauk SSSR, Ural. Fil.* (1964) **5**, 154; *Chem. Abstr.* (1965) **62**, 8472.
54c. Guillou, M., *Rev. Gen. Elec.* (1967) **76**, 58.
54d. Stricker, D. W., Carlson, W. G., *J. Am. Ceram. Soc.* (1965) **48**, 286.
54e. Tannenberger, H., Schachner, H., Kovacs, P., *Rev. Energ. Primaire* (1966) **2**, 19.
54f. Möbius, H.-H., Proeve, G., *Z. Chem.* (1965) **5**, 431.
55. Schmalzried, H., *Z. Electrochem., Ber. Bunsenges. Phys. Chem.* (1962) **66**, 572.
56. Yanagida, H., reported in Yuan, D., Kröger, F. A., *J. Electrochem Soc.* (1969) **116**, 594.
57. Etsell, T. H., Flengas, S. N., *J. Electrochem. Soc.* (1972) **119**, 1.
58. Komatsu, S., Yonehara, M., Kouzuka, Z., *Bull. Metall. Soc. Jpn.* (1972) **36**, 674.
59. Patterson, J. W., Bogren, E. C., Rapp, R. A., *J. Electrochem. Soc.* (1967) **114**, 752.
60. Steele, B. C. H., Alcock, C. B., *Trans. Metall. Soc. AIME* (1965) **233**, 1359.
61. Shores, D. A., Rapp, R. A., *J. Electrochem. Soc.* (1971) **118**, 1107.
62. VanHandel, G. J., Blumenthal, R. N., *J. Electrochem. Soc.* (1974) **121**, 1198.
63. Tuller, H. L., Nowick, A. S., *J. Electrochem. Soc.* (1975) **122**, 836.
64. Panlener, R. J., Blumenthal, R. N., Garnier, J. E., *J. Phys. Chem. Solids* (1975) **36**, 1213.
65. Kevane, C. J., Holvesson, E. L., Armstrong, J. B., *Proc. Rare Earth Res. Conf., 5th Ames, Iowa, August, 1965*, Book **4**, 43.
66. Noddack, W., Walch, H., *v. Phys. Chem.* (1959) **211**, 194.
67. Rudolph, J., *Z. Naturforsch.* (1959) **14A**, 727.
68. Greener, E. H., Wimmer, J. M., Hirthe, W. M., in "Rare Earth Research II," K. S. Vorres, Ed., Gordon and Breach, New York, 1964.
69. Blumenthal, R. N., Pinz, B. A., *J. Appl. Phys.* (1967) **38**, 2376.
70. Blumenthal, R. N., Lee, P. W., Panlener, R. J., *J. Electrochem. Soc.* (1971) **118**, 123.
71. Kofstad, P., Hed, A. Z., *J. Am. Ceram. Soc.* (1967) **50**, 681.
72. Blumenthal, R. N., Hofmaier, R. L., *J. Electrochem. Soc.* (1974) **121**, 126.
73. Croatto, U., Mayer, A., *Gazz. Shim. Ital.* (1943) **73**, 199.
74. Takahashi, T., Ito, K., Iwahara, H., *Proc. J. Int. Etude Piles Combust.* (1965) **1-III**, 42.
75. Takahashi, T., Iwahara, H., *Denki Kagaku* (1966) **34**, 254.
76. Singman, D., *J. Electrochem. Soc.* (1966) **113**, 502.

77. Kudo, T., Obayashi, H., *J. Electrochem. Soc.* (1975) **122**, 142.
78. Blumenthal, R. N., Brugner, F. S., Garnier, J. E., *J. Electrochem. Soc.* (1973) **120**, 2340.
79. Seitz, M. A., Holliday, T. B., *J. Electrochem. Soc.* (1974) **121**, 122.
80. Tuller, H. L., Nowick, A. S., *J. Electrochem. Soc.* (1975) **122**, 255.
81. Pal'guev, S. F., Volchenkova, Z. S., *Tr. Inst. Elektrokhim., Akad. Nauk SSSR, Ural. Fil.* (1961) **2**, 157; *Chem. Abs.* (1963) **59**, 12267.
82. Kudo, T., Obayashi, H., to be published in *J. Electrochem. Soc.* (1976) **123**.
83. Takahashi, T., Ito, K., Iwahara, H., *Electrochim. Acta* (1967) **12**, 21.
84. Alcock, C. B., Steele, B. C. H., in "Science of Ceramics," Vol. II, G. H. Stewart, Ed., Academic, London, 1965; *Trans. Br. Ceram. Soc.* (1964) 397.
85. Laskar, M. F., Rapp, R. A., *Z. Phys. Chem.* (1966) **49**, 211.
86. Choudhury, N. S., Patterson, J. W., *J. Am. Ceram. Soc.* (1974) **57**, 90.
87. Diness, A. M., Roy, R., *J. Mater. Sci.* (1969) **4**, 613.
88. Mehrotra, A. K., Maiti, H. S., Subbarao, E. C., *Mater. Res. Bull.* (1973) **8**, 899.
89. Etsell, T. H., *Z. Naturforsch. Teil A* (1972) **27A**, 1138.
90. Volchenkova, Z. S., Pal'guev, S. F., *Electrochem. Molten Solid Electrolyte* (1964) **2**, 53.
91. Möbius, H. -H., Witzmann, H., Witte, W., *Z. Chem.* (1964) **4**, 152.
92. Swinkels, D. A. J., *J. Electrochem Soc.* (1970) **117**, 1267.
93. Hardaway, J. B., III, Patterson, J. W., Wilder, D. R., Schieltz, J. D., *J. Am. Ceram. Soc.* (1971) **54**, 94.
94. Tretyakov, Y. D., Muan, A., *J. Am. Ceram. Soc.* (1969) **116**, 331.
95. Wells, A. F., "Structural Inorganic Chemistry," 3rd Ed., pp. 668–670, Clarendon, Oxford, 1962.
96. Takahashi, T., Iwahara, H., Nagai, Y., *Proc. Annu. Meet. Electrochem. Soc. (Jpn.), 37th, Tokyo, May, 1970*, Paper A209.
97. Takahashi, T., Iwahara, H., Arao, T., *Proc. Annu. Meet. Electrochem. Soc. (Jpn.), 38th, Osaka, May, 1971*, Paper D111.
98. Takahashi, T., Iwahara, H., *Proc. Annu. Meet. Electrochem. Soc. (Jpn.), 39th, Tokyo, March, 1972*, Paper C305.
99. Goodenough, J. B., Longo, J. M., *Landolt Börnstein New Series* III/4a, p. 126, Springer-Verlag, Berlin, 1970.
100. Galasso, F. S., "Structure, Properties and Preparation of Perovskite-Type Compounds," *Int. Ser. of Monographs in Solid State Physics*, **5**, Pergamon, Oxford, 1969.
101. Takahashi, T., Iwahara, H., *Denki Kagaku* (1967) **35**, 433.
102. Takahashi, T., Iwahara, H., Imaichi, T., *Denki Kagaku* (1969) **37**, 857.
103. Takahashi, T., Iwahara, H., Imaichi, T., Aoyama, H., *Annu. Rep. Asahi Glass Found. Contrib. Ind. Technol.* (1969) **15**, 237.
104. Matsuo, Y., Sasaki, H., *Jpn. J. Appl. Phys.* (1964) **3**, 799.
105. Schwarz, D. B., Anderson, H. U., *J. Electrochem. Soc.* (1975) **12**, 707.
106. Stephenson, C. V., Flanagan, C. E., *J. Chem. Phys.* (1961) **34**, 2203.
107. Roth, R. S., *J. Res. Nat. Bur. Stand.* (1956) **56**, 17.
108. Aleskin, E., Roy, R., *J. Am. Ceram. Soc.* (1962) **45**, 18.
109. Mazelsky, R., Kramer, W. E., *J. Electrochem. Soc.* (1964) **111**, 528.
110. Wells, A. F., "Structural Inorganic Chemistry," 3rd Ed., p. 465, Clarendon, Oxford, 1962.
111. Etsell, T. H., Flengas, S. N., *J. Electrochem. Soc.* (1969) **116**, 771.
112. Tedmon, C. S., Jr., Spacil, M. S., Mitoff, S. P., *J. Electrochem. Soc.* (1969) **116**, 1170.
113. Yanagida, H., Brook, R. J., Kröger, F. A., *J. Electrochem. Soc.* (1970) **117**, 593.
114. Etsell, T. H., Flengas, S. N., *J. Electrochem. Soc.* (1961) **118**, 1890.

115. Kröger, F. A., *J. Electrochem. Soc.* (1973) **120**, 75.
116. Yanagida, H., Brook, R. J., Kröger, F. A., *J. Electrochem. Soc.* (1970) **117**, 593.
117. Hirsch, H. H., White, D. W., Ref. 4 *in* Tedmon, C. S., Jr., Spacil, H. S. Mitoff, S. P., *J. Electrochem. Soc.* (1969) **116**, 1170.
118. Tannenberger, H., Siegert, H., "Abstracts of Papers," 154th National Meeting, ACS, Sept. 1967, FUEL 042.
119. Tannenberger, H., Siegert, H., *in* "Fuel Cell Systems—II," B. S. Baker, Ed., *Advan. Chem. Ser.* **90**, 281.
120. Goodenough, J. B., "Metallic Oxides," *in Prog. Solid State Chem.* (1972) **5**, 145.
121. Verwey, E. J. W., Haaijman, P. W., Romeijn, F. C., Van Oosterhout, G. W., *Philips Res. Rep.* (1950) **5**, 173.
122. van Houten, S., *J. Phys. Chem. Solids* (1960) **17**, 7.
123. Takahashi, T., Suzuki, Y., Ita, K., Hasegawa, H., *Denki Kagaku* (1967) **35**, 201.
124. Takahashi, T., Iwahara, H., Suzuki, Y., *Proc. J. Int. Etude Piles Combust.* (1969) **3**, 113.
125. Böhm, R., Kleinschmager, H., *Z. Naturforsch.* (1971) **26A**, 780.
126. Heikes, R. R., Miller, R. C., Mazelsky, R., *Physica* (1964) **30**, 1600.
127. Goodenough, J. B., Raccah, P. M., *J. Appl. Phys.* (1963) **36**, 1031.
128. Sverdrup, E. F., Archer, D. H., Glasser, A. D., *in* "Fuel Cell Systems—II," B. S. Baker, Ed., *Advan. Chem. Ser.* (1969) **90**, 301.
129. Bosman, A. J., Crevecoeur, C., *Phys. Rev.* (1966) **144**, 763.
130. Shirai, K., Ihara, S., Sato, H., *Bull. Electrotech. Lab.*, Tokyo (1974) **38**, 378.
131. Weiher, R. L., Ley, R. P., *J. Appl. Phys.* (1963) **34**, 1833.
132. Kudo, T., Obayashi, H., Gejo, T., *J. Electrochem. Soc.* (1975) **122**, 159.
133. Obayashi, H., Kudo, T., Gejo, T., *Jpn. J. Appl. Phys.* (1974) **13**, 1.
134. Similar results were obtained for $LaCoO_3$ by G. H. J. Broers, et al., private communication.
135. Obayashi, H., Kudo, T., *Jpn. J. Appl. Phys.* (1975) **14**, 330.
136. Obayashi, H., et al., unpublished data.
137. Eysel, H. H., "Extended Abstracts," No. 36, *Electrochem. Soc. Meeting*, Atlantic City, Oct., 1970.
138. Tedmon, C. S., Jr., Spacil, H. S., Mitoff, S. P., G. E. Report No. **69-C-056** (1969).
139. Rohland, B., Möbius, H. H., *Naturwissenschaften* (1968) **55**, 227.
140. Takahashi, T., Suzuki, Y., Ito, K., Yamanaka, H., *Denki Kagaku* (1968) **36**, 345.
141. Sun, C. C., Hawk, E. W., Sverdrup, E. F., *J. Electrochem. Soc.* (1972) **119**, 1433.
142. Schmalzried, H., *Ber. Bunsenges. Phys. Chem.* (1963) **67**, 93.
143. Kleinschmager, H., Reich, A., *Z. Naturforsch.* (1972) **27A**, 363.
144. Boll, R. H., Bhada, R. K., *Energy Convers.* (1968) **8**, 3.
145. Kudo, T., Obayashi, H., *Energy Convers.* (1976) **15**, 121.
146. Fischer, W., Kleinschmager, H., Rohr, F. J., Steiner, R., Eysel, H. H., *Chem. Ing. Tech.* (1972) **44**, 726.
147. Eysel, H. H., Kleinschmager, H., *BBC-Nachr.* (1972) **54**, 13.
148. Spacil, H. S., Tedmon, C. S., Jr., G. E. Report No. **69-C-176** (1969).
149. *Chem. Eng. News* (1968) **46**, 48.
150. Allison, H. J., Hughes, W. L., *Intersoc. Energy Convers. Eng. Conf., Rec., 10th, 1975* (1975) 104.
151. Browall, K. W., Hanneman, R. E., G. E. Report No. **75CRD012** (1975).
152. Takahashi, T., Iwahara, H., *Energy Convers.* (1971) **11**, 105.

RECEIVED July 26, 1976.

INDEX

INDEX

A

Absorbers, selective-black	151, 159
Acid–base hypothesis	230
Acrylonitrile	314
Adenine	307
Algae	94, 98
anaerobic fermentation of	98
conversion into	107
Alkali-ion transport	179, 193
Alloy(s)	
aluminum	225
chromium–nickel	229
nickel–chromium	229
Alumina	229
β-Alumina layer structure, schematic of	192
β''-Alumina ceramic electrolyte	207
Aluminum oxide	226
AN	214
Anabena cylindrica	100
Anaerobic fermentation of algae	98
Anion polarizability, effect on cation mobility	201
Anisole	310
Anodes	352
Anodic current at *n*-type semiconductors, photoeffect on	44
Anodic dissolution of photon absorber by TiO_2 coatings, prevention of	67
Anti-Beevers–Ross sites	192
Azobenzene	310
Azolla	100

B

Bacteria, photosynthetic	98
Bacteriochlorophyll	105
Bacteriopheophytin	105
Band gap	116, 120
energy	75
Batteries, secondary, energy storage in	2
Batteries, secondary, high-specific-energy	179
Batteries, sodium–sulfur high-specific-energy	180
Beevers–Ross sites	192
Beta phase	259
Bismuth	202
Bismuth-containing oxides	202
Born equation	43
Bottleneck barrier	200
Boudouard reaction	17
Bronzes	
molybdenum	168
oxides	165, 167
tungsten	167
vanadium	169

C

Cadmium selenide	79
-based photoelectrochemical cells, power conversion efficiency using	88
photoelectrodes	71
stabilization of	77
Cadmium sulfide	77, 79
Calvin–Benson cycle	97
Carbon, oxidation of	6
Carbon monoxide	15
hydrogenation	17
oxidation of	6
Catalyst(s)	
deactivation	23
Fischer–Tropsch	19
metal	17, 24
nickel	29
nickel/alumina	19
palladium	28
Zeigler–Natta	298
Cathodes	344
Cation-exchange capacity	301
Cation mobility, effect of anion polarizability on	201
C.e.c.	301
Cell(s)	
electrolytic	179
silicon solar	134
solar, cost of	127
technology	355
Ceramic electrolyte	205
β''-alumina	207
Cernegieite	182
Chalcogenides, transition-metal	304
Chlorobenzene	310
Chlorophyll, antenna	101
Chlorophyll, light harvesting	101
Chromium–nickel alloys	229
Chromium oxide, attack by sodium sulfate	230
Coal	1, 15
liquification	3
Coatings, solar selective	162
Coatings, TiO_2 prevention of anodic dissolution of photon absorber by	67

367

Collector, solar, flat-plate 150
Conservation 284
Cooling, solar 271, 273
Corrosion, hot 225
  definition 226
  phenomenology of 229
Corrosive agents 227
Cost of solar cells 127
Crystallographic principles 180
Current–voltage properties 81
CVD 134
Cyclododecatrienes 299

D

Data, ionic conductivity 196
Deacon equilibrium 6
Debye–Waller factor 184
Decomposition of water by
  thermochemical cycles 5
4,4'-Diamino-*trans*-stilbene 309
Dibenzene titanium 299
Dickite 301, 304
Dihydrides 285
Diodes
  metal–insulator–semiconductor . 125
  Schottky barrier 124
  semiconductor 109
  as solar cells 113
Diphenylethylene 309
Double-jump transport process 193

E

Efficiency
  fuel-cell energy 332
  power conversion 86
  quantum 48
  storage, maximum theoretical ... 75
Electric generators, wind-powered . 94
Electrode(s) 344
  reactions at a semiconductor–
    solution interface, photo-
    effect on 36
Electrolysis cells 3, 316, 356
Electrolyte(s)
  β"-alumina ceramic 207
  ceramic 205
  new solid 181
  O^{2+}-ion solid 319
  polysulfide 77
  solid 242
    properties of 182
  stability of *n*-type photoelectrode
    in aqueous polysulfide 78
Electrolytic cell(s) 2, 179
Electrolytic domain 323
Energy
  band gap 75
  efficiency, fuel-cell 332
  farming 98
  requirements, future 33
  reserves in the United States ... 94

Energy (*continued*)
  solar 93
    calculated hydrogen production
      rate using photo-driven
      cells 54
  sources 15
  storage in secondary batteries ... 2
Enthalpy 37
Entropies of binary hydrides 291
Epsilon phase 259
Europium 285
Evaporation 134
Extractive metallurgy 230

F

Farming, energy 98
Fermentation of algae, anaerobic .. 98
Ferroelectrics 201
Film, transparent heat mirror 150
Films, heteroepitaxial 134
Fischer–Tropsch catalysts 19
Fischer–Tropsch synthesis 16
Fluorite-related materials 240
  in energy-winning roles 242
Fluorite-related ordered phases ... 247
Fluorite-related oxides for the
  nuclear industry 245
Fluorites 323
Fossil fuel(s) 1, 93
Free energies of formation 290
Fuel
  automotive, methanol for 13
  cells 316, 355
    criteria for 344
    energy efficiency of 332
  conversion of organic waste,
    plants, and algae into 107
  fossil 1, 93
  nuclear 1
Fused oxides 230

G

Gallium phosphide 35
Germanium 34
Greenhouse effect 95

H

Halogenides 7
Hatch–Slack pathway 97
Heat engine 94
Heat mirror(s)
  film, transparent 150
  thermal stability of 150
  transparent 152
Heating, solar 94, 95, 271
Helium 7
Heterodiodes 124
Heteroepitaxial films 134
Hollandite structure 188
Hot corrosion, definition 226
Hybridization, ease of 202

INDEX

Hydride(s)
  entropies of binary 291
  iron–titanium 273, 276, 277
  new alloy 278
  solid metal 271
  stability 288
Hydrocarbons, catalytic synthesis
  of 15
Hydrogen 1, 15
  cost 3
  distribution 9
  production 2, 3
    in a photoelectrochemical cell . 36
  photosynthetic 98
  rate from solar energy using
    photo-driven cells 54
  safety problems 12
  storage 9
  uses 3
Hydrogenase 99
Hydrogenation, carbon monoxide . 17
Hyoxanthine 307
Hypothesis, acid–base 230

I

Intercalate, tetrahydrofuran 304
Intercalates, sheet-silicate 301
Interconnectors 344, 352
Intermetallic compounds 285
Ionic conductivity data 196
Iota phase, calculated images of .. 255
Iota phase, observed images of ... 257
Iron-titanium hydride 273, 276, 277
Isostructural series 195

K

Kaolinite 301, 304

L

Laser hosts 247
Leucothionine 104, 105
Liquification of coal 3
LPE 134
Lutetium 285

M

Madelung energy 186
Magnesium
  compounds 229
  sulfate 229
  vanadates 229
Metal catalysts 17, 24
Metal-hydrogen systems 289
Metal–insulator–semiconductor
  diodes 125
Metal titanates, photoelectro-
  chemical reactions on 65
Metallurgy, extractive 230
Methacrylonitrile 314

Methanation 19
Methanol for automotive fuel 13
MgO/Au film 160
Mirror, heat, thermal stability of .. 150
Mirrors, transparent heat 152
MIS 125
Molybdenum bronzes 168
Montmorillonite 301–313
Montmorillonites, transition-metal
  exchanged 304
Mullite 235

N

Naphtha 3, 16
Nernst glowers 242
Nernst mass 319
Network formed by linked
  octahedra 183, 189
Network formed by linked
  tetrahedra 182, 189
Nickel/alumina catalysts 19
Nickel catalysts 29
Nickel–chromium alloys 229
Niobates 201
Niobium 285
Nitrogenase 99
Nuclear
  control materials 247
  fuel 1
  industry, fluorite-related oxides
    for the 245

O

O^{2+}-ion solid electrolytes 319
Ocean thermal gradients 94, 95
Octahedra, linked, network formed
  by 183
Octahedra and tetrahedra, linked,
  network formed by 189
OHP 42
Oil 1
Oil shale 1, 15
OPEC 284
Open circuit photopotential 81
Optical Born energy 42
Ordered phases, fluorite-related ... 247
Organic waste, conversion into fuel 107
Oxidation of carbon 6
Oxidation of carbon monoxide 6
Oxide bronzes 165
  structures of 167
Oxides, bismuth-containing 202

P

Peltier effect 135, 146
Perovskite(s) 340
  cubic 341
  structure 166
Phase transformations 264
Phenomenology of hot corrosion .. 229
Phosphors 247

Photocurrent	122
Photoeffect on the anodic current at n-type semiconductors	44
Photoeffect on electrode reactions at a semiconductor–solution interface	36
Photoelectric devices	94, 102
Photoelectrochemical cell(s)	72
CdSe-based, power conversion efficiency using	88
CdSe-based, quantum efficiency for electron flow for	89
hydrogenation production in	36
properties of TiO_2 film	62
reaction on TiO_2 single crystal	58
reactions on metal titanates	65
Photoelectrode, n-type, in aqueous polysulfide electrolytes stability	78
Photoelectrodes, cadmium selenide	71
Photoelectrolysis of water	71
Photon absorber by TiO_2 coatings, prevention of anodic dissolution of	67
Photon absorption	38
Photopotential, open circuit	81
Photosynthesis	93
Photosynthetic bacteria	98
hydrogen production	94, 98
systems	94
Photovoltage	115
Plants, conversion into fuel	107
Polarizability	202
anion effect on cation mobility	201
Polybutadiene	299
Polysulfide electrolytes	77
stability of n-type photoelectrode in aqueous	78
Power conversion efficiency	86
using CdSe-based photoelectrochemical cells	88
Power, solar	109
Priderite	188
Pseudophases, definition	249
Pyrochlore(s)	343
defect	186
structure, ideal formula for	196

Q

Quantum efficiency	48
for electron flow for CdSe-based photoelectrochemical cells	89

R

Radiation losses	149
Raney nickel	19
Rare earth oxides, texture in the binary	249
Redox cycles	8
Refining, zone	135
Rule of reversed stability	279

S

Safety problems with hydrogen	12
Samarium	285
Schottky barrier diodes	124
Selective-black absorbers	151, 159
SEM	210
Semiconductor(s) diodes	109
investigated photoelectrochemically	34
photoeffect on the anodic current at n-type	44
–solution interface, photoeffect on electrode reactions at	36
p-type	38
SiC	226
Si_3N_4	226
SiO_2	226
Silicate intercalates, sheet	301
Silicates, sheet, structural characteristics of	301
Silicon solar cells	134
Sn-doped In_2O_3 film	156
SNG	19
Sodium chromate	235
–sodium cell(s)	209
schematic of	208
sulfate	227
attack of chromium oxide by	230
–sulfur battery, application of	207
–sulfur cell(s)	212
after testing, examination of	216
materials costs estimate for	222
schematic of	206
–sulfur high-specific-energy batteries	180
Solar cells	109
cost of	127
diodes operating as	113
silicon	134
collector, flat-plate	150
cooling	271, 273
energy	93
using photo-driven cells, calculated hydrogen production rate from	54
heating	94, 95, 271
power	109
radiation, normalized	151
selective coatings	162
Solid electrolytes	181, 242
properties of	182
Specific-energy secondary batteries, high-	179
Spinach	100
Sputtering	134

Stability
  hydride 288
  reversed, rule of 279
  of n-type photoelectrode in
    aqueous polysulfide electro-
    lytes 78
Stabilization of CdSe 77
Steam–carbon process 6
Steel production 3
Storage
  efficiency, maximum theoretical . 75
  energy, in secondary batteries .. 2
  of hydrogen 9
Structure 249
Sulfation attack, definition 226
Sulfur electrode 205
Synthesis of hydrocarbons,
  catalytic 15

T

Tantalates 201
Tantalum 285
Tar sands 15
Tetrahedra, linked, network
  formed by 182
Tetrahedra, network formed by
  linked octahedra and 189
Tetrahydrofuran intercalate 304
Texture in the binary rare earth
  oxides 249
Thermal
  gradients, ocean 95
  stability of heat mirror 150
  storage 272
Thermodynamic cycles, classifi-
  cations of 6
Thermoelectric generation **177**
Thionine 104, 105
Thoria–rare earth oxide systems .. 244
Titanates, metal, photoelectrochemical reactions 65
Titanium dioxide 34, 35
  coatings, prevention of anodic
    dissolution of photon
    absorber by 67
  film, photoelectrochemical
    properties 62
  single crystal, photoelectro-
    chemical reaction on 58
TOT 301

Transition-metal
  chalcogenides 304
  exchanged montmorillonites 304
  oxides 7
Transparent heat mirrors 152
Transport, alkali-ion 179, 193
Transport process, double-jump ... 193
Triphenylamine 309
Tungsten bronzes 167
Turbine, wind-powered 94
p-Type semiconductor 38

U

United States, energy reserves in .. 94
Uranium dioxide–rare earth oxide
  systems 245

V

Vanadium 285
  bronzes 169
  dihydride 276
Vermiculite 301
Vinylcyclohexenes 299

W

Water, decomposition by thermo-
  chemical cycles 5
Water photoelectrolysis 71
Welsbach gas mantles 242
Wigner–Seitz 288
Wind 94, 107
Wind-powered electric generators . 94
Wind-powered turbine 94

Y

Ytterbium 285

Z

Zeigler–Natta catalysts 298
Zeta phase 259
Zirconia(s) 4
  –Hafnia-based materials 242
  stabilized 320
  yttria-stabilized 319
Zone refining 135

The text of this book is set in 10 point Caledonia with two points of leading. The chapter numerals are set in 30 point Garamond; the chapter titles are set in 18 point Garamond Bold.

The book is printed offset on Text White Opaque, 50-pound. The cover is Joanna Book-Binding blue linen.

Jacket design by Ger Quinn.
Editing and production by Joan Comstock.

The book was composed by the Service Composition Co., Baltimore, Md., printed and bound by The Maple Press Co., York, Pa.

/541.0421S686>C1/

DATE DUE